Advances in Applied Mathematics

QUADRATIC PROGRAMMING WITH COMPUTER PROGRAMS

Michael J. Best

University of Waterloo
Ontario, Canada

CRC Press
Taylor & Francis Group
Boca Raton London New York

CRC Press is an imprint of the
Taylor & Francis Group, an **informa** business
A CHAPMAN & HALL BOOK

CRC Press
Taylor & Francis Group
6000 Broken Sound Parkway NW, Suite 300
Boca Raton, FL 33487-2742

First issued in paperback 2022

ISBN 13: 978-1-03-247694-0 (pbk)
ISBN 13: 978-1-4987-3575-9 (hbk)

DOI: 10.1201/9781315120881

Visit the Taylor & Francis Web site at
http://www.taylorandfrancis.com

and the CRC Press Web site at
http://www.crcpress.com

QUADRATIC
PROGRAMMING
WITH COMPUTER
PROGRAMS

Advances in Applied Mathematics

Series Editor: Daniel Zwillinger

Published Titles

Advanced Engineering Mathematics with MATLAB, Fourth Edition
Dean G. Duffy

CRC Standard Curves and Surfaces with Mathematica®, Third Edition
David H. von Seggern

Dynamical Systems for Biological Modeling: An Introduction
Fred Brauer and Christopher Kribs

Fast Solvers for Mesh-Based Computations *Maciej Paszyński*

Green's Functions with Applications, Second Edition *Dean G. Duffy*

Handbook of Peridynamic Modeling *Floriin Bobaru, John T. Foster,
Philippe H. Geubelle, and Stewart A. Silling*

Introduction to Financial Mathematics *Kevin J. Hastings*

Linear and Complex Analysis for Applications *John P. D'Angelo*

Linear and Integer Optimization: Theory and Practice, Third Edition
Gerard Sierksma and Yori Zwols

Markov Processes *James R. Kirkwood*

Pocket Book of Integrals and Mathematical Formulas, 5th Edition
Ronald J. Tallarida

Stochastic Partial Differential Equations, Second Edition *Pao-Liu Chow*

Quadratic Programming with Computer Programs *Michael J. Best*

To: My wife Patti

and my daughters

Jiameng, Zhengzheng, Jing, Yuanyuan, Xiaohui, Lan, Lei, Qun

and with special thanks to Patricia M. Best for excellent proof reading.

Contents

Preface

This book is about Quadratic Programming (QP), Parametric Quadratic Programming (PQP), the theory of these two optimization problems, and solution algorithms for them. Throughout this text, we rely heavily on vector and matrix notation. Prime ($'$) will be used to denote matrix transposition. All unprimed vectors will be column vectors. The i–th component of the vector x_0 will be denoted by $(x_0)_i$. In a numerical example, we sometimes use the symbol times ($\times$) to denote multiplication. For 2–dimensional examples, we generally use the problem variables x_1 and x_2 with $x = (x_1, x_2)'$. In other situations, x_1 and x_2 may denote two n–vectors. The meaning should be clear from the context. The end of a proof is signified by an unfilled box ($\square$) and the end of an example is signified by an unfilled diamond ($\lozenge$).

In Chapter 1, we consider several 2–dimensional examples of QP's. We teach the reader how to draw these problems and how to deduce an optimal solution using graphical arguments. We also deduce from the graphical representation algebraic conditions which are both necessary and sufficient for optimality.

Chapter 2 is devoted to the development of portfolio optimization. Portfolio optimization is concerned with combining a number of assets, each with known expected return and correlation with the other assets into a single portfolio. For the combined portfolio, with any level of expected return, no other portfolio should have a smaller variance. Such a portfolio is called efficient. Finding such a portfolio is equivalent to solving a certain QP. Finding all such portfolios is equivalent to solving a PQP. Thus Chapter 2 provides many examples of problems which are naturally QP's or PQP's.

Chapter 3 begins by looking at the problem of unconstrained quadratic minimization theory then development of a solution algorithm for it using conjugate directions (Algorithm 1). Although this problem can be solved directly, we use a conjugate direction method to solve it because it will be a building block for full QP algorithms developed subsequently. Chapter 3 continues by addressing the problem of quadratic minimization subject to linear equality constraints. Theoretical properties and a solution algorithm (Algorithm 2) are developed as generalizations of those for unconstrained quadratic minimization.

Whenever we develop an algorithm in Chapter 3 or subsequent chapters, we first give the informal development of the algorithm. Then we give a complete and detailed formulation and prove its properties. We also give several numerical examples which are illustrated with detailed calculations. Then in a later section, a Matlab program is given to implement the algorithm. These programs are included on the accompanying CD. This computer program is run on the same data as the example to validate its results. In some cases, we illustrate the computer program by solving some problems which are too large for hand computation.

Chapter 4 is devoted to the theory of QP. Optimality conditions which are both necessary and sufficient for optimality are developed. Duality theory for QP may be regarded as a generalization of duality theory for LP. Indeed, the two types of duality share many of the same properties. All of the QP algorithms automatically compute an optimal solution for the dual. In some situations, obtaining the optimal solution for the dual may be equally important as finding the optimal solution for the given problem (the primal).

Sometimes the data for a QP or PQP is taken from physical measurements and is only known to a certain accuracy. In this case it is important to know how sensitive the (approximate) optimal solution is to small changes in the problem data. This area is called sensitivity analysis and is also developed in Chapter 4.

In Chapter 5, we generalize Algorithm 2 to solve a QP having both linear inequality and equality constraints. The resulting algorithm (Algorithm 3) will solve a QP provided a certain nondegeneracy assumption is satisfied. Algorithm 3 will also find an optimal solution for the dual QP. An initial feasible point is required to start Algorithm 3. We solve this problem by formulating an initial point problem (an LP) whose optimal solution will be feasible for the original problem. We also specialize Algorithm 3 to solve an LP (LP Algorithm 3) and this can be used to efficiently solve the initial point problem.

Chapter 5 continues by noting that when a new constraint becomes active, all of the conjugate direction built up by Algorithm 3 are destroyed and must be rebuilt. A Householder transformation is introduced which when applied to the existing conjugate directions will result in a new set of conjugate directions which will not be destroyed (Algorithm 4) and this is a very efficient and fast QP algorithm.

Both algorithms require a nondegeneracy assumption to guarantee finite termination. We complete Chapter 5 by giving a generalization of Bland's rules which ensures finite termination if degeneracy is encountered. Computer programs for Algorithms 3 and 4 are given.

Most QP algorithms are based on satisfying primal feasibility and complementary slackness at each iteration and iteratively working toward satisfying dual feasibility. In Chapter 6, we present a dual QP algorithm (Algorithm 5) which satisfies dual feasibility and complementary slackness at each iteration and iteratively works to satisfy primal feasibility. An example of where this is appropriate is the following. Suppose a QP has just been solved. But then it is realized that additional primal constraints are necessary. So the previously optimal solution satisfies dual feasibility and complementary slackness and so can be efficiently solved using a dual QP method.

In Chapter 7, we formulate general QP algorithms (Algorithm 6) and PQP algorithms (Algorithm 7). These methods are general in the sense that the method of solving certain linear equations is left unspecified. The resulting method just contains the QP and PQP logic. The linear equations may be sparse or contain structure like flow constraints. This structure may be utilized to formulate a more efficient solution method. One way to formulate an implementable version of these methods is to solve the relevant linear equations using the inverse of the coefficient matrix. Then there are 3 possible types of updates: add a row and column, delete a row and column and exchange a row and column. These 3 symmetric inverse updates are developed in detail.

The Simplex Method (Algorithm 8) and Parametric Simplex Method (Algorithm 9) are developed in Chapter 8. They both assume that the problem constraint structures are of the form $\{ x \mid Ax = b,\ x \geq 0\}$, which is identical to that of the Simplex Method for LP. The efficiency of this method is a consequence of that if the intermediate points constructed have many variables at zero, then the linear equations that must be solved are very small in number. This would be the case if the Hessian matrix for the problem has low rank.

In Chapter 9, we consider the problem of nonconvex quadratic programming. This type of problem may possess many local minima, a global minimum or be unbounded from below. We show that the first order Karush–Kuhn–Tucker conditions are necessary for a local minimum but not in general sufficient. We then develop the second order sufficiency conditions which with the first order necessary conditions will guarantee that a point is a strong local minimum.

With additional steps in Algorithm 4, we formulate a new method (Algorithm 10) which will determine a strong or weak local minimum (if such a point exists), or determine that the problem is unbounded from below.

Using the Matlab Programs

Every Matlab® program shown in this book is included in the file QPProgs.zip which is available for download on the author's website http://www.math.uwaterloo.ca/~mjbest/ . Make a fresh directory on your computer and name it "MYQP". Download QPProgs.zip into "MYQP" and unzip all files into this same directory. The following instructions apply to MATLAB R2016a but other releases of MATLAB should be similar. Start Matlab and press the "HOME" button and then the "Set Path" button on the "HOME" page toolbar. This brings up a window entitled "Set Path". Press "Add Folder..." and keep browsing until you locate the "MYQP" folder. Highlight "MYQP" and then press the "Select Folder". This will close the "Add Folder..." window and return you to the "Set Path" window. At the top of this window you will see the correct path to "MYQP". Now press "Save" and then "Close". All of this sets Matlab's path pointing to "MYQP". This only needs to be done once. Matlab will remember this setting when you restart Matlab.

Next, suppose you want to run eg3p3.m (Figure 3.8). Press the "EDITOR" button to show the drop down menu. Then press the "OPEN" button and then the second "Open" button. This will produce a new window containing all of the Matlab programs in the book. Scroll through the names until you come to eg3p3. Highlight this line and press "Open". The program eg3p3.m will now appear in the "Editor" window. The program may now be run by pressing the "EDITOR" button and pressing the "Run" button (green triangle icon). The output will appear in the "Command Window". Note that eg3p3.m references the functions checkdata, Alg1 and checkopt1. These will be loaded automatically because they are in the "MYQP" directory.

Next suppose we want to modify the data for eg3p3. It is good practice to make a copy of eg3p3.m. The original version of eg3p3.m thus remains intact and a copy of it can be made as follows. Press "EDITOR", then "New" then "Script". This will open a new empty window entitled "Untitledj" where j is some number. Press the tab eg3p3.m which will open the file eg3p3.m, highlight the contents (press Ctrl plus a), press Ctrl plus Ins, then open "Untitledj" and press Shift plus Ins. A copy of eg3p3.m will then appear in the window

"Untitledj". Suppose we want to see the effect of changing $c(2)$ from its present value of -8 to +8. In line 3 of the "Untitledj" window, just change the -8 to 8. Press the "EDITOR" button, choose "Save" then "Save As..." from the drop down menu. Enter some name like eg3p3mod.m. Now the modified example can be run by choosing "Editor", then "Run" as above.

Another useful thing is to show the intermediate calculations of an algorithm. Suppose we wish to see the intermediate values of "x" in the iterations of Algorithm 1. Open Alg1.m and observe that "x" is updated on line 49. Remove the semi colon (;) from the end of that line, save the modified Alg1, open an example for Alg1.m and run it. Every time line 49 is executed, the new value of "x" will be printed. This can be done for any variable in Alg3.m.

MATLAB is a registered trademark of The Math Works, Inc. For product information, please contact:

The MathWorks, Inc.
3 Apple Hill Drive
Natick, MA 01760-2098 USA
Tel: 5086477000
Fax: 508-647-7001
E-mail: info@mathworks.com
Web: www.mathworks.com

Chapter 1

Geometrical Examples

The purpose of this chapter is to introduce the most important properties of quadratic programming by means of geometrical examples. In Section 1.1 we show how to determine an optimal solution geometrically. A major difference between linear and quadratic programming problems is the number of constraints active at an optimal solution and this is illustrated by means of several geometrical examples. Optimality conditions are derived from geometrical considerations in Section 1.2. The geometry of quadratic functions is developed in terms of eigenvectors and eigenvalues in Section 1.3. Nonconvex quadratic programming problems are introduced geometrically in Section 1.4 where it is shown that such problems may possess many local minima.

1.1 Geometry of a QP: Examples

We begin our analysis of a quadratic programming problem by observing properties of an optimal solution in several examples.

Example 1.1

$$
\begin{aligned}
\text{minimize:} \quad & -4x_1 - 10x_2 + x_1^2 + x_2^2 \\
\text{subject to:} \quad -{} & x_1 + x_2 \le 2, & (1) \\
& x_1 + x_2 \le 6, & (2) \\
& x_1 \phantom{{}+ x_2} \le 5, & (3) \\
& \phantom{x_1 +{}} -x_2 \le 0, & (4) \\
-{} & x_1 \phantom{{}+ x_2} \le -1. & (5)
\end{aligned}
$$

The *objective function*[1] for this problem is $f(x) = -4x_1 - 10x_2 + x_1^2 + x_2^2$. The *feasible region*, denoted by R, is the set of points

$$x = \begin{bmatrix} x_1 \\ x_2 \end{bmatrix}$$

which simultaneously satisfy constraints (1) to (5). An *optimal solution* is a feasible point for which the objective function is smallest among those points in R. The feasible region has the same form as that for a linear programming problem. The difference between a quadratic and a linear programming problem is that the former has a quadratic objective function while the latter has a linear objective function. A linear programming problem is thus a special case of a quadratic programming problem in which the coefficients of all the quadratic terms in the objective function are zero.

Because this example has only two variables, we can obtain a graphical solution for it. The feasible region is shown in Figure 1.1. The behavior of the objective function can be determined by observing that the set of points for which $f(x) = $ a constant, is a circle centered at $(2, 5)'$[2] and that points inside the circle give a smaller value of f. Figure 1.1 shows the circle $f(x) = -25$. Each such circle (more generally, an ellipse) is called a *level set* of f.

Points inside this circle and in R, shown in the shaded region of Figure 1.1, give a better objective function value. Also shown is the analogous situation for $f(x) = -27$. Points in the heavily shaded area are more attractive than those on the circle segment $f(x) = -27$. Intuitively, we can obtain an optimal solution by decreasing the radius of the circle until it intersects R at a single point. Doing so gives the optimal solution $x_0 = (2, 4)'$ with $f(x_0) = -28$.
$\Diamond$

Note that the optimal solution for Example 1.1 is an *extreme point* (geometrically a corner of the feasible region) just as in linear programming. This is not always the case, however, as we show in the following two examples.

Example 1.2

$$
\begin{array}{llrcrcl}
\text{minimize:} & & -10x_1 & - & 6x_2 & + & x_1^2 + x_2^2 \\
\text{subject to:} & & -x_1 & + & x_2 & \leq & 2, \quad (1) \\
& & x_1 & + & x_2 & \leq & 6, \quad (2) \\
& & x_1 & & & \leq & 5, \quad (3) \\
& & & - & x_2 & \leq & 0, \quad (4)
\end{array}
$$

---

[1]The *italicized* terms are introduced informally in this chapter. Precise definitions are given in subsequent chapters.

[2]Prime ($'$) denotes transposition. All nonprimed vectors are column vectors.

$$- \quad x_1 \qquad \leq -1. \quad (5)$$

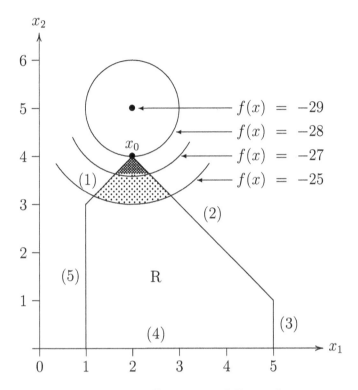

Figure 1.1. Geometry of Example 1.1.

The feasible region for this example is identical to that of Example 1.1. The linear terms in the new objective function $f(x) = -10x_1 - 6x_2 + x_1^2 + x_2^2$ have been changed so that the level sets are concentric circles centered at $(5, 3)'$. Figure 1.2 shows the feasible region and level set $f(x) = -29$. Points inside this level set and in R give a better objective function value. "Shrinking" the level sets as in Example 1.1 gives the optimal solution $x_0 = (4, 2)'$. $\Diamond$

Note that at the optimal solution for Example 1.2, the active constraint is tangent to the level set of f. Also observe that x_0 is **not** an extreme point: only one constraint [namely constraint (2)] is active at x_0. It is also possible for no constraints to be active at the optimal solution as shown in our third example.

Example 1.3

$$\begin{aligned} \text{minimize:} \quad & - 6x_1 - 4x_2 + x_1^2 + x_2^2 \\ \text{subject to:} \quad & - x_1 + x_2 \leq 2, \quad (1) \end{aligned}$$

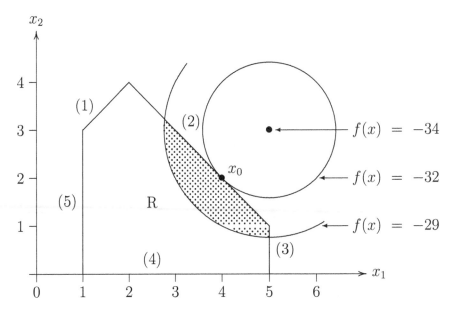

Figure 1.2.   Geometry of Example 1.2.

$$\begin{aligned}
x_1 + x_2 &\leq 6, & (2) \\
x_1 &\leq 5, & (3) \\
- x_2 &\leq 0, & (4) \\
- x_1 &\leq -1. & (5)
\end{aligned}$$

The constraints are identical to those of the previous two examples. The level sets of the new objective function are circles centered at $(3 , 2)'$. Figure 1.3 shows the level set for $f(x) = -12$. Since the center of this level set lies within R, it follows that the optimal solution x_0 is precisely this point. $\Diamond$

Based on Examples 1.1, 1.2, and 1.3, we can make

Observation 1.

Any number of constraints can be active at the optimal solution of a quadratic programming problem.

It is this observation that distinguishes a quadratic programming problem from a linear one. Although the algorithms for the solution of a quadratic programming problem are quite similar to those for linear programming problems, the principal difference is the method of accounting for a varying number of active constraints.

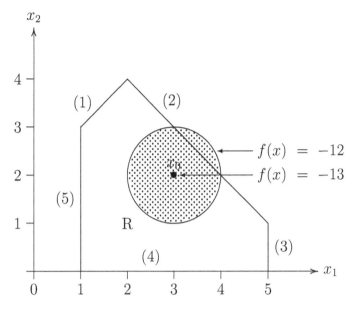

Figure 1.3. Geometry of Example 1.3.

1.2 Optimality Conditions

An optimal solution for a linear programming problem is characterized by the condition that the negative gradient of the objective function lies in the cone spanned by the gradients of those constraints active at it. Under a simple assumption, the same characterization holds for a quadratic programming problem. Recalling that the *gradient* of a function is just the vector of first partial derivatives, the gradient of the objective functions for the previous three examples is

$$g(x) \equiv \left[\begin{array}{c} \dfrac{\partial f(x)}{\partial x_1} \\ \dfrac{\partial f(x)}{\partial x_2} \end{array} \right].$$

Geometrically, the gradient of a function at a point x_0 is orthogonal to the tangent of the level set at x_0.

Figure 1.4 shows the negative gradient of the objective function for Example 1.1 at several points;[3] x_0, x_1, x_2, and x_3.

---

[3]There is a possibility of confusion with this notation. Here x_0, x_1, x_2, and x_3 refer to points in the plane and are thus 2-vectors. However, we have also used scalars x_1 and x_2 to denote the coordinates of a general point in the plane. We could avoid this difficulty by utilizing additional subscripts. We prefer not to do so. It is only necessary in this chapter to refer to the specific coordinates of a point. The reader should infer from context whether a scalar or vector is being discussed.

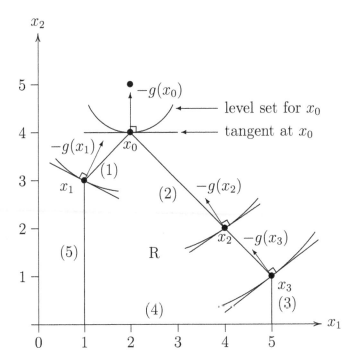

Figure 1.4.   Gradient vectors for objective function of Example 1.1.

The negative gradient is drawn at x_0, for example, as follows. The level set at x_0 is first drawn. This is the circle centered at $(2 , 5)'$ which passes through x_0. Next, the tangent to this level set is drawn at x_0. Since $-g(x_0)$ is orthogonal to this tangent, it must point either straight up or straight down. In general, the gradient points in the direction of maximum local increase so $-g(x_0)$ must point in the direction of maximum local decrease; that is, towards the center of the circle. The remaining negative gradients are constructed using a similar procedure.

Let a_i, $i = 1, \ldots, 5$, denote the gradient of the ith constraint function for Example 1.1. Then

$$a_1 = \begin{bmatrix} -1 \\ 1 \end{bmatrix}, \quad a_2 = \begin{bmatrix} 1 \\ 1 \end{bmatrix}, \ldots, \quad a_5 = \begin{bmatrix} -1 \\ 0 \end{bmatrix}.$$

The a_i's can be drawn using the same procedure as for $g(x)$. Because the constraint functions are linear, their gradients are constant whereas the objective function is quadratic and therefore its gradient is linear. The gradients of the constraints are orthogonal to the boundaries of R and point out and away from R.

Figure 1.5 shows $-g(x)$ and the cone spanned by the gradients of the constraints active at x for each of the points $x = x_0, x_1, x_2,$ and x_3. From this, we observe that at the optimal solution x_0, $-g(x_0)$ does indeed lie in the

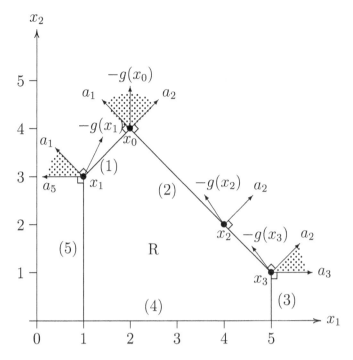

Figure 1.5. Optimality conditions for Example 1.1.

cone spanned by the gradients of the active constraints a_1 and a_2. Furthermore, that condition is **not** satisfied at any of the nonoptimal points x_1, x_2, and x_3. At x_1, for example, there are two active constraints, namely (1) and (5), and $-g(x_1)$ lies outside the cone spanned by a_1 and a_5. At x_2, only one constraint is active [constraint (2)] and the cone spanned by a_2 is the set of vectors pointing in the same direction as a_2. Since $-g(x_2)$ is not parallel to a_2, it is not in this cone.

Observation 2.

A point is optimal if and only if it is feasible and the negative gradient of the objective function at the point lies in the cone spanned by the gradients of the active constraints.[4]

We have given a geometrical verification that for Example 1.1 x_0 satisfies the optimality conditions formulated in Observation 2. We next do the same thing algebraically. For this example we have

$$g(x) = \begin{bmatrix} -4 + 2x_1 \\ -10 + 2x_2 \end{bmatrix},$$

---

[4]We will show (Theorem 4.3) that this statement is valid generally provided the objective function is a convex function (see Section 1.4).

so that $g(x_0) = (0, -2)'$. The cone part of the optimality conditions requires that there are nonnegative numbers u_1 and u_2 satisfying

$$-g(x_0) = u_1 a_1 + u_2 a_2.$$

These are precisely the 2 linear equations in 2 unknowns

$$- u_1 + u_2 = 0,$$
$$u_1 + u_2 = 2.$$

These have solution $u_1 = u_2 = 1$. Since both u_1 and u_2 are nonnegative, we have algebraically verified that the optimality conditions are satisfied.

The optimality conditions for Example 1.2 can be analyzed in a similar manner. Recall from Figure 1.2 that the optimal solution is $x_0 = (4, 2)'$ and that only constraint (2) is active at x_0. The cone spanned by the gradient of this single constraint is precisely the set of positive multipliers of a_2. From Figure 1.6, it is clear that $-g(x_0)$ points in the same direction as a_2 so that the optimality conditions are indeed satisfied at x_0.

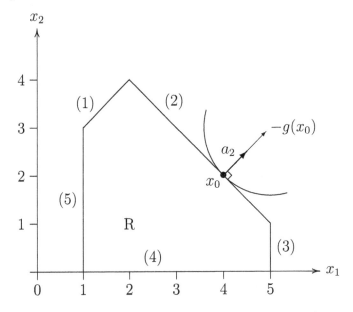

Figure 1.6.   Optimality conditions for Example 1.2.

To provide algebraic verification of the optimality conditions for Example 1.2, we observe that

$$g(x) = \begin{bmatrix} -10 + 2x_1 \\ -6 + 2x_2 \end{bmatrix}$$

so that $g(x_0) = (-2, -2)'$. The cone part of the optimality conditions requires that there is a nonnegative number u_2 with

$$-g(x_0) = u_2 a_2.$$

These are precisely the 2 linear equations in 1 unknown

$$u_2 = 2,$$
$$u_2 = 2,$$

which have the obvious solution $u_2 = 2$. Since u_2 is nonnegative, we have algebraically verified that the optimality conditions are satisfied.

Finally, the optimal solution $x_0 = (3, 2)'$ for Example 1.3 lies in the interior of R. Since no constraints are active at x_0, the cone part of the optimality conditions requires that $-g(x_0) = 0$. Since

$$g(x) = \begin{bmatrix} -6 + 2x_1 \\ -4 + 2x_2 \end{bmatrix},$$

it follows that $-g(x_0) = 0$, and we have algebraically verified that the optimality conditions are satisfied.

These three examples are the basis for Observations 1 and 2. The number of active constraints at the optimal solution for Examples 1.1, 1.2, and 1.3 is 2, 1, and 0, respectively. In each example, the optimal solution is characterized by the condition that it is feasible and that the gradient of the objective function lies in the cone spanned by the gradients of the active constraints.

1.3 Geometry of Quadratic Functions

The objective functions for Examples 1.1, 1.2, and 1.3 all have the two properties that the coefficients of x_1^2 and x_2^2 are identical and that there are no terms involving $x_1 x_2$. Consequently, the level sets for each objective function are circles which can be drawn by inspection. However, for the function

$$f(x) = -67x_1 - 59x_2 + \tfrac{13}{2}x_1^2 + 5x_1 x_2 + \tfrac{13}{2}x_2^2,$$

it is not immediately obvious how the level sets should be drawn. The purpose of this section is to show a method by which the level sets can easily be drawn. The relevant mathematical tools are eigenvectors and eigenvalues. The reader unfamiliar with these concepts may wish to consult an appropriate reference ([13], for example). However, it is our intention that this section be reasonably self-contained.

Consider a general quadratic function of two variables

$$f(x) = c'x + \tfrac{1}{2}x'Cx, \tag{1.1}$$

where x and c are 2-vectors and C is a $(2, 2)$ symmetric matrix. Let S denote the $(2, 2)$ matrix whose columns are the eigenvectors of C and let D

denote the $(2, 2)$ diagonal matrix of corresponding eigenvalues. The defining property of S and D is

$$CS = SD .\qquad(1.2)$$

An elementary property of eigenvectors is that they are orthogonal. Provided that they are also scaled to have unit norm,

$$S'S = I ,\qquad(1.3)$$

where I denotes the $(2, 2)$ identity matrix. Multiplying (1.2) on the left by S' and using (1.3) gives

$$S'CS = D .\qquad(1.4)$$

Let x_0 be the point where $f(x)$ is minimized. Since this is an unconstrained quadratic minimization problem, the optimality conditions of the previous section imply that the gradient of f at x_0 must be equal to zero. Let $g(x)$ denote the gradient of f. By writing $f(x)$ explicitly in terms of the components of x, it is easy to see that

$$g(x) = c + Cx .$$

Because $g(x_0) = 0$, x_0 is the solution of the linear equations

$$Cx_0 = -c .\qquad(1.5)$$

We next introduce a change of variable y related to x by

$$x = Sy + x_0 .\qquad(1.6)$$

Substituting this expression for x into (1.1) gives the identity

$$f(x) = f(x_0) + g'(x_0)Sy + \tfrac{1}{2}y'S'CSy .\qquad(1.7)$$

We have chosen x_0 such that $g(x_0) = 0$. With this and (1.7) simplifies to

$$f(x) = f(x_0) + \tfrac{1}{2}y'Dy .\qquad(1.8)$$

We are now ready to make the key point. Because D is diagonal, (1.8) expresses f in y coordinates solely in terms of y_1^2 and y_2^2. In particular, (1.8) includes no linear terms in y and no cross product terms involving y_1y_2. Using (1.8), it is easy to draw level sets for f in y coordinates.

Summarizing, level sets for f may be drawn by utilizing the following steps.

1. Obtain S and D by solving the eigenvalue problem for C.
2. Solve $Cx_0 = -c$.
3. Draw the y_1, y_2 axes in the x_1, x_2 space by drawing the column vectors of S centered at x_0.
4. Draw the level sets of f in y_1, y_2 space using (1.8).

The steps of the procedure are illustrated in the following example.

Example 1.4
Draw the level sets for

$$f(x) = -67x_1 - 59x_2 + \tfrac{13}{2}x_1^2 + 5x_1x_2 + \tfrac{13}{2}x_2^2 .$$

Here

$$c = \begin{bmatrix} -67 \\ -59 \end{bmatrix} \text{ and } C = \begin{bmatrix} 13 & 5 \\ 5 & 13 \end{bmatrix} .$$

By definition, an eigenvector $s = (s_1 , s_2)'$ of C and its corresponding eigenvalue λ satisfy

$$Cs = \lambda s , \text{ or } (C - \lambda I)s = 0 .$$

This last is a system of 2 linear equations in 2 unknowns. Because the right-hand side is zero, a necessary condition for the system to have nontrivial solutions is $\det(C - \lambda I) = 0$; that is

$$(13 - \lambda)^2 - 25 = 0 .$$

This has roots $\lambda = 18$ and $\lambda = 8$ so

$$D = \begin{bmatrix} 18 & 0 \\ 0 & 8 \end{bmatrix} .$$

The eigenvector corresponding to $\lambda = 18$ is obtained by substituting into $(C - \lambda I)s = 0$. This gives $s_1 = s_2$ so that

$$s = s_2 \begin{bmatrix} 1 \\ 1 \end{bmatrix} .$$

Because (1.3) requires $s's = 1$, we choose $s_2 = 1/\sqrt{2}$ giving

$$s = \frac{1}{\sqrt{2}} \begin{bmatrix} 1 \\ 1 \end{bmatrix} .$$

Proceeding in a similar manner with $\lambda = 8$ gives the second eigenvector

$$s = \frac{1}{\sqrt{2}} \begin{bmatrix} -1 \\ 1 \end{bmatrix} .$$

Thus

$$S \; = \; \frac{1}{\sqrt{2}} \begin{bmatrix} 1 & -1 \\ 1 & 1 \end{bmatrix} ,$$

and this completes Step 1.

For Step 2, we solve the linear equations

$$\begin{bmatrix} 13 & 5 \\ 5 & 13 \end{bmatrix} x_0 \; = \; \begin{bmatrix} 67 \\ 59 \end{bmatrix} .$$

These have solution $x_0 \; = \; (4 , 3)'$.

For Step 3, the origin of the y coordinate system is at x_0 in the x coordinate system. The y_1 axis extends from x_0 in the direction $2^{-1/2}(1 , 1)'$ and the y_2 axis extends from x_0 in the direction $2^{-1/2}(-1 , 1)'$. This is shown in Figure 1.7.

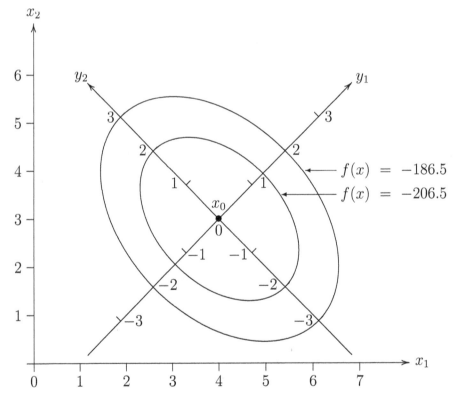

Figure 1.7.   Level sets for Example 1.4.

For Step 4, we observe that from (1.8) with $f(x_0) \; = \; -222.5$, the level sets of f in y coordinates are

$$f(x) \; = \; -222.5 + 9y_1^2 + 4y_2^2 .$$

The level set for $f(x) = -186.5$ is thus

$$\frac{y_1^2}{4} + \frac{y_2^2}{9} = 1 .$$

This is an ellipse with intercepts at $(\pm 2 , 0)'$ and $(0 , \pm 3)'$, and is shown in Figure 1.7. Similarly, the level set for $f(x) = -206.5$ is

$$\frac{y_1^2}{4} + \frac{y_2^2}{9} = \frac{4}{9} ,$$

which is "parallel" to the previous one (see Figure 1.7).   ◊

In Step 2, the solution of $Cx_0 = -c$ may not be uniquely determined as shown in

Example 1.5
Draw the level sets for

$$f(x) = -8x_1 - 8x_2 + x_1^2 + 2x_1x_2 + x_2^2 .$$

Here

$$c = \begin{bmatrix} -8 \\ -8 \end{bmatrix} \quad \text{and} \quad C = \begin{bmatrix} 2 & 2 \\ 2 & 2 \end{bmatrix} .$$

Using the same analysis as before, we find that the eigenvalues are 4 and 0 with corresponding eigenvectors

$$\frac{1}{\sqrt{2}} \begin{bmatrix} 1 \\ 1 \end{bmatrix} \quad \text{and} \quad \frac{1}{\sqrt{2}} \begin{bmatrix} -1 \\ 1 \end{bmatrix} ,$$

respectively. Thus

$$S = \frac{1}{\sqrt{2}} \begin{bmatrix} 1 & -1 \\ 1 & 1 \end{bmatrix} \quad \text{and} \quad D = \begin{bmatrix} 4 & 0 \\ 0 & 0 \end{bmatrix} .$$

The equations for x_0 are

$$\begin{bmatrix} 2 & 2 \\ 2 & 2 \end{bmatrix} x_0 = \begin{bmatrix} 8 \\ 8 \end{bmatrix}$$

and are satisfied by any point on the line $x_1 + x_2 = 4$. Any such point will do and, for concreteness, we choose $x_0 = (2 , 2)'$.
From (1.8) with $f(x_0) = -16$, the level sets of f in y coordinates are

$$f(x) = -16 + 2y_1^2 .$$

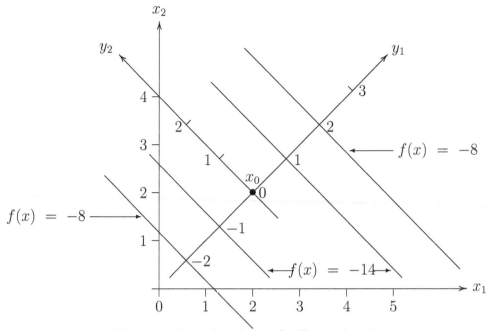

Figure 1.8.   Level sets for Example 1.5.

For $f(x) = -14$, the level set is simply $y_1^2 = 1$. This is just the pair of parallel lines $y_1 = 1$ and $y_1 = -1$ (see Figure 1.8). For $f(x) = -8$, the level set is the pair of parallel lines $y_1 = 2$ and $y_1 = -2$. $\Diamond$

We have seen examples where the equations $Cx_0 = -c$ have a unique solution (Example 1.4) and a multiplicity of solutions (Example 1.5). The final possibility is that of having no solutions and is illustrated in

Example 1.6

Draw the level sets for

$$f(x) = -10x_1 - 8x_2 + x_1^2 + 2x_1x_2 + x_2^2 .$$

Here

$$c = \begin{bmatrix} -10 \\ -8 \end{bmatrix} \quad \text{and} \quad C = \begin{bmatrix} 2 & 2 \\ 2 & 2 \end{bmatrix} .$$

Note that $f(x)$ differs from that of Example 1.5 only in that the coefficient of x_1 has been changed from -8 to -10. Since C is unchanged we have

$$S = \frac{1}{\sqrt{2}} \begin{bmatrix} 1 & -1 \\ 1 & 1 \end{bmatrix} \quad \text{and} \quad D = \begin{bmatrix} 4 & 0 \\ 0 & 0 \end{bmatrix} ,$$

as before. The equations for $Cx_0 = -c$ become

$$\begin{bmatrix} 2 & 2 \\ 2 & 2 \end{bmatrix} x_0 = \begin{bmatrix} 10 \\ 8 \end{bmatrix},$$

which are obviously inconsistent. In deriving (1.8) from (1.7), we chose x_0 such that $g(x_0) = 0$. The reason for doing this was to make the linear term in (1.7), namely $g'(x_0)Sy$, vanish. Although we cannot make the coefficients of both y_1 and y_2 vanish, we can make one vanish. Since

$$[\, g'(x_0)S\,]' = \frac{1}{\sqrt{2}} \begin{bmatrix} (4\,,\,4)\, x_0 - 18 \\ 2 \end{bmatrix},$$

the coefficient of y_1 vanishes for any x_0 satisfying

$$(4\,,\,4)\, x_0 = 18\,.$$

In particular, $x_0 = (2\,,\,5/2)'$ satisfies the above.

Since $f(x_0) = -79/4$, (1.7) gives

$$f(x) = \tfrac{-79}{4} + \sqrt{2}y_2 + 2y_1^2\,.$$

In y coordinates, the level set for $f(x) = -79/4$ is

$$y_2 = -\sqrt{2}y_1^2\,,$$

which is a parabola. Similarly, using $x_0 = (7/2\,,\,1)'$ the level set for $f(x) = -91/4$ is

$$y_2 = -3/\sqrt{2} - \sqrt{2}y_1^2\,.$$

These two level sets are shown in Figure 1.9. ◇

1.4 Nonconvex QP's

In this section, we distinguish between convex and nonconvex quadratic programming problems and discuss some of the difficulties associated with the latter problem.

We refine the definition of an optimal solution given in Section 1.1 as follows. For the model problem min $\{\, f(x) \mid x \in R \,\}$, we say that x_0 is a *global optimal solution* if $x_0 \in R$ and $f(x_0) \leq f(x)$ for all $x \in R$. In addition, x_0 is a *local optimal solution* if $x_0 \in R$ and $f(x_0) \leq f(x)$ for all $x \in R$ which are also in some neighborhood of x_0. In each of Examples 1.1, 1.2, and 1.3, the optimal solution is also the global optimal solution. This

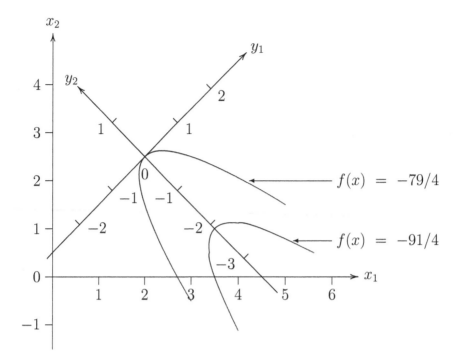

Figure 1.9.   Level sets for Example 1.6.

is a consequence of the objective function being *convex*, a property we shall make precise in Chapter 2. Functions which are *nonconvex* may result in local optimal solutions which are different than the global optimal solution. We illustrate this in

Example 1.7

$$
\begin{array}{llrll}
\text{minimize:} & 8x_1 & + & 6x_2 & - & x_1^2 & - & x_2^2 & \\
\text{subject to:} & x_1 & & & \le & 5, & & (1) \\
& & & x_2 & \le & 4, & & (2) \\
& - & x_1 & & \le & -1, & & (3) \\
& & - & x_2 & \le & -1. & & (4)
\end{array}
$$

We can analyze this problem using the graphical technique of Section 1.1. First observe that from Figure 1.10, the level sets of the objective function are circles centered at $(4 , 3)'$. However, the **largest** value of f occurs at the center. This is in contrast to the objective functions of Examples 1.1, 1.2, and 1.3 where the **smallest** value of the objective function occurs at the center of the circular level sets.

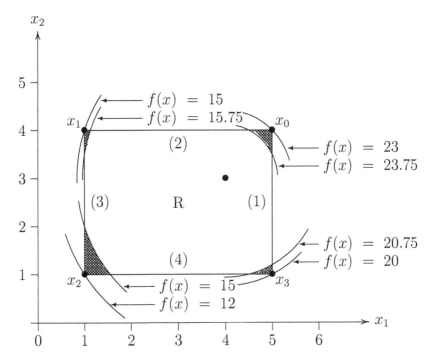

Figure 1.10.   Local optimal solutions for Example 1.7.

Consider the situation at x_0 (Figure 1.10). The shaded area adjacent to x_0 is part of R. Furthermore, for all of x in this neighborhood, $23.75 \geq f(x) \geq 23 = f(x_0)$. Consequently, x_0 is a local optimal solution for this problem. Similarly, for all x in the shaded region adjacent to x_1, $15.75 \geq f(x) \geq 15 = f(x_1)$. Thus x_1 is also a local optimal solution. A similar argument shows that both x_2 and x_3 are local optimal solutions. Each local optimal solution has a different objective function value, and of the four, x_2 gives the smallest objective function value. The global optimal solution is thus x_2.

Figure 1.11 shows graphically that the optimality conditions of Section 1.2 are satisfied at each of $x_0, \ldots, x_3$. Furthermore, they are also satisfied at each of $x_4, \ldots, x_7$ although none of these last four points is a local optimal solution. ◊

It is possible to formulate an algorithm which will find a local optimal solution for a nonconvex quadratic programming problem. Indeed, we shall do so in Chapter 9. Such an algorithm must have the ability to determine that points like $x_4, \ldots, x_7$, which do satisfy the optimality conditions, are not local optimal solutions.

The problem of determining a global optimal solution for a nonconvex quadratic programming problem is extremely difficult. The essential problem

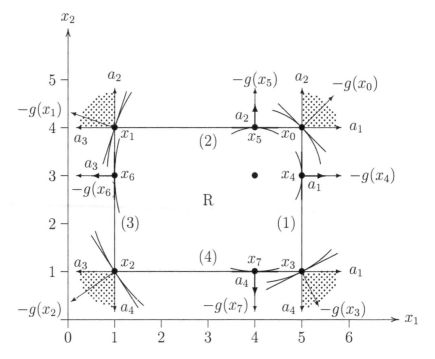

Figure 1.11.   Optimality conditions for Example 1.7.

is to be able to correctly deduce whether a local optimal solution is a global optimal solution, and if not, to find a point with smaller objective function value.

If the objective function for a quadratic programming problem is convex, then every point which satisfies the optimality conditions is a global optimal solution (Theorem 4.3) and the problem is much simpler. Throughout this text, we shall restrict ourselves primarily to convex problems.

1.5   Exercises

1.1   (a) Solve the following problem by drawing a graph of the feasible region and level sets for the objective function.

$$
\begin{aligned}
\text{minimize:} \quad & -6x_1 - 4x_2 + x_1^2 + x_2^2 \\
\text{subject to:} \quad & x_1 + x_2 \le 11, \quad (1) \\
& -x_1 - x_2 \le -7, \quad (2) \\
& x_1 - x_2 \le 7, \quad (3) \\
& x_1 - x_2 \le 3. \quad (4)
\end{aligned}
$$

(b) Repeat part (a) with the objective function replaced by

$$-18x_1 - 8x_2 + x_1^2 + x_2^2 \ .$$

(c) Repeat part (a) with the objective function replaced by

$$-14x_1 - 4x_2 + x_1^2 + x_2^2 \ .$$

1.2 Verify geometrically and algebraically that the optimality conditions are satisfied at each of the optimal solutions for the problems of Exercise 1.1.

1.3 Use the optimality conditions to solve the problem

$$\text{minimize}: \quad 2x_1^2 + 3x_2^2 + 4x_3^2$$
$$\text{subject to}: \quad 2x_1 + 3x_2 + 4x_3 = 9.$$

1.4 Let $f(x) = -196x_1 - 103x_2 + \frac{1}{2}(37x_1^2 + 13x_2^2 + 32x_1x_2)$.

(a) Use the eigenvalues and eigenvectors of C to sketch the level sets of f.

(b) Give a graphical solution to the problem of minimizing $f(x)$ subject to

$$
\begin{aligned}
- \ x_1 + x_2 &\leq 0, & (1) \\
x_1 \qquad &\leq 3, & (2) \\
- \ x_2 &\leq 0. & (3)
\end{aligned}
$$

(c) Show both geometrically and algebraically that the optimality conditions are satisfied at the point obtained in part (b).

1.5 Consider the quadratic programming problem

$$
\begin{aligned}
\text{minimize}: \quad - \ 8x_1 - 8x_2 &+ x_1^2 + x_2^2 + 2x_1x_2 \\
\text{subject to}: \quad 2x_1 + 2x_2 &\leq 5, & (1) \\
- \ x_1 \qquad &\leq 0, & (2) \\
- \ x_2 &\leq 0. & (3)
\end{aligned}
$$

(a) Determine **all** optimal solutions by graphing the feasible region and drawing level sets for the objective function.

(b) Verify geometrically and algebraically that the optimality conditions are satisfied at each optimal solution.

1.6 Consider the quadratic programming problem

$$\text{minimize:} \quad -10x_1 - 8x_2 + x_1^2 + x_2^2 + 2x_1x_2$$

$$\text{subject to:} \quad \begin{aligned} -x_2 &\leq 0, & (1) \\ -x_1 \qquad\quad &\leq -1, & (2) \\ x_2 &\leq 2, & (3) \\ 2x_1 - 2x_2 &\leq 5. & (4) \end{aligned}$$

(a) Determine the optimal solution x_0 by graphing the feasible region and drawing level sets for the objective function.

(b) Verify geometrically and algebraically that the optimality conditions are satisfied at x_0.

1.7 Consider the quadratic programming problem

$$\text{minimize:} \quad x_1x_2$$

$$\text{subject to:} \quad \begin{aligned} x_1 - x_2 &\leq 2, & (1) \\ -x_1 + x_2 &\leq 1, & (2) \\ x_1 + x_2 &\leq 1, & (3) \\ -x_1 - x_2 &\leq 2. & (4) \end{aligned}$$

Determine all local minima and the global minimum by graphing the feasible region and sketching level sets for the objective function. Verify algebraically that the optimality conditions are satisfied at each local minimum. Does $x_0 = (0, 0)'$ satisfy the optimality conditions? Is x_0 a local minimum?

Chapter 2

Portfolio Optimization

Portfolio optimization is the study of finding an optimal portfolio of risky assets. The subject was pioneered by H. M. Markowitz in [11]. Later on, W. F. Sharpe [14] made significant theoretical contributions to the subject. In 1990, Markowitz shared the Nobel prize for Financial Economics with W. F. Sharpe and M. H. Miller. In this chapter we will develop some of this theory based on Best [3]. The relevance here is that much of the theory can be formulated in terms of various types of QP's. The reader can thus see how QP's arise quite naturally.

QP's also arise in the Civil Engineering area of deformation analysis [12] and [2].

2.1   The Efficient Frontier

Consider a universe of n assets. Let μ_i denote the expected return of asset i and μ denote the $n-$vector of these expected returns: i.e.,

$$\mu = (\mu_1, \mu_2, \ldots, \mu_n)'.$$

Let σ_{ij} denote the covariance between assets i and j and let

$$\Sigma = [\, \sigma_{ij} \,].$$

μ is called the expected return vector for the n assets and Σ is called the covariance matrix for the n assets. Note that any covariance matrix is positive semi definite. Also note that although mathematicians usually reserve the symbol Σ for summation, finance people use Σ to denote a covariance matrix. Throughout this chapter we will follow the financial convention.

We need to decide what proportion of our wealth to invest in each asset. To that end, let x_i denote the proportion of wealth invested in asset i, $i = 1, 2, \ldots, n$ and let

$$x = (x_1, x_2, \ldots, x_n)'.$$

The expected return on the portfolio, μ_p, is given by

$$\mu_p = \mu_1 x_1 + \mu_2 x_2 + \cdots + \mu_n x_n = \mu' x.$$

The variance of the portfolio, σ_p^2 is given by

$$\sigma_p^2 = x' \Sigma x.$$

The components of x are proportions and so must add up to 1. This constraint is formulated as

$$x_1 + x_2 + \ldots + x_n = 1.$$

This latter is called the *budget constraint* and can be written more compactly as

$$l' x = 1,$$

where $l = (1, 1, \ldots, 1)'$ is an $n-$vector of ones.

There are three definitions of an efficient portfolio. They are all equivalent (for suitable choices of their defining parameters) and each gives a different insight into an efficient portfolio.

First we have the following. If an investor has his choice of two portfolios, both having the same expected return, but one has a large variance and the other has a small variance then he will choose the one with the small variance because for the same expected return that is the one with less risk. This leads to the following definition. A portfolio is variance-efficient if for fixed level of expected return, no other portfolio gives a smaller variance. The variance efficient portfolios are thus the optimal solutions to

$$\min\{x'\Sigma x \mid \mu'x = \mu_p, \ l'x = 1\}. \tag{2.1}$$

In the definition of variance efficient, the quantity μ_p acts like a parameter which can be chosen by the user. Each new value of μ_p produces a new variance efficient portfolio.

Note that (2.1) is a QP for fixed μ_p and a parametric QP if μ_p is allowed to vary.

Our second definition is somewhat like the reverse of the definition of variance efficient. Suppose an investor has a choice of two portfolios. Both portfolios have the same variance. However one has a large expected return while the other has a small expected return. Obviously, the portfolio with the higher expected return is more attractive since both have the same risk. This leads to the following definition. A portfolio is expected return-efficient if for fixed level of variance, no other portfolio gives a higher expected return. The expected return efficient portfolios are thus the optimal solutions to

$$\max\{\mu'x \mid x'\Sigma x = \sigma_p^2, \ l'x = 1\}. \tag{2.2}$$

In the definition of expected return efficient, the quantity σ_p^2 acts like a parameter which can be chosen by the user. Each new value of σ_p^2 produces a new expected return efficient portfolio. Note that (2.2) has a linear objective function, a single linear constraint and a single quadratic constraint. It does not fit into the category of an LP or QP.

There is a third formulation for an efficient portfolio which we now develop. An efficient (or optimal) portfolio should have μ_p large and σ_p^2 small. But these are conflicting goals. Let t be a nonnegative scalar which reflects how much weight we put on making μ_p large. Since making $t\mu_p$ large is equivalent to making $-t\mu_p$ small, the problem we need to solve is

$$\min\{ -t\mu'x + \frac{1}{2}x'\Sigma x \mid l'x = 1\}. \tag{2.3}$$

An optimal solutions for (2.3) is called a parameter-efficient portfolio.

When t is small, the σ_p^2 term dominates and (2.3) focusses on minimizing the variance at the expense of maximizing the expected return. When $t = 0$, (2.3) finds the (global) minimum variance portfolio. When t is large, (2.3) tends to maximize the expected return at the expense of keeping the variance small.

For fixed t, (2.3) is a special case of (3.50) which minimizes a quadratic function subject to a number of linear equality constraints. When t is considered as a parameter and allowed to vary from 0 to $+\infty$, (2.3) becomes a parametric QP and is a special case (7.16).

Each of (2.1), (2.2) and (2.3) generates a family of optimal solutions as their respective parameters (μ_p, σ_p^2 and t) are allowed to vary. An important result is that the efficient portfolios for each of these three problems are identical for suitable values of their respective parameters provided μ is not a multiple of l[1]. Thus minimum variance-efficient, maximum expected-return efficient and parameter-efficient are all identical and we refer to them all as simply efficient portfolios.

Because (2.3) has such a simple structure, we can obtain an explicit optimal solution for it. According to Theorem 3.11, necessary and sufficient conditions for x_0 to be optimal for (2.3) are

$$l'x = 1, \tag{2.4}$$
$$t\mu - \Sigma x = ul. \tag{2.5}$$

The first equation, (2.4), is simply the budget constraint. The left hand-side of (2.5) is the negative gradient of the objective function for (2.3). The scalar u is the multiplier for the budget constraint.

---

[1]If μ is a multiple of l, then $\mu = \theta l$ for some scalar θ and this implies that all of the expected returns are identical.

We make the assumption that Σ is positive definite. From Exercise 3.9(a), this implies Σ is nonsingular. Multiplying both sides of (2.5) on the left by Σ^{-1} and rearranging gives

$$x = -u\Sigma^{-1}l + t\,\Sigma^{-1}\mu \tag{2.6}$$

Substituting this last expression into the budget constraint (2.4) gives

$$1 = l'x = -ul'\Sigma^{-1}l + t\,l'\Sigma^{-1}\mu$$

This last is a single linear equation which can be solved for the multiplier u. Doing so gives

$$u = \frac{-1}{l'\Sigma^{-1}l} + \frac{l'\Sigma^{-1}\mu}{l'\Sigma^{-1}l}\,t\,. \tag{2.7}$$

Note that because we have assumed Σ is positive definite, Σ is also non-singular (Exercise 3.9(a)). Furthermore, $l \neq 0$ and by Exercise 3.9(b), Σ^{-1} is positive definite so $l'\Sigma^{-1}l > 0$. Thus, division by $l'\Sigma^{-1}l$ does not cause problems. Substituting (2.7) into (2.6) gives an explicit formula for the optimal solution for (2.3) as follows:

$$x \equiv x(t) = \frac{\Sigma^{-1}l}{l'\Sigma^{-1}l} + t(\Sigma^{-1}\mu - \frac{l'\Sigma^{-1}\mu}{l'\Sigma^{-1}l}\Sigma^{-1}l). \tag{2.8}$$

Notice that in (2.8), $l'\Sigma^{-1}l$ and $l'\Sigma^{-1}\mu$ are both scalars and both $\Sigma^{-1}l$ and $\Sigma^{-1}\mu$ are vectors. For ease of notation, let

$$h_0 = \frac{\Sigma^{-1}l}{l'\Sigma^{-1}l} \quad \text{and} \quad h_1 = \Sigma^{-1}\mu - \frac{l'\Sigma^{-1}\mu}{l'\Sigma^{-1}l}\Sigma^{-1}l. \tag{2.9}$$

The efficient portfolios can then be written

$$x(t) = h_0 + th_1. \tag{2.10}$$

We can use (2.10) to express the expected return and variance of the portfolio in terms of t as follows.

$$\mu_p = \mu'x(t) = \mu'h_0 + t\mu'h_1 \tag{2.11}$$

and

$$\begin{aligned}\sigma_p^2 &= (h_0 + th_1)'\Sigma(h_0 + th_1) \\ &= h_0'\Sigma h_0 + 2th_1'\Sigma h_0 + t^2 h_1'\Sigma h_1.\end{aligned} \tag{2.12}$$

In (2.12), we have used the fact that $h_1'\Sigma h_0 = h_0'\Sigma h_1$ because each of these last terms is a scalar, Σ is symmetric and the transpose of a scalar is just the scalar itself.

We define the constants

$$\alpha_0 = \mu' h_0 \quad \text{and} \quad \alpha_1 = \mu' h_1, \tag{2.13}$$

and

$$\beta_0 = h_0' \Sigma h_0, \quad \beta_1 = h_1' \Sigma h_0, \quad \text{and} \quad \beta_2 = h_1' \Sigma h_1. \tag{2.14}$$

With this notation, we have

$$\mu_p = \alpha_0 + \alpha_1 t \tag{2.15}$$
$$\sigma_p^2 = \beta_0 + 2\beta_1 t + \beta_2 t^2, \tag{2.16}$$

where α_0, α_1, β_0, β_1 and β_2 are just the constants defined by (2.13) and (2.14), respectively. There are two critical properties of these coefficients, namely

$$\alpha_1 = \beta_2, \tag{2.17}$$
$$\beta_1 = 0. \tag{2.18}$$

The proofs of these two properties are left as exercises (Exercises 2.4 and 2.5, respectively).

With (2.17) and (2.18) substituted into (2.15) and (2.16) we get

$$\mu_p = \alpha_0 + \beta_2 t \tag{2.19}$$
$$\sigma_p^2 = \beta_0 + \beta_2 t^2, \tag{2.20}$$

Solving each of these for t^2 and equating gives

$$[(\mu_p - \alpha_0)/\beta_2]^2 = [\sigma_p^2 - \beta_0]/\beta_2.$$

Cancelling a β_2 on each side and rearranging gives

$$\sigma_p^2 - \beta_0 = (\mu_p - \alpha_0)^2/\beta_2. \tag{2.21}$$

Equation (2.21) is one of the most important in financial theory. It shows the relationship between expected return (μ_p) and variance (σ_p^2) for efficient portfolios. This relationship is called the efficient frontier. If we regard σ_p^2 as a variable in itself then (2.21) defines a parabola in (σ_p^2, μ_p) space opening up to the right.

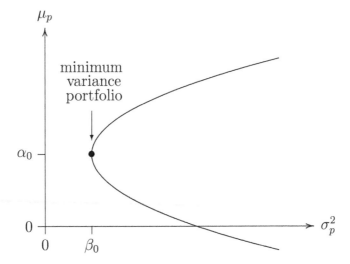

Figure 2.1.   The efficient frontier.

Figure 2.1 shows the graph of a typical efficient frontier. The "nose" of
the parabola at the point (β_0, α_0) corresponds to $t = 0$ in (2.19) and (2.20).
As t increases from 0, the investor moves up and along the efficient frontier
achieving a higher expected return at the expense of a riskier portfolio. The
dashed lower half of the efficient frontier corresponds to $t \leq 0$ and is called
the *inefficient frontier*.

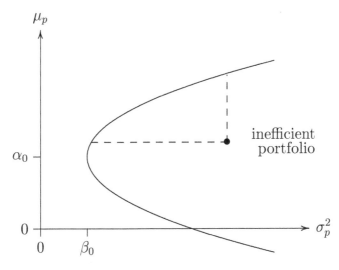

Figure 2.2.   Inefficient portfolio.

The heavily dotted point in Figure 2.2 labeled "Inefficient Portfolio" is
indeed inefficient. To see this, imagine moving this portfolio straight up. Any
such point below or on the efficient frontier will give a portfolio with higher
expected return, but the same variance. Thus this point violates the definition
of an expected return-efficient portfolio. Furthermore, moving this portfolio to

the left but to the left of the efficient frontier gives a portfolio with the same expected return but with a reduced variance, thus violating the definition of a variance-efficient portfolio.

One might think that solving (2.1), (2.2) or (2.3) will solve practical portfolio optimization problems. But practitioners need to add constraints to these models so that the resulting efficient portfolios satisfy known constraints. A common type of restriction is "no short sales" which means all of the components of x must be nonnegative. Short sales would mean the investor is allowed to borrow against some assets to purchase more of attractive assets and this is regulated. The version of (2.3) with no short sales allowed is

$$\min\{ -t\mu'x + \frac{1}{2}x'\Sigma x \mid l'x = 1, \ x \geq 0\}, \qquad (2.22)$$

where the no short sales restrictions are formulated as the n inequality constraints $x \geq 0$. These constraints may also be written $x_i \geq 0, \ i = 1, 2, \ldots, n$.

We consider a specific version of (2.22) in

Example 2.1

We consider the version of (2.22) with data given in Figure 2.6 of Section 2.3. The problem has $n = 20$ variables, $m = 20$ inequality constraints (the nonnegativity constraints) and $r = 1$ equality constraint (the budget constraint). All of these are defined in line 2 of the program eg2p1.m. Line 3 defines μ as a random vector and line 4 defines t. Line 5 computes $-t\mu$ and line 6 sets a program tolerance. Line 7 sets D^{-1} to the (n, n) identity matrix. Line 8 defines a column vector of n ones (gradient of budget constraint) which is then used to replace the last column of D' and the inverse of the modified matrix is computed using Procedure Phi (see Chapter 3). Lines 10 through 16 define the constraint matrix A and right hand-side b. Line 17 defines the starting point for the QP algorithm to be $x_i = 1/n, \ i = 1, 2, \ldots, n$.

The covariance matrix is set up in line 18 to be $diag(10^{-1})$. An index set required by the QP algorithm is defined in lines 19-20. The function checkdata is invoked in line 21 which checks that Sigma is indeed positive semidefinite. Line 22 invokes the function checkconstraints which verifies that the given initial feasible point does indeed satisfy the constraints. Line 23 which is continued to line 24, invokes Algorithm 4 (Section 5.3) which solves the specified QP. Finally, in line 25, the routine checkKKT is called. This routine prints a summary of the optimization results, the optimal (primal) solution and the optimal dual solution (see Section 4.2).

The reader may be only able to follow a portion of the material in this section. We mean the material here to be an overview of things to come. After reading subsequent chapters, where everything is developed in detail, it may be useful to reread this chapter. Our intent in presenting this example here is to formulate an example of a QP having more variables (here $n = 20$) than the 2-dimensional examples of Chapter 1.

To that end, Figure 2.7 shows the optimal solution to this (primal) problem. It also shows the optimal solution for the dual problem (see Section 4.2). Note that for a problem of the form (2.22), at an optimal solution $\hat{x}$, any component $\hat{x}_i$ must satisfy either $\hat{x}_i = 0$ (i-th nonnegativity constraint is *active* at $\hat{x}_i$) or $\hat{x}_i > 0$ (i-th nonnegativity constraint is *inactive* at $\hat{x}_i$). Any QP algorithm which solves (2.22) must be able to identify those nonnegativity constraints which are active at the optimal solution. Each such constraint may be either active or inactive and so there are $2^n = 2^{20}$ possibilities. Fortunately, QP algorithms are generally able to solve problems in a low order polynomial of n number of iterations, where n is the number of variables. $\Diamond$

Most modern portfolio analysts insist on provision for transactions costs. These are usually incurred relative to some target portfolio, say x_0. This may be the presently held portfolio or some industry standard such as the S&P 500 in the US. The idea is that the portfolio model should include costs for buying or selling assets away from the current holdings. This will have two desirable effects. First, if the transaction cost is too high, the transaction will not take place. Second, very small but mathematically correct transactions will be precluded. This situation can be mathematically modeled as follows. A key idea is to treat separately what we purchase and what we sell. Let x^+ denote the vector of purchases and x^- denote the vector of sales. Then the holdings vector x may be represented as

$$x = x_0 + x^+ - x^-, \quad x^+ \geq 0, \quad x^- \geq 0.$$

Let p and q denote the vectors of purchase and sales transactions costs, respectively. Having separated out the sales and purchases, the total transactions cost is

$$p'x^+ + q'x^-.$$

It could be that the mathematics suggests quite large changes in a portfolio. In practice, most money managers will not accept such large changes. Changes can be controlled by introducing upper bounds on amount of assets purchased and sold. Let d and e denote upper bounds on the amount sold and purchased, respectively. Assembling these results, the analog of (2.3) is

$$\left.\begin{array}{ll} \text{minimize}: & -t(\mu'x - p'x^+ - q'x^-) + \frac{1}{2}x'\Sigma x \\[2ex] \text{subject to}: & l'x = 1, \\ & x - x^+ + x^- = x_0, \\ & 0 \le x^- \le d, \\ & 0 \le x^+ \le e. \end{array}\right\} \qquad (2.23)$$

Note that the Hessian matrix (see Section 3.1) for the objective function of (2.23) is

$$\begin{bmatrix} \Sigma & 0 & 0 \\ 0 & 0 & 0 \\ 0 & 0 & 0 \end{bmatrix}.$$

Even if Σ is positive definite, the above matrix will always be positive semidefinite and as such, any QP algorithm for its solution must account for a positive semidefinite Hessian.

In addition to the budget constraint in (2.23), one could include other linear equality constraints such as constraints on the portfolio's beta and sector constraints. In our experience, such constraints tend to be small in number.

The problem (2.23) is a quadratic programming problem (for fixed t). Methods to solve this are developed in Chapters 5, 6, 7 and 8. If (2.23) is to be solved for all t, then (2.23) is a parametric quadratic programming problem. Solution methods for it are developed in Chapters 7, 8 and 9.

2.2 The Capital Market Line

This section will deal with a variation of the model of the previous section. In addition to the n risky assets, we will now suppose there is an additional asset with special properties. This asset will be risk free and as such will have a zero variance and a zero covariance with the remaining n risky assets. A good example of a risk free asset is Treasury bills or T$-$bills.

Let x_{n+1} denote the proportion of wealth invested in the risk free asset and let r denote its return. The expected return of this portfolio is

$$\begin{aligned} \mu_p &= \mu_1 x_1 + \mu_2 x_2 + \cdots + \mu_n x_n + r x_{n+1} \\ &= (\mu', r) \begin{bmatrix} x \\ x_{n+1} \end{bmatrix}. \end{aligned}$$

Its variance is

$$\sigma_p^2 = x'\Sigma x = \begin{bmatrix} x \\ x_{n+1} \end{bmatrix}' \begin{bmatrix} \Sigma & 0 \\ 0' & 0 \end{bmatrix} \begin{bmatrix} x \\ x_{n+1} \end{bmatrix}.$$

The covariance matrix for this $n + 1$ dimensional problem is

$$\begin{bmatrix} \Sigma & 0 \\ 0' & 0 \end{bmatrix},$$

for which the last row and column contain all zeros corresponding to the risk free asset. This matrix is positive semidefinite, whereas Σ, the covariance matrix for the n risky assets, is assumed to be positive definite. The efficient portfolios are the optimal solutions for (2.3). For the case at hand, (2.3) becomes

$$\left.\begin{aligned} \text{minimize}: \quad & -t(\mu', r)\begin{bmatrix} x \\ x_{n+1} \end{bmatrix} + \tfrac{1}{2}\begin{bmatrix} x \\ x_{n+1} \end{bmatrix}'\begin{bmatrix} \Sigma & 0 \\ 0' & 0 \end{bmatrix}\begin{bmatrix} x \\ x_{n+1} \end{bmatrix} \\ \text{subject to}: \quad & l'x + x_{n+1} = 1 \end{aligned}\right\} . (2.24)$$

In partitioned matrix form, the optimality conditions for (2.24) are

$$t\begin{bmatrix} \mu \\ r \end{bmatrix} - \begin{bmatrix} \Sigma & 0 \\ 0' & 0 \end{bmatrix}\begin{bmatrix} x \\ x_{n+1} \end{bmatrix} = u\begin{bmatrix} l \\ 1 \end{bmatrix}, \tag{2.25}$$

and

$$l'x + x_{n+1} = 1. \tag{2.26}$$

The second partition of (2.25) gives the scalar multiplier u:

$$u = tr.$$

Using this, the first partition of (2.25) becomes

$$\Sigma x = t(\mu - rl). \tag{2.27}$$

Because Σ is positive definite, it is also nonsingular (see Exercise 3.9(a)). Thus, we can obtain the efficient risky assets as

$$x \equiv x(t) = t\Sigma^{-1}(\mu - rl) \tag{2.28}$$

and the efficient risk free asset

$$x_{n+1} \equiv x_{n+1}(t) = 1 - tl'\Sigma^{-1}(\mu - rl). \tag{2.29}$$

Using (2.28) and (2.29) we can calculate an efficient portfolio's variance and expected return as follows:

$$\sigma_p^2 = t^2(\mu - rl)'\Sigma^{-1}(\mu - rl), \tag{2.30}$$

$$\begin{aligned} \mu_p &= t\mu'\Sigma^{-1}(\mu - rl) + r - trl'\Sigma^{-1}(\mu - rl) \\ &= r + t[\mu'\Sigma^{-1}(\mu - rl) - rl'\Sigma^{-1}(\mu - rl)] \\ &= r + t[(\mu - rl)'\Sigma^{-1}(\mu - rl)]. \end{aligned} \qquad (2.31)$$

Note that since Σ is positive definite, so too is Σ^{-1} (see Exercise 3.9(b)) and thus

$$(\mu - rl)'\Sigma^{-1}(\mu - rl) > 0$$

provided $\mu \neq rl$. With this restriction, it follows from (2.30) and (2.31) that σ_p^2 and μ_p are both strictly increasing functions of t.

Observe from (2.29) that $x_{n+1}(t)$ is strictly decreasing if and only if $l'\Sigma^{-1}(\mu - rl) > 0$; i.e.,

$$r < \frac{l'\Sigma^{-1}\mu}{l'\Sigma^{-1}l}. \qquad (2.32)$$

Because Σ^{-1} is symmetric (Exercise 3.9(b)) and the transpose of a scalar is just the scalar,

$$l'\Sigma^{-1}\mu = (l'\Sigma^{-1}\mu)' = \mu'\Sigma^{-1}l$$

and (2.32) becomes

$$r < \mu'\left[\frac{\Sigma^{-1}l}{l'\Sigma^{-1}l}\right]. \qquad (2.33)$$

From (2.9) and (2.13), the quantity on the right−hand side of this inequality is α_0, the expected return of the minimum variance portfolio. Thus we have shown that the allocation in the risk free asset will be a strictly decreasing function of t if and only if the risk free rate is strictly less than the expected return of the minimum variance portfolio. We will assume that $r < \alpha_0$ throughout the remainder of this section.

Because $x_{n+1}(t)$ is a strictly decreasing function of t, as t is increased eventually $x_{n+1}(t)$ is reduced to zero. This occurs when

$$t = t_m \equiv \frac{1}{l'\Sigma^{-1}(\mu - rl)}. \qquad (2.34)$$

The corresponding portfolio of risky assets is

$$x(t_m) \equiv x_m = t_m\Sigma^{-1}(\mu - rl). \qquad (2.35)$$

x_m is called *the market portfolio* (hence the m subscript). It is that efficient portfolio for which there is zero holdings in the risk free asset. Using (2.28), (2.34) and (2.35) we can write the efficient risky assets as

$$x(t) = \frac{t}{t_m}x_m. \qquad (2.36)$$

Using (2.30) and (2.31) we can obtain the equation of the efficient frontier for this model:

$$\mu_p - r = \sigma_p[(\mu - rl)'\Sigma^{-1}(\mu - rl)]^{\frac{1}{2}}. \tag{2.37}$$

In mean–standard deviation space, the efficient frontier is a line. It is called the *Capital Market Line* (CML) and is illustrated in Figure 2.3. Investors move up and down the Capital Market Line according to their aversion to risk. According to (2.28) and (2.29), for $t = 0$ all wealth is invested in the risk free asset and none is invested in the risky assets. As t increases from 0, the amount invested in the risk free asset is reduced whereas the holdings in the risky assets increases. All investors must lie on the CML. From (2.36), the sole thing that distinguishes individual investors is the proportion of the market portfolio they hold.

The CML can be written in a perhaps more revealing way as follows. Recall from (2.28) that part of the optimality conditions for (2.24) are

$$x_m = t_m\Sigma^{-1}(\mu - rl). \tag{2.38}$$

Taking the inner product of both sides with $(\mu - rl)$ gives

$$(\mu - rl)'x_m = t_m(\mu - rl)'\Sigma^{-1}(\mu - rl). \tag{2.39}$$

At the market portfolio, the holdings in the risk free asset have been reduced to zero. Consequently,

$$l'x_m = 1 \quad \text{and} \quad \mu_m = \mu'x_m. \tag{2.40}$$

Using (2.40) in (2.39), taking square roots and rearranging gives

$$\left[\frac{\mu_m - r}{t_m}\right]^{\frac{1}{2}} = [(\mu - rl)'\Sigma^{-1}(\mu - rl)]^{\frac{1}{2}}. \tag{2.41}$$

Now, taking the inner product of both sides of (2.38) with Σx_m gives

$$x_m'\Sigma x_m = t_m(\mu - rl)'x_m = t_m(\mu_m - r),$$

so that

$$\sigma_m = [t_m(\mu_m - r)]^{\frac{1}{2}},$$

or,

$$t_m^{\frac{1}{2}} = \left[\frac{\sigma_m}{(\mu_m - r)^{\frac{1}{2}}}\right]. \tag{2.42}$$

Substitution of (2.42) into (2.41), rearranging and simplifying gives

$$[(\mu - rl)'\Sigma^{-1}(\mu - rl)]^{\frac{1}{2}} = \frac{(\mu_m - r)}{\sigma_m}. \tag{2.43}$$

Substituting the quantity in the left−hand side of (2.37) shows that the Capital Market Line can also be expressed as

$$\mu_p = r + \left[\frac{(\mu_m - r)}{\sigma_m}\right]\sigma_p. \tag{2.44}$$

Equation (2.44) shows that the slope of the CML is

$$\frac{(\mu_m - r)}{\sigma_m}.$$

This is in agreement with direct observation in Figure 2.3.

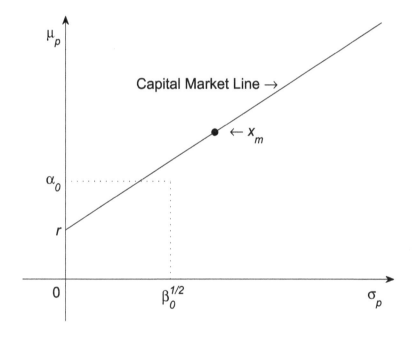

Figure 2.3.   Capital Market Line.

Consider the following implication of this model. Suppose in an economy that everyone is mean-variance efficient and everyone "sees" the same covariance matrix Σ and the same vector of expected returns μ. Let two typical investors be Mr. X and Ms Y and suppose their t values are t_x and t_y, respectively. Note that t_x and t_y could be obtained from the amounts Mr. X and

Ms. Y invest in the risk free asset by means of (2.29). Let Mr. X and Ms. Y's holdings in the risky assets be denoted by x and y, respectively. From (2.36),

$$x = \frac{t_x}{t_m}x_m, \quad \text{and,} \quad y = \frac{t_y}{t_m}x_m.$$

This implies that their holdings in risky assets are *proportional*; i.e., $x = (t_x/t_y)y$. For example, suppose $n = 4$ and Mr. X's risky holdings are $x = (0.1, 0.11, 0.12, 0.13)'$. Suppose we knew Ms. Y's holdings in just the first risky asset were 0.2. But then her holdings in the remaining risky assets must be in the same proportion (namely 2) as Mr. X's. Thus we would know the remainder of her holdings: $y = (0.2, 0.22, 0.24, 0.26)'$.

We next turn to the question of what happens when t exceeds t_m. By definition, if the Capital Market Line were to continue, holdings in the risk free asset would become negative. This would allow short selling the risk free asset which we will not allow in this model. For all $t \geq t_m$, we shall impose the condition $x_{n+1}(t) = 0$. Substituting $x_{n+1} = 0$ everywhere in (2.24), the problem (2.24) reduces to (2.3), the n risky asset problem of the previous section for $t \geq t_m$. Figure 2.4 shows the Capital Market Line and the efficient frontier for the n risky asset problem superimposed, in mean-standard deviation space.

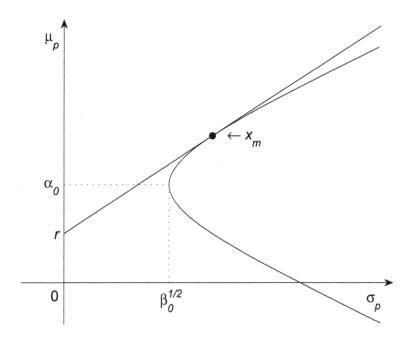

Figure 2.4.   Capital Market Line and efficient frontier for risky assets.

Implicit in the model problem (2.24) is the assumption that the risk free asset must be nonnegative. The key result for this problem is that the efficient frontier for it is composed of two pieces (see Figure 2.5). The first piece is the Capital Market Line going from r to the market portfolio x_m. The second piece is a part of the hyperbola (2.21), the efficient frontier for the risky assets. The two pieces of the efficient frontier are shown as the solid line in Figure 2.5.

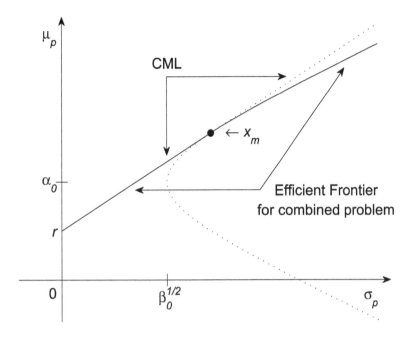

Figure 2.5. Two pieces of the Efficient Frontier.

Figure 2.5 is drawn in a special way, and not by accident. The mean and standard deviation at x_m on the CML is identical to that for the risky asset part of the frontier. This is always the case. Indeed, the two corresponding portfolios are both x_m. To see this, from (2.8) and (2.34) we can calculate

$$
\begin{aligned}
x(t_m) &= \frac{\Sigma^{-1}l}{l'\Sigma^{-1}l} + \frac{1}{l'\Sigma^{-1}(\mu - rl)}\left[\Sigma^{-1}\mu - \frac{l'\Sigma^{-1}\mu}{l'\Sigma^{-1}l}\Sigma^{-1}l\right] \\
&= \left[1 - \frac{l'\Sigma^{-1}\mu}{l'\Sigma^{-1}(\mu - rl)}\right]\frac{\Sigma^{-1}l}{l'\Sigma^{-1}l} + \frac{1}{l'\Sigma^{-1}(\mu - rl)}\Sigma^{-1}\mu \\
&= \frac{-rl'\Sigma^{-1}l}{l'\Sigma^{-1}(\mu - rl)}\frac{\Sigma^{-1}l}{l'\Sigma^{-1}l} + \frac{\Sigma^{-1}\mu}{l'\Sigma^{-1}(\mu - rl)} \\
&= \frac{\Sigma^{-1}(\mu - rl)}{l'\Sigma^{-1}(\mu - rl)} \\
&= x_m.
\end{aligned}
$$

Since the two portfolios are identical, their means are equal and so too are their standard deviations. Thus, the efficient frontier is continuous.

Not only is the efficient frontier continuous, it is also differentiable. One way to think of this is that in mean−standard deviation space, the CML is

tangent to the efficient frontier for the risky assets. This is shown in Figure
2.5. From (2.37), the slope of the CML is

$$((\mu - rl)'\Sigma^{-1}(\mu - rl))^{\frac{1}{2}}. \tag{2.45}$$

To verify the differentiability assertion, we need only show that the slope of
the efficient frontier for risky assets at x_m has this same value. A convenient
way to do this is to use (2.15) and (2.16), namely

$$\mu_p = \alpha_0 + \alpha_1 t \text{ and } \sigma_p = (\beta_0 + 2\beta_1 t + \beta_2 t^2)^{\frac{1}{2}}.$$

Using the chain rule of calculus, $\beta_1 = 0$ (2.18) and $\beta_2 = \alpha_1$ (2.17), we determine

$$
\begin{aligned}
\frac{d\mu_p}{d\sigma_p} &= \frac{d\mu_p}{dt}\frac{dt}{d\sigma_p} \\
&= \alpha_1 \left[\frac{2\beta_2 t}{2\sigma_p}\right]^{-1} \\
&= \frac{\sigma_p}{t}.
\end{aligned}
\tag{2.46}
$$

When $t = t_m$, (2.30) with (2.35) imply

$$\sigma_p = \left[\frac{(\mu - rl)'\Sigma^{-1}(\mu - rl)}{(l'\Sigma^{-1}(\mu - rl))^2}\right]^{\frac{1}{2}}, \tag{2.47}$$

so that with (2.34) we have

$$\frac{d\mu_p}{d\sigma_p} = ((\mu - rl)'\Sigma^{-1}(\mu - rl))^{\frac{1}{2}}, \tag{2.48}$$

and this is identical to (2.45). Thus the slope of the CML is equal to the slope
of the efficient frontier for risky assets at x_m and the total efficient frontier is
indeed differentiable.

Example 2.2
Find the equation of the CML using (2.37), t_m and the market portfolio x_m
for the problem with $n = 3$, $\mu = (1.1, 1.2, 1.3)'$, $\Sigma = \text{diag}(10^{-2}, 5 \times 10^{-2}, 7 \times 10^{-2})$ and a risk free return of $r = 1.02$. Verify the equation of the CML by
recalculating it using (2.44).

The required quantities all depend on $\Sigma^{-1}(\mu - rl)$ so we calculate this first:

$$\Sigma^{-1}(\mu - rl) = 100 \begin{bmatrix} 1 & 0 & 0 \\ 0 & \frac{1}{5} & 0 \\ 0 & 0 & \frac{1}{7} \end{bmatrix} \begin{bmatrix} 0.08 \\ 0.18 \\ 0.28 \end{bmatrix} = \begin{bmatrix} 8 \\ 3.6 \\ 4 \end{bmatrix}. \tag{2.49}$$

Thus

$$(\mu - rl)'\Sigma^{-1}(\mu - rl) = \begin{bmatrix} 0.08 & 0.18 & 0.28 \end{bmatrix} \begin{bmatrix} 8 \\ 3.6 \\ 4 \end{bmatrix} = 2.4080$$

and since $\sqrt{2.4080} = 1.5518$, it follows from (2.37) that the equation of the CML is

$$\mu_p = 1.02 + 1.5518\,\sigma_p. \tag{2.50}$$

From (2.34), we have

$$t_m = \frac{1}{l'\Sigma^{-1}(\mu - rl)}.$$

From (2.49),

$$l'\Sigma^{-1}(\mu - rl) = \begin{bmatrix} 1 & 1 & 1 \end{bmatrix} \begin{bmatrix} 8 \\ 3.6 \\ 4 \end{bmatrix} = 15.6,$$

so that

$$t_m = 0.0641.$$

Finally, we have from (2.35) and (2.49) that the market portfolio is

$$\begin{aligned} x_m &= t_m\Sigma^{-1}(\mu - rl) \\ &= 0.0641 \begin{bmatrix} 8 \\ 3.6 \\ 4 \end{bmatrix} \\ &= \begin{bmatrix} 0.5128 \\ 0.2308 \\ 0.2564 \end{bmatrix}. \end{aligned}$$

We next calculate the equation of the CML using (2.44) as follows. Using x_m above, we calculate

$$\mu_m = \mu'x_m = 1.1744, \quad \sigma_m^2 = x_m'\Sigma x_m = 0.0099,$$

and

$$\sigma_m = 0.0995, \quad \frac{\mu_m - r}{\sigma_m} = \frac{1.1744 - 1.02}{0.0995} = 1.5518.$$

From (2.44), the equation of the CML is

$$\mu_p = 1.02 + 1.5518\,\sigma_p,$$

in agreement with (2.50).   ◇

2.3   Computer Programs

In this section, we show the Matlab program which constructs the data for Example 2.1 and calls Alg4 to solve it (Figure 2.6). We also show the optimal solution and related quantities obtained by Algorithm 4 (Alg 4, Chapter 5) in Figure 2.7.

```
1   %eg2p1
2   n = 20 ; m = n; r = 1;
3   mu = rand(n,1) - 0.8;
4   t = 0.054;
5   mtmu = - t*mu;
6   tol = 1.e-6;
7   Dinv = eye(n);
8   d = zeros(n,1) + 1.;
9   [Dinv] = Phi(Dinv,d,n,n);
10  A = zeros(n+1,n);
11  for i=1:n
12      A(i,i) = -1.;
13      A(n+1,i) = 1.;
14  end
15  b = zeros(m+r,1);
16  b(m+r) = 1.;
17  x = zeros(n,1) + 1./n;
18  Sigma = eye(n)/10.e1;
19  J = zeros(1,n);
20  J(n) = m + r;
21  checkdata(Sigma);
22  checkconstraints(A,b,x,m,r,n);
23  [x,J,fx,Dinv,msg,dual,j,alg] = ...
24      Alg4(mtmu,Sigma,n,n,1,tol,x,J,Dinv,A,b);
25  checkKKT(A,b,x,n,m,r,dual,mtmu,Sigma,fx,j,msg,alg);
```

Figure 2.6.   QP Inequality Constraints: eg2p1.m.

```
1   eg2p1
2   Optimization Summary
3   Optimal solution obtained
4   Algorithm: Algorithm 4
5   Optimal objective value    -0.00776344
6   Iterations    19
7   Maximum Primal Error        0.00000000
8   Maximum Dual Error          0.00000000
9
10
11    Optimal Primal Solution
12       1    0.00000000
```

```
13    2    0.00000000
14    3    0.00000000
15    4    0.00000000
16    5    0.00000000
17    6    0.00000000
18    7    0.00000000
19    8    0.00000000
20    9    0.17611758
21   10    0.21597876
22   11    0.00000000
23   12    0.24678169
24   13    0.17428219
25   14    0.00000000
26   15    0.00000000
27   16    0.00000000
28   17    0.00000000
29   18    0.00000000
30   19    0.00000000
31   20    0.18683977
32
33
34   Optimal Dual Solution
35    1    0.00594911
36    2    0.00103143
37    3    0.04308691
38    4    0.00062190
39    5    0.01579679
40    6    0.04467701
41    7    0.03490529
42    8    0.02041259
43    9    0.00000000
44   10    0.00000000
45   11    0.04143309
46   12    0.00000000
47   13    0.00000000
48   14    0.02373391
49   15    0.00672905
50   16    0.04228233
51   17    0.02716908
52   18    0.00049447
53   19    0.00716500
54   20    0.00000000
55   21    0.00674419
```

Figure 2.7. Solution of Example 2.1.

2.4   Exercises

2.1 Consider a Portfolio Optimization problem with data $n = 3$, $\mu = (1.1, 1.15, 1.2)'$, and $\Sigma = \text{diag}(10^{-4}, 10^{-3}, 10^{-2})$. Find h_0, h_1, $x(t)$, α_0, α_1, β_0, β_1, and β_2, (as in Examples 2.1) and sketch the efficient frontier.

2.2 Consider a Portfolio Optimization problem with data $n = 2$, $\mu = (1.1, 1.2)'$, and $\Sigma = \text{diag}(10^{-2}, 10^{-1})$. Augment this problem with a risk free asset having expected return $r = 1.05$. Determine the efficient portfolios and the two pieces of the efficient frontier.

2.3 Show $\beta_2 = 0$ if and only if μ is a multiple of l.

2.4 Verify (2.17) ($\alpha_1 = \beta_2$).

2.5 Verify (2.18) ($\beta_1 = 0$).

2.6 Suppose $\Sigma = \text{diag}(\sigma_i)$. Show the efficient portfolios have components

$$x_i = [\theta_1 + t(\mu_i - \theta_2)]/\sigma_i, \quad i = 1, \ldots, n,$$

where

$$\theta_1 = 1/(\sigma_1^{-1} + \cdots + \sigma_n^{-1}) \quad \text{and} \quad \theta_2 = \theta_1(\mu_1/\sigma_1 + \cdots + \mu_n/\sigma_n).$$

Also, show the multiplier for the budget constraint is $u = -\theta_1 + t\theta_2$.

Chapter 3

QP Subject to Linear Equality Constraints

Our goal in this chapter is to develop necessary and sufficient conditions for optimality for a quadratic minimization problem subject to linear equality constraints and to formulate an algorithm for the solution of this problem. Section 3.1 develops some basic concepts of quadratic functions which will be used throughout this book including Taylor's Theorem and a matrix updating formula. Section 3.2 presents the notion of convexity, and gives necessary and sufficient conditions for an optimal solution of an unconstrained quadratic minimization problem. An algorithm for the solution of this problem using conjugate directions is given in Section 3.3. Sections 3.4 and 3.5 extend the results of Sections 3.2 and 3.3, respectively, for quadratic minimization subject to linear equality constraints.

3.1 QP Preliminaries

Taylor's series is an essential tool for analyzing a quadratic function. It expresses a quadratic function at a new point in terms of its value at another point and changed functional quantities. We develop it in the following.

Let x be an n-vector with $x = (x_1, \ldots, x_n)'$. A general quadratic function of n variables, $f(x)$, may be written

$$f(x) = \sum_{i=1}^{n} c_i x_i + \tfrac{1}{2} \sum_{i=1}^{n} \sum_{j=1}^{n} \gamma_{ij} x_i x_j, \tag{3.1}$$

where the c_i and γ_{ij} are constants. We can assume without loss of generality that $\gamma_{ij} = \gamma_{ji}$, for if not, both quantities may be replaced with $(\gamma_{ij} + \gamma_{ji})/2$

while leaving the function unchanged.[1] Let

$$
c = \begin{bmatrix} c_1 \\ c_2 \\ \vdots \\ c_n \end{bmatrix} \quad \text{and} \quad C = \begin{bmatrix} \gamma_{11} & \gamma_{12} & \cdots & \gamma_{1n} \\ \gamma_{21} & \gamma_{22} & \cdots & \gamma_{2n} \\ \vdots & \vdots & & \vdots \\ \gamma_{n1} & \gamma_{n2} & \cdots & \gamma_{nn} \end{bmatrix} .
$$

Then c is an n-vector, C is an (n, n) symmetric matrix, and (3.1) may be written in the more compact form

$$
f(x) = c'x + \tfrac{1}{2}x'Cx . \tag{3.2}
$$

The <u>gradient</u>[2] of f at x is the n-vector

$$
\begin{bmatrix} \dfrac{\partial f(x)}{\partial x_1} \\ \dfrac{\partial f(x)}{\partial x_2} \\ \vdots \\ \dfrac{\partial f(x)}{\partial x_n} \end{bmatrix}
$$

and is denoted by $g(x)$. It is straightforward to show (Exercise 3.1) that

$$
g(x) = c + Cx. \tag{3.3}
$$

The <u>Hessian</u> <u>matrix</u> of f at x is the (n, n) matrix whose (i, j)th element is

$$
\frac{\partial^2 f(x)}{\partial x_i \partial x_j} ,
$$

and is denoted by $H(x)$. Again, it is straightforward to show (Exercise 3.1) that

$$
H(x) = C . \tag{3.4}
$$

The reader may recall from elementary calculus that if h is a function of a single variable x, the Taylor series of h about a given point x_0 is

$$
h(x) = h(x_0) + \frac{1}{1!}\frac{dh(x_0)}{dx}(x - x_0) + \frac{1}{2!}\frac{d^2 h(x_0)}{dx^2}(x - x_0)^2 +
$$
$$
\frac{1}{3!}\frac{d^3 h(x_0)}{dx^3}(x - x_0)^3 \ldots \tag{3.5}
$$

---

[1]For example, $4x_2 x_1 + 6x_1 x_2$ and $5x_1 x_2 + 5x_2 x_1$ both yield the same result, namely $10x_1 x_2$. For the first expression $\gamma_{12} = 4 \neq \gamma_{21} = 6$ whereas for the second expression $\gamma_{12} = \gamma_{21} = 5$.

[2]Terms which are <u>underlined</u> are formal definitions.

Using the vector quantities we have just developed, there is an analogous Taylor series for $f(x)$ as in (3.2). Let x_0 be a given n-vector. The <u>Taylor series</u> for f about x_0 is

$$f(x) = f(x_0) + g'(x_0)(x - x_0) + \tfrac{1}{2}(x - x_0)'C(x - x_0) . \qquad (3.6)$$

Throughout this text, we shall frequently use the expression "by Taylor's series" and it is to (3.6) that we refer. Note the similarity between (3.5) and (3.6). The latter includes only terms up to quadratic, however, and this is because f is a quadratic function. It is straightforward to verify (3.6) by using the definition of $g(x_0)$ and simplifying $(x - x_0)'C(x - x_0)$ (Exercise 3.2).

Closely related to (3.6) is the identity

$$g(x) = g(x_0) + C(x - x_0) \qquad (3.7)$$

which we also refer to by the phrase "by Taylor's series."

We illustrate these concepts in

Example 3.1
Let

$$f(x) = x_1 + 3x_2 + 3x_1^2 + 7x_1x_2 + 5x_2^2 .$$

Here $c = (1 , 3)'$ and, using either (3.1) or (3.4),

$$C = \begin{bmatrix} 6 & 7 \\ 7 & 10 \end{bmatrix} .$$

Continuing the example, we illustrate the Taylor's series for f by taking $x_0 = (1 , -1)'$. Then from (3.3),

$$g(x_0) = \begin{bmatrix} 1 \\ 3 \end{bmatrix} + \begin{bmatrix} 6 & 7 \\ 7 & 10 \end{bmatrix} \begin{bmatrix} 1 \\ -1 \end{bmatrix} = \begin{bmatrix} 0 \\ 0 \end{bmatrix}$$

so that (3.6) becomes

$$f(x) = -1 + 3(x_1 - 1)^2 + 7(x_1 - 1)(x_2 + 1) + 5(x_2 + 1)^2 .$$

$\diamond$

The following situation arises frequently. A matrix and its inverse are known. The matrix is then changed in some way. It is then desired to find the inverse of the modified matrix in terms of the inverse of the original matrix and the data defining the change.

Our first updating procedure concerns the following situation. A matrix and its inverse are known. The matrix is then modified by changing a single row and it is desired to find the inverse of the modified matrix. An efficient way to do this is to express the new inverse in terms of the old inverse and the modified row. This update is formulated as

Procedure Φ.

Let $D' = [\, d_1, \ldots, d_n \,]$ be an $(n \, , \, n)$ nonsingular matrix and let $D^{-1} = [\, c_1, \ldots, c_n \,]$. Let d and k be a given n-vector and an index, respectively, such that $d'c_k \neq 0$. Define

$$\hat{D}' = [\, d_1, \ldots, d_{k-1} \, , \quad d \, , \quad d_{k+1}, \ldots, d_n \,] \; ;$$

that is, $\hat{D}'$ is obtained from D' by replacing column k with d. We use the notation $\Phi(D^{-1}, d, k)$ to denote $\hat{D}^{-1}$. One way of computing $\hat{D}^{-1} = [\hat{c}_1, \ldots, \hat{c}_n \,]$ is

$$\hat{c}_i = c_i - \frac{d'c_i}{d'c_k} \, c_k \, , \quad \text{for all } i = 1, \ldots, n \, , \quad i \neq k \, , \quad \text{and}$$

$$\hat{c}_k = \frac{1}{d'c_k} \, c_k \, .$$

The procedure for modifying the inverse of a matrix due to a change in a row or column was first formulated by Sherman and Morrison [15] and Woodbury [17].

We illustrate Procedure Φ as follows.

Example 3.2

Let

$$D' = \begin{bmatrix} 1 & 0 & 0 \\ 2 & 1 & 0 \\ 3 & 0 & 1 \end{bmatrix} , \quad \text{then } D^{-1} = \begin{bmatrix} 1 & -2 & -3 \\ 0 & 1 & 0 \\ 0 & 0 & 0 \end{bmatrix} .$$

Suppose that we obtain $\hat{D}'$ from D' by replacing column 3 with $(-1 \, , \, 1 \, , \, -1)'$. We thus compute

$$\hat{D}^{-1} = \Phi \left(D^{-1} \, , \begin{bmatrix} -1 \\ 1 \\ -1 \end{bmatrix} , \, 3 \right) = [\, \hat{c}_1 \, , \quad \hat{c}_2 \, , \quad \hat{c}_3 \,] \, ,$$

where

$$
\hat{c}_1 = \begin{bmatrix} 1 \\ 0 \\ 0 \end{bmatrix} - \frac{(-1)}{2} \begin{bmatrix} -3 \\ 0 \\ 1 \end{bmatrix} = \begin{bmatrix} -1/2 \\ 0 \\ 1/2 \end{bmatrix},
$$

$$
\hat{c}_2 = \begin{bmatrix} -2 \\ 1 \\ 0 \end{bmatrix} - \frac{3}{2} \begin{bmatrix} -3 \\ 0 \\ 1 \end{bmatrix} = \begin{bmatrix} 5/2 \\ 1 \\ -3/2 \end{bmatrix}, \text{ and}
$$

$$
\hat{c}_3 = \frac{1}{2} \begin{bmatrix} -3 \\ 0 \\ 1 \end{bmatrix}.
$$

Thus

$$
\hat{D}^{-1} = \begin{bmatrix} -1/2 & 5/2 & -3/2 \\ 0 & 1 & 0 \\ 1/2 & -3/2 & 1/2 \end{bmatrix}.
$$

$\diamondsuit$

In the analysis and algorithms presented in this book, we shall make frequent use of the relationship between a matrix and its inverse. This is simply a convenient and compact way of expressing relationships between certain naturally occurring vectors. We next formulate some of these in detail. Let D be a nonsingular (n, n) matrix. Let

$$
D' = [\, d_1, d_2, \ldots, d_n \,] \quad \text{and} \quad D^{-1} = [\, c_1, c_2, \ldots, c_n \,] . \tag{3.8}
$$

Each d_i and c_i is an n-vector. The ith column of D' is d_i. Equivalently, the ith row of D is d_i'. The ith column of D^{-1} is c_i. The definition of the inverse matrix states that $DD^{-1} = I$, where I denotes the (n, n) identity matrix. Expanding this equation gives

$$
\begin{bmatrix} d_1' \\ d_2' \\ \vdots \\ d_n' \end{bmatrix} [\, c_1, c_2, \ldots, c_n \,] = \begin{bmatrix} d_1'c_1 & d_1'c_2 & \cdots & d_1'c_n \\ d_2'c_1 & d_2'c_2 & \cdots & d_2'c_n \\ \vdots & \vdots & & \vdots \\ d_n'c_1 & d_n'c_2 & \cdots & d_n' \end{bmatrix} = \begin{bmatrix} 1 & 0 & \cdots & 0 \\ 0 & 1 & \cdots & 0 \\ \vdots & \vdots & & \vdots \\ 0 & 0 & \cdots & 1 \end{bmatrix} .
$$

Comparing terms in the last part gives

$$
d_i'c_j = \begin{cases} 1, & \text{if } i = j, \\ 0, & \text{if } i \neq j. \end{cases} \tag{3.9}
$$

We will refer to (3.9) as a consequence of (3.8) by the phrase "by definition of the inverse matrix."

There is another identity which we will sometimes use. Let D and D^{-1} be as in (3.8) and suppose that d is an arbitrary given n-vector. It is occasionally useful to express d as a linear combination of the columns of D'. Let $x = (x_1, \ldots, x_n)'$ denote the coefficients of this linear combination. The components of x can be expressed in terms of the columns of D^{-1} and d itself as follows. By definition of x,

$$d = x_1 d_1 + x_2 d_2 + \cdots + x_n d_n . \tag{3.10}$$

Equivalently,

$$D'x = d , \quad \text{or,} \quad x'D = d' .$$

Multiplying on the right by D^{-1} in this last equation gives $x' = d'D^{-1}$. Expanding this gives

$$(x_1, \ldots, x_n) = d'[c_1, c_2, \ldots, c_n] = (d'c_1, d'c_2, \ldots, d'c_n) .$$

Comparing like components gives

$$x_i = d'c_i , \quad i = 1, \ldots, n .$$

Substituting into (3.10) gives the identity

$$d = (d'c_1)d_1 + (d'c_2)d_2 + \cdots + (d'c_n)d_n .$$

We summarize this result in

Lemma 3.1 *Let D be a nonsingular (n, n) matrix with $D' = [d_1, d_2, \ldots, d_n]$ and $D^{-1} = [c_1, c_2, \ldots, c_n]$. Let d be any n-vector. Then*

$$d = (d'c_1)d_1 + (d'c_2)d_2 + \cdots + (d'c_n)d_n .$$

In Chapters 3 and 4, several arguments are considerably simplified by the use of a certain orthogonal decomposition. It is likely that the reader will be familiar with the result. However we set out the result here for easy reference.

Theorem 3.1 *(Orthogonal Decomposition Theorem) Let $a_1, \ldots, a_r$ be given n-vectors. Then any vector g can be written as $g = g_1 + g_2$, where $g_1 \in$ span $\{a_1, \ldots, a_r\}$ and g_2 is orthogonal to $\{a_1, \ldots, a_r\}$. Furthermore, g_1 and g_2 are uniquely determined.*

Proof: Let $b_1, \ldots, b_k$ be a basis for span $\{a_1, \ldots, a_r\}$. Let Q denote the orthogonal complement of span $\{a_1, \ldots, a_r\}$; that is,

$$Q = \{x \mid a_i'x = 0 , \quad i = 1, \ldots, r\} .$$

The dimension of Q is $n - k$ and if $b_{k+1}, \ldots, b_n$ is a basis for it, then $b_1, \ldots, b_n$ spans all of n-space. Thus for any g there are scalars $\theta_1, \ldots, \theta_n$ such that

$$g = \theta_1 b_1 + \cdots + \theta_n b_n .$$

Defining $g_1 = \sum_{i=1}^{k} \theta_i b_i$ and $g_2 = \sum_{i=k+1}^{n} \theta_i b_i$ completes the proof. $\square$

3.2 QP Unconstrained: Theory

We consider the unconstrained quadratic minimization problem

$$\min \left\{ c'x + \tfrac{1}{2}x'Cx \right\} , \tag{3.11}$$

where c and x are n-vectors, and C is an (n , n) symmetric matrix. $f(x) = c'x + \tfrac{1}{2}x'Cx$ is called the objective function for (3.11). A point x_0 is called a global minimizer or an optimal solution for (3.11) if $f(x_0) \leq f(x)$ for all x. The objective function for (3.11) is unbounded from below (or more briefly, problem (3.11) is unbounded from below] if there are n-vectors x_0 and s_0 such that $f(x_0 - \sigma s_0) \to -\infty$ as $\sigma \to +\infty$.

Let x_0 be an optimal solution. We first obtain a necessary condition for optimality. Let s be any n-vector and let σ be a scalar. Keeping s fixed for the moment and letting σ vary around zero, $x_0 - \sigma s$ is the point obtained from x_0 by moving in the direction $-s$. The distance along $-s$ is controlled by σ. Since x_0 is optimal, the smallest value of $f(x_0 - \sigma s)$ must occur when $\sigma = 0$. By Taylor's series,

$$f(x_0 - \sigma s) = f(x_0) - \sigma g'(x_0)s + \tfrac{1}{2}\sigma^2 s'Cs.$$

This gives $f(x_0 - \sigma s)$ as an explicit quadratic function of σ: the coefficient of the linear term is $-g'(x_0)s$ and that of the quadratic term is $s'Cs/2$. Since $f(x_0 - \sigma s)$ is minimized when $\sigma = 0$, it follows from elementary calculus that its derivative with respect to σ must vanish when $\sigma = 0$. That is,

$$g'(x_0)s = 0 . \tag{3.12}$$

Now (3.12) must hold for all n-vectors s. In particular, it must hold for each of the n unit vectors e_i.[3] Thus

$$g'(x_0)e_i = \frac{\partial f(x_0)}{\partial x_i} = 0 , \quad i = 1, \ldots, n .$$

Therefore $g(x_0) = 0$ and we have proven

---

[3] The ith unit vector e_i has zeros in all components except the ith which has value 1.

Theorem 3.2 *Let x_0 be an optimal solution for* min $\{ c'x + \frac{1}{2}x'Cx \}$. *Then* $g(x_0) = 0$.

Theorem 3.2 gives a necessary condition for optimality. However, it is not in general sufficient. Consider the function of two variables

$$f(x) = x_1 x_2 .$$

Here $g(x) = (x_2 , x_1)'$ and $g(x_0) = 0$ for $x_0 = (0 , 0)'$. Also, $f(x_0) = 0$. However, the point $(\epsilon , -\epsilon)'$ can be made arbitrarily close to x_0 by taking the scalar ϵ sufficiently small. Furthermore, the objective function value at this point is $-\epsilon^2$ and is strictly less than $f(x_0)$ provided that $\epsilon \neq 0$. Thus x_0 is not an optimal solution.

Suppose now that x_0 is such that $g(x_0) = 0$. We seek a condition that will imply that x_0 is an optimal solution. Any point x can be written as $x_0 - s$ for a suitably chosen n-vector s. By Taylor's series with $g(x_0) = 0$,

$$f(x_0 - s) = f(x_0) + \frac{1}{2}s'Cs .$$

Thus x_0 is optimal provided that $s'Cs \geq 0$. Furthermore, $s'Cs \geq 0$ must be satisfied for all n-vectors s.

Quadratic functions which possess this last property are quite important and are given a special name. A function $f(x) = c'x + \frac{1}{2}x'Cx$ is convex if $s'Cs \geq 0$ for all n-vectors s and is strictly convex if $s'Cs > 0$ for all $s \neq 0$.

With Theorem 3.2 and the previous discussion, we have shown

Theorem 3.3 *Let $f(x) = c'x + \frac{1}{2}x'Cx$ be convex. Then x_0 is an optimal solution for* min $\{ c'x + \frac{1}{2}x'Cx \}$ *if and only if $g(x_0) = 0$.*

The reader may have previously seen the conditions in the definition of convex and strictly convex functions. A symmetric matrix C is said to be positive semidefinite if $s'Cs \geq 0$ for all s and positive definite if $s'Cs > 0$ for all $s \neq 0$. Thus a quadratic function is convex if its Hessian matrix is positive semidefinite and is strictly convex if its Hessian matrix is positive definite.

We next give some examples of convex and strictly convex functions. A linear function $f(x) = c'x$ is a special case of a quadratic function with $C = 0$. It follows from definition that a linear function is convex but not strictly convex. Next consider a quadratic function $f(x) = c'x + \frac{1}{2}x'Cx$ with

$$C = \text{diagonal} (\gamma_i) .$$

First suppose that some diagonal element is strictly negative. In particular, suppose that $\gamma_k < 0$. With $s = e_k$ (the kth unit vector),

$$s'Cs = \gamma_k < 0$$

so that f is not convex. Next suppose that all of the γ_i are nonnegative. With $s = (s_1, s_2, \ldots, s_n)'$,

$$s'Cs = \gamma_1 s_1^2 + \gamma_2 s_2^2 + \cdots + \gamma_n s_n^2 ,$$

which is nonnegative. In this case $f(x)$ is convex. Finally, suppose that each γ_i is strictly positive. If $s \neq 0$, then at least one component is nonzero and thus at least one term in the previous sum must be strictly positive. Consequently, $s'Cs > 0$ and $f(x)$ is strictly convex.

Another example of a convex function $f(x) = c'x + \frac{1}{2}x'Cx$ is one for which the Hessian matrix can be written as $C = H'H$ where H is some (k, n) matrix. Then

$$s'Cs = s'H'Hs = (Hs)'(Hs)$$

and $s'Cs$ being the sum of squares of the components of Hs is necessarily nonnegative. Thus f is convex. In addition, f will be strictly convex if $Hs \neq 0$ whenever $s \neq 0$. This would be the case if H were square and nonsingular.

With the notion of strict convexity, we can now give a condition under which an optimal solution is uniquely determined.

Theorem 3.4 *Let x_0 be an optimal solution for* $\min \{ c'x + \frac{1}{2}x'Cx \}$. *If $f(x) = c'x + \frac{1}{2}x'Cx$ is strictly convex, then x_0 is the unique optimal solution.*

Proof: Suppose to the contrary that x_1 is another optimal solution and that $x_1 \neq x_0$. From Theorem (3.3),

$$g(x_0) = c + Cx_0 = 0$$

and

$$g(x_1) = c + Cx_1 = 0 .$$

Subtracting gives $C(x_1 - x_0) = 0$. Premultiplying this last by $(x_1 - x_0)'$ gives

$$(x_1 - x_0)'C(x_1 - x_0) = 0 .$$

The definition of strict convexity implies that $(x_1 - x_0) = 0$, or, that $x_1 = x_0$. The assumption that there is a second optimal solution leads to a contradiction and is therefore false. □

There are other characterizations of convex quadratic functions. One is formulated in terms of the gradient of the function. Let x and x_0 be any two points. Then by Taylor's series,

$$f(x) - f(x_0) - g'(x_0)(x - x_0) = \tfrac{1}{2}(x - x_0)'C(x - x_0) . \qquad (3.13)$$

Now f is convex if and only if the right-hand side of (3.13) is nonnegative for all points x and x_0. Similarly, f is strictly convex if and only if the right-hand side of (3.13) is strictly positive for all points x and x_0 with $x \neq x_0$. We have thus proven

Theorem 3.5 *The following hold.*

    (a) $f(x) = c'x + \tfrac{1}{2}x'Cx$ is convex if and only if $f(x) \geq f(x_0) + g'(x_0)(x - x_0)$ for all x, x_0.

    (b) $f(x) = c'x + \tfrac{1}{2}x'Cx$ is strictly convex if and only if $f(x) > f(x_0) + g'(x_0)(x - x_0)$ for all x, x_0 with $x \neq x_0$.

Theorem 3.5 has the following geometric interpretation. The function $f(x_0) + g'(x_0)(x - x_0)$ is obtained from the Taylor's series for f at x_0 by omitting the quadratic term $(1/2)(x - x_0)'C(x - x_0)$. As such, it may be thought of as the linear approximation to f at x_0. Note that the linear approximation agrees with f for $x = x_0$. Theorem 3.5 states that f is convex if and only if its linear approximation always underestimates f (see Figure 3.1).

Let x_1 and x_2 be any two points. For σ between 0 and 1, $\sigma x_1 + (1 - \sigma)x_2$ is on the line segment joining x_1 and x_2. Such a point is called a *convex combination* of x_1 and x_2. The following characterization of a convex function relates the function value at a point to the corresponding convex combination of function values (see Figure 3.2).

Theorem 3.6 *The following hold.*

    (a) $f(x)$ is convex if and only if for any x_1 and x_2 $f(\sigma x_1 + (1 - \sigma)x_2) \leq \sigma f(x_1) + (1 - \sigma)f(x_2)$ for all σ with $0 \leq \sigma \leq 1$.

    (b) $f(x)$ is strictly convex if and only if for any x_1 and x_2 with $x_1 \neq x_2$, $f(\sigma x_1 + (1 - \sigma)x_2) < \sigma f(x_1) + (1 - \sigma)f(x_2)$ for all σ with $0 < \sigma < 1$.

Proof: (a) Suppose first that f is convex. Let x_1 and x_2 be any points and choose σ with $0 \leq \sigma \leq 1$. For brevity, define $w = \sigma x_1 + (1 - \sigma)x_2$. Applying Theorem 3.5(a) with x and x_0 replaced by x_1 and w, respectively, gives

$$f(x_1) \geq f(w) + g'(w)(x_1 - w) . \qquad (3.14)$$

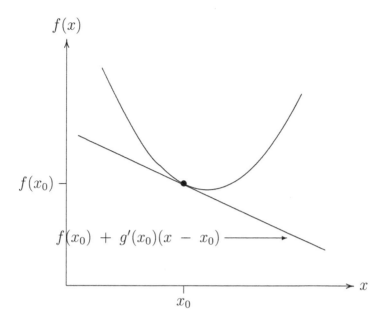

Figure 3.1.   Linear approximation to a convex function (Theorem 3.5).

Again applying Theorem 3.5(a) with x and x_0 replaced by x_2 and w, respectively, gives

$$f(x_2) \geq f(w) + g'(w)(x_2 - w). \qquad (3.15)$$

Multiplying (3.14) and (3.15)) by nonnegative σ and $(1 - \sigma)$, respectively, and adding gives

$$\sigma f(x_1) + (1 - \sigma)f(x_2) \geq (\sigma + 1 - \sigma)f(w) + g'(w)(\sigma x_1 + (1 - \sigma)x_2 - w).$$

Upon simplification, this becomes

$$f(\sigma x_1 + (1 - \sigma)x_2) \leq \sigma f(x_1) + (1 - \sigma)f(x_2) ,$$

as required.

Conversely, suppose that the inequality in the statement of the theorem holds for all x_1 and x_2, and σ with $0 \leq \sigma \leq 1$. Let x_1 be arbitrary, let s be any n-vector, and let σ be such that $0 < \sigma < 1$. Replacing x_1 and x_2 in the statement of the inequality with $x_1 + s$ and x_1, respectively, gives

$$f(x_1 + \sigma s) \leq \sigma f(x_1 + s) + (1 - \sigma)f(x_1) .$$

Expanding both $f(x_1 + \sigma s)$ and $f(x_1 + s)$ using Taylor's series gives

$$f(x_1) + \sigma g'(x_1)s + \tfrac{1}{2}\sigma^2 s'Cs \leq \sigma f(x_1) + \sigma g'(x_1)s + \tfrac{\sigma}{2}s'Cs + (1 - \sigma)f(x_1) .$$

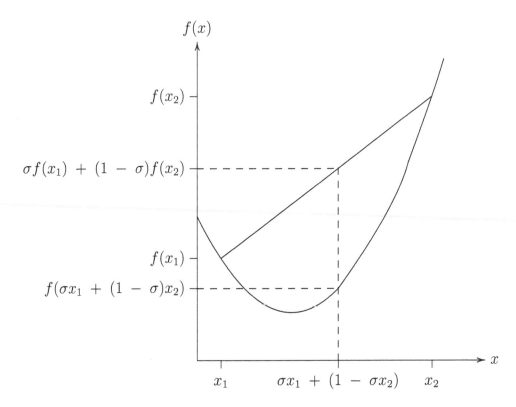

Figure 3.2.   Characterization of a convex function (Theorem 3.6).

Simplifying results in $\sigma(1 - \sigma)s'Cs \geq 0$. Since $0 < \sigma < 1$, this shows that f is indeed convex.

(b) Using $x_1 \neq x_2$, $s \neq 0$, σ with $0 < \sigma < 1$, and Theorem 3.14(b), the arguments are identical to those of part (a).   □

Many presentations of the theory of convex functions use the conditions of Theorem 3.6(a) and (b) to define a convex and a strictly convex function, respectively. Theorem 3.5 and the condition of our definition are then deduced as consequences. We have used the condition $s'Cs \geq 0$ to define a convex function because it occurs more naturally than the condition of Theorem 3.6(a).

Thus far we have assumed the existence of an optimal solution. We next show that if the objective function is bounded from below, then an optimal solution exists.

Theorem 3.7 *Suppose that there is a γ with $f(x) = c'x + \frac{1}{2}x'Cx \geq \gamma$ for all x. Then there exists an optimal solution for $\min \{ c'x + \frac{1}{2}x'Cx \}$.*

Proof: From Theorem 3.1, we may write c as $c = c_1 + c_2$, where c_1 lies in the space spanned by the columns of C and c_2 is orthogonal to that space. Then

$$c_1 = Cw , \qquad (3.16)$$

for some n-vector w, and

$$c_2'C = 0 . \qquad (3.17)$$

Then $c_2'c_1 = c_2'Cw = 0$ so that

$$c_2'c = c_2'(c_1 + c_2) = c_2'c_2 . \qquad (3.18)$$

Let σ be a scalar. From (3.17) and (3.18),

$$f(-\sigma c_2) = -\sigma c_2'c_2 . \qquad (3.19)$$

Because $c_2'c_2$ is the sum of the squares of the components of c_2, $c_2'c_2 \geq 0$. Furthermore, $c_2'c_2 = 0$ if and only if $c_2 = 0$. Now if $c_2 \neq 0$, $c_2'c_2 > 0$ and by taking σ arbitrarily large, (3.19) implies that f is not bounded below. This contradicts the hypothesis of the theorem and is therefore false. Consequently, $c_2 = 0$ and thus $c = c_1$. Setting $x_0 = -w$ in (3.16), we have established the existence of a point satisfying

$$g(x_0) = 0 . \qquad (3.20)$$

If there were some n-vector s satisfying $s'Cs < 0$, then Taylor's series with (3.20) would imply that

$$f(x_0 - \sigma s) = f(x_0) + \tfrac{1}{2}\sigma^2 s'Cs .$$

By taking the scalar σ sufficiently large, $f(x_0 - \sigma s)$ can be reduced without bound in contradiction to the hypothesis of the theorem. Consequently,

$$s'Cs \geq 0 , \quad \text{for all } s .$$

Thus f is necessarily convex. With (3.20), Theorem 3.3 asserts that x_0 is an optimal solution. □

3.3 QP Unconstrained: Algorithm 1

In this section, we present an algorithm for the solution of the unconstrained quadratic minimization problem

$$\min \{ c'x + \tfrac{1}{2}x'Cx \}. \qquad (3.21)$$

We assume that the objective function is convex. If the dimension of x is n, the algorithm will terminate in n steps or less with either an optimal solution or the information that (3.21) is unbounded from below.

According to Theorem 3.3, solving (3.21) is equivalent to solving the n simultaneous linear equations $Cx_0 = -c$. Although these equations may be solved directly, we will solve (3.21) using a different approach; one which can more easily be modified to account for linear inequality constraints.

Suppose that we have an initial estimate, x_0, of an optimal solution for (3.21). As in linear programming, we consider a new point of the form $x_0 - \sigma s_0$, where σ is a scalar and s_0 is some known vector which we interpret geometrically as a search direction. Let g_0 denote $g(x_0)$. Suppose that $g_0' s_0 \neq 0$. By replacing s_0 with $-s_0$ if necessary, we can assume that $g_0' s_0 > 0$. We first address the question of a suitable choice for σ. The most natural choice is to choose σ_0 to minimize $f(x_0 - \sigma s_0)$. By Taylor's series,

$$f(x_0 - \sigma s_0) = f(x_0) - \sigma g_0' s_0 + \tfrac{1}{2}\sigma^2 s_0' C s_0 . \tag{3.22}$$

This gives $f(x_0 - \sigma s_0)$ as an explicit quadratic function of σ. Because f is convex, $s_0' C s_0 \geq 0$. If $s_0' C s_0 = 0$, then because $g_0' s_0 > 0$, $f(x_0 - \sigma s_0) \to -\infty$ as $\sigma \to +\infty$; that is, (3.21) is unbounded from below. The remaining possibility is that $s_0' C s_0 > 0$. Since σ_0 minimizes $f(x_0 - \sigma s_0)$, the derivative of (3.22) with respect to σ, evaluated at σ_0, must be equal to zero; that is,

$$-g_0' s_0 + \sigma_0 s_0' C s_0 = 0.$$

Solving for σ_0 gives

$$\sigma_0 = \frac{g_0' s_0}{s_0' C s_0}. \tag{3.23}$$

This choice for σ_0 is called the <u>optimal</u> <u>step</u> <u>size</u> (see Figure 3.3).

Because we have adopted the convention that $g_0' s_0 > 0$, it follows from (3.23) that σ_0 is positive.

The next approximation to an optimal solution is

$$x_1 = x_0 - \sigma_0 s_0 .$$

From Taylor's series, $g_1 = g_0 - \sigma_0 C s_0$. Taking the inner product of both sides of this with s_0 and using (3.22) gives

$$g_1' s_0 = g_0' s_0 - \sigma_0 s_0' C s_0 = g_0' s_0 - g_0' s_0 = 0.$$

Thus

$$g_1' s_0 = 0; \tag{3.24}$$

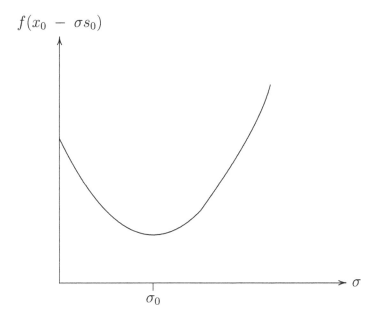

Figure 3.3.   Optimal step size.

that is, using the optimal step size results in the new gradient being orthogonal to the search direction.

Suppose that we continue in a similar manner giving $x_2 = x_1 - \sigma_1 s_1$ with $\sigma_1 = g_1' s_1 / s_1' C s_1$. Then as with (3.24)

$$g_2' s_1 = 0. \tag{3.25}$$

Continuing would give

$$x_{j+1} = x_j - \sigma_j s_j \tag{3.26}$$

with

$$\sigma_j = \frac{g_j' s_j}{s_j' C s_j} \tag{3.27}$$

and consequently

$$g_{j+1}' s_j = 0 , \tag{3.28}$$

for $j = 0 , 1 , \ldots$.

With no restrictions on s_j, this sequence could continue indefinitely. We next look for a relationship among the s_j which assures termination in a finite number of steps. To motivate the concept, suppose for the moment that

$n = 2$. After two iterations, (3.25) states that $g_2's_1 = 0$. If in addition, $g_2's_0 = 0$ then

$$g_2'[\, s_0, s_1 \,] = 0.$$

If s_0 and s_1 are linearly independent, $[\, s_0, s_1 \,]$ is nonsingular and multiplying on the right by its inverse results in $g_2 = 0$. Theorem 3.3 then asserts that x_2 is optimal.

The critical condition is thus $g_2's_0 = 0$. The optimal step size makes g_2 orthogonal to s_1 and some additional relationship between s_0 and s_1 is necessary to make g_2 orthogonal to s_0. Because $g_2 = g_1 - \sigma_1 C s_1$, it follows from (3.24) that

$$g_2's_0 = -\sigma_1 s_1' C s_0 \,.$$

Consequently, $g_2's_0 = 0$ and the iterative process finds an optimal solution in 2 steps provided that $s_1' C s_0 = 0$. Search directions s_0 and s_1 which satisfy this are said to be *conjugate*.

Returning now to the n-dimensional problem (3.21), the n-vectors s_0, $s_1, \ldots, s_k$ are <u>conjugate directions</u> if

$$s_i' C s_i > 0 \,, \quad i = 0, \ldots, k \,, \quad \text{and}$$
$$s_i' C s_j = 0 \,, \quad i = 0, \ldots, k \,, \quad j = 0, \ldots, k \,, \quad i \neq j \,.$$

In addition, if $s_i' C s_i = 1$ for $i = 0, \ldots, k$, then $s_0, \ldots, s_k$ are <u>normalized conjugate directions</u>. It is easy to see that a set of conjugate directions may be normalized by replacing each s_i with $(s_i' C s_i)^{-1/2} s_i$.

We next present two important properties of conjugate directions. Following this, we will present a method to construct them.

Lemma 3.2 *Let $s_0, \ldots, s_k$ be conjugate directions. Then $s_0, \ldots, s_k$ are linearly independent.*

Proof: Assume to the contrary that $s_0, \ldots, s_k$ are linearly dependent. Then there are scalars $\lambda_0, \ldots, \lambda_k$, not all zero, with

$$\lambda_0 s_0 + \lambda_1 s_1 + \cdots + \lambda_k s_k = 0 \,.$$

Suppose that $\lambda_i \neq 0$. Multiplying the last equation by $s_i' C$ gives

$$\lambda_0 s_i' C s_0 + \lambda_1 s_i' C s_1 + \cdots + \lambda_k s_i' C s_k = 0 \,.$$

Because $s_0, \ldots, s_k$ are conjugate directions, all terms in this summation vanish except for the one having subscript i. Thus

$$\lambda_i s_i' C s_i = 0 \,.$$

But $s_i'Cs_i > 0$, so that $\lambda_i = 0$. This contradiction establishes that $s_0, \ldots, s_k$ are linearly independent. $\qquad\square$

As a consequence of Lemma 3.2, there can be no more than n conjugate directions for an n-dimensional problem. The next result shows that using conjugate search directions will locate the minimum of a strictly convex quadratic function in n steps.

Theorem 3.8 *Assume that $f(x)$ is strictly convex and let $s_0, s_1, \ldots, s_{n-1}$ be conjugate directions. Let x_0 be arbitrary and let $x_1, \ldots, x_n$ be constructed using (3.16) and (3.17). Then*

(a) $g_{j+1}'s_i = 0$ *for $j = 0, 1, \ldots, n-1$ and $i = 0, 1, \ldots, j$.*

(b) x_n *is the optimal solution for* $\min\{ c'x + \frac{1}{2}x'Cx \}$.

Proof: (a) Let j satisfy $0 \leq j \leq n-1$ and let i be such that $i < j$. By Taylor's series,

$$g_{j+1} = g_{i+1} + C(x_{j+1} - x_{i+1}) .$$

From (3.28), $g_{i+1}'s_i = 0$ so that

$$g_{j+1}'s_i = (x_{j+1} - x_{i+1})'Cs_i$$
$$= -\sigma_{i+1}s_{i+1}'Cs_i - \cdots - \sigma_j s_j'Cs_i .$$

Because $s_i, \ldots, s_j$ are conjugate directions, each term in this last sum vanishes. Therefore

$$g_{j+1}'s_i = 0 , \quad i = 0, 1, \ldots, j-1 .$$

In addition, (3.28) asserts that $g_{j+1}'s_j = 0$ which completes the proof of part (a).

(b) Applying part (a) with $j = n-1$ gives

$$g_n's_i = 0 , \quad i = 0, 1, \ldots, n-1 .$$

From Lemma 3.2, g_n is orthogonal to n linearly independent vectors. Therefore $g_n = 0$, and Theorem 3.2 implies that x_n is optimal. $\qquad\square$

Theorem 3.8(a) is interesting because it offers some insight into using conjugate directions as search directions. After the first iteration, g_1 is orthogonal to s_0. After the second, g_2 is orthogonal to both s_0 and s_1. In general, the new gradient is orthogonal to all previously searched conjugate directions. After iteration $n-1$, g_n is orthogonal to n conjugate directions and thus vanishes.

Some other properties of conjugate directions are developed as exercises (Exercises 3.12 to 3.14).

Theorem 3.8 applies only to a strictly convex objective function. We now formulate a method which uses conjugate directions to solve (3.21) when f is assumed only to be convex. In order to formulate such a method, it is helpful to conceptualize an intermediate step. Suppose that a certain number of conjugate directions have already been constructed and searched. We seek the appropriate information to construct an additional conjugate direction. Suppose that for some j with $0 \leq j \leq n - 1$ there is a nonsingular (n, n) matrix D which is related to its inverse as follows:

$$\left. \begin{array}{l} D' = [\, Cc_1\,,\ Cc_2, \ldots, Cc_j\,,\ d_{j+1}, \ldots, d_n\,]\,, \\ D^{-1} = [\ \ c_1\,,\ \ c_2, \ldots,\ \ c_j\,,\ c_{j+1}, \ldots, c_n\,]\,. \end{array} \right\} \tag{3.29}$$

The columns of D^{-1} are $c_1, \ldots, c_n$, and those of D' are $Cc_1, \ldots, Cc_j$, $d_{j+1}, \ldots, d_n$.

The relationship between D and D^{-1} in (3.29) will play a critical role in the algorithm presented here and in subsequent chapters. First observe that by definition of the inverse matrix (see (3.9)),

$$c_i' Cc_k = 0\,,\ \ 1 \leq i\,,\ k \leq j\,,\ i \neq k\,,\ \text{and} \tag{3.30}$$

$$c_i' Cc_i = 1\,,\ \ i = 1, \ldots, j\,. \tag{3.31}$$

Therefore $c_1, \ldots, c_j$ is a set of normalized conjugate directions.

Now suppose that we are in the process of solving (3.21) using the general iterative process defined by (3.26) and (3.27). Suppose that $s_0, \ldots, s_{j-1}$ are parallel[4] to $c_1, \ldots, c_j$, respectively, so that

$$c_1 = (s_0' Cs_0)^{-1/2} s_0, \ldots, c_j = (s_{j-1}' Cs_{j-1})^{-1/2} s_{j-1}\,, \tag{3.32}$$

and $s_0, \ldots, s_{j-1}$ are (unnormalized) conjugate directions. Let $g_j = g(x_j)$. As is suggested by Theorem 3.8(a) assume that

$$g_j' s_i = 0\,,\ \ i = 0, \ldots, j - 1\,,$$

so that from (3.32),

$$g_j' c_i = 0\,,\ \ i = 1, \ldots, j\,. \tag{3.33}$$

In other words, assume that the gradient of the objective function at the current iterate is orthogonal to all of the normalized conjugate direction columns of D^{-1}.

---

[4]We say that two vectors are parallel if one is a nonzero multiple of the other.

In order to continue, we require a search direction s_j such that $s_0, \ldots, s_j$ is a set of conjugate directions. Since $s_0, \ldots, s_{j-1}$ is a set of conjugate directions, s_j need only satisfy

$$s_j' C s_i \;=\; 0 \,, \quad i \;=\; 0, \ldots, j-1, \tag{3.34}$$

$$s_j' C s_j \;>\; 0. \tag{3.35}$$

From (3.31), (3.32), and the definition of the inverse matrix, choosing s_j parallel to any of $c_{j+1}, \ldots, c_n$ will satisfy (3.34). In other words, each of $c_{j+1}, \ldots, c_n$ is a candidate for the next conjugate direction. For simplicity, we choose c_{j+1}. Assume that $g_j' c_{j+1} \neq 0$. In order to satisfy our requirement of $g_j' s_j > 0$, we set

$$s_j \;=\; \begin{cases} c_{j+1} \,, & \text{if } g_j' c_{j+1} > 0 \,, \\ -c_{j+1} \,, & \text{otherwise} \,. \end{cases}$$

Since f is convex, $s_j' C s_j \geq 0$. If $s_j' C s_j = 0$, then by Taylor's series

$$f(x_j - \sigma s_j) \;=\; f(x_j) - \sigma g_j' s_j.$$

This implies that $f(x_j - \sigma s_j) \to -\infty$ as $\sigma \to +\infty$ and (3.21) is unbounded from below. The remaining possibility is that $s_j' C s_j > 0$. In this case, both (3.34) and (3.35) are satisfied and $s_0, \ldots, s_j$ are indeed a set of conjugate directions. Using the optimal step size σ_j, we obtain the next iterate x_{j+1}.

From Taylor's series, $g_{j+1} = g_j - \sigma_j C s_j$. Let i be such that $1 \leq i \leq j$. Then

$$g_{j+1}' c_i \;=\; g_j' c_i - \sigma_j c_i' C s_j \,. \tag{3.36}$$

From (3.33), $g_j' c_i = 0$. Furthermore, by construction s_j is parallel to c_{j+1} so that from (3.29) and the definition of the inverse matrix,

$$c_1' C s_j \;=\; c_2' C s_j \;=\; \cdots \;=\; c_j' C s_j \;=\; 0 \,.$$

With (3.36), this implies that

$$g_{j+1}' c_i \;=\; 0 \,, \quad i \;=\; 1, \ldots, j. \tag{3.37}$$

Finally, because σ_j is the optimal step size and s_j is parallel to c_{j+1}, (3.28) asserts that

$$g_{j+1}' c_{j+1} \;=\; 0 \,. \tag{3.38}$$

Summarizing (3.37) and (3.38);

$$g_{j+1}' c_i \;=\; 0 \,, \quad i \;=\; 1, \ldots, j+1, \tag{3.39}$$

which shows that (3.33) remains satisfied with j replaced by $j + 1$.

In order to continue, D must be modified. Since $c_1, \ldots, c_j$ together with $(c'_{j+1}Cc_{j+1})^{-1/2}c_{j+1}$ form a set of $j + 1$ normalized conjugate directions, it is reasonable to update D' by replacing d_{j+1} with $(c'_{j+1}Cc_{j+1})^{-1/2}Cc_{j+1}$. Let $\hat{D}'$ denote this new matrix. Then

$$\hat{D}' = \left[\, Cc_1, \ldots, Cc_j \, , \, (c'_{j+1}Cc_{j+1})^{-1/2}Cc_{j+1} \, , \, d_{j+2}, \ldots, d_n \, \right] \, .$$

With no additional information, there is no reason to expect that the first $j + 1$ columns of $\hat{D}^{-1}$ will be $c_1, \ldots, c_j, (c'_{j+1}Cc_{j+1})^{-1/2}c_{j+1}$. However, this will indeed be the case because of (3.40) as we will now show. Let

$$\hat{D}^{-1} = [\hat{c}_1, \ldots, \hat{c}_n] \, .$$

Since $\hat{D}'$ differs from D' only in column $j + 1$, Procedure Φ (see Section 3.1) gives the $\hat{c}_i$ in terms of the c_i as follows:

$$\hat{c}_i = c_i - \left(\frac{(c'_{j+1}Cc_{j+1})^{-1/2}c'_{j+1}Cc_i}{(c'_{j+1}Cc_{j+1})^{-1/2}c'_{j+1}Cc_{j+1}} \right) c_{j+1} \, , \quad i = 1, \ldots, n \, ,$$

$$i \neq j + 1 \, , \text{ and}$$

$$\hat{c}_{j+1} = \frac{1}{(c'_{j+1}Cc_{j+1})^{-1/2}c'_{j+1}Cc_{j+1}} c_{j+1} \, .$$

From (3.29) and the definition of the inverse matrix,

$$c'_{j+1}Cc_1 = c'_{j+1}Cc_2 = \cdots = c'_{j+1}Cc_j = 0.$$

Accounting for this and cancellation, the updating formulae become

$$\hat{c}_i = c_i \, , \quad i = 1, \ldots, j, \tag{3.40}$$

$$\hat{c}_{j+1} = (c'_{j+1}Cc_{j+1})^{-1/2}c_{j+1} \, , \quad \text{and} \tag{3.41}$$

$$\hat{c}_i = c_i - \left(\frac{c'_{j+1}Cc_i}{c'_{j+1}Cc_{j+1}} \right) c_{j+1} \, , \quad i = j + 2, \ldots, n. \tag{3.42}$$

A critical consequence of this is that the first j columns of D^{-1} are unchanged by the updating. Consequently, the relationship between $\hat{D}'$ and $\hat{D}^{-1}$ can be summarized as

$$\left. \begin{array}{l} \hat{D}' = \left[\, Cc_1 \, , \, Cc_2, \ldots, \, Cc_j \, , \, C\hat{c}_{j+1} \, , \, d_{j+2}, \ldots, d_n \, \right] \, , \\ \hat{D}^{-1} = \left[\quad c_1 \, , \quad c_2, \ldots, \quad c_j \, , \quad \hat{c}_{j+1} \, , \, \hat{c}_{j+2}, \ldots, \hat{c}_n \, \right] \, , \end{array} \right\} \tag{3.43}$$

which has the same form as (3.29) with j replaced by $j + 1$. Thus the algorithm may be continued to the next iteration setting

$$s_{j+1} = \begin{cases} \hat{c}_{j+2}, & \text{if } g'_{j+1}\hat{c}_{j+1} > 0, \\ -\hat{c}_{j+2}, & \text{otherwise}. \end{cases}$$

In the development thus far, the normalized conjugate direction columns of D^{-1} were assumed to be the **first** j and s_j was chosen parallel to c_{j+1}. This ordering assumption was for notational convenience only. It is equally appropriate to choose s_j parallel to any one of $c_{j+1}, \ldots, c_n$. Intuitively, it is attractive to choose s_j parallel to a c_i for which the reduction in f is substantial. Identifying the c_i giving the greatest reduction in f would require the computation of the optimal step size and the corresponding new point for each possible search direction and this would be computationally expensive (see Exercise 3.19). A less expensive criterion is to choose that c_i which gives the greatest **rate** of decrease. From Taylor's series,

$$f(x_j - \sigma s_j) = f(x_j) - \sigma g'_j s_j + \tfrac{1}{2}\sigma^2 s'_j C s_j . \tag{3.44}$$

The rate of decrease of f at $\sigma = 0$ is just the derivative with respect to σ of $f(x_j - \sigma s_j)$ evaluated at $\sigma = 0$. From (3.44), this is $-g'_j s_j$. Accordingly, we compute k such that

$$| g'_j c_k | = \max \{ | g'_j c_i | \, | \, i = j + 1, \ldots, n \} \tag{3.45}$$

and set

$$s_j = \begin{cases} c_k, & \text{if } g'_j c_k > 0, \\ -c_k, & \text{otherwise}. \end{cases}$$

There is an additional benefit to choosing s_j according to (3.45). It may occur that $g'_j c_k = 0$ and in this case, we claim that x_j is optimal. To see this, observe that from Lemma 3.1

$$g_j = (g'_j c_1)C c_1 + \cdots + (g'_j c_j)C c_j + (g'_j c_{j+1})d_{j+1} + \cdots + (g'_j c_n)d_n .$$

Equation (3.33) asserts that the coefficients of the first j terms all vanish. In addition, $g'_j c_k = 0$ implies that

$$g'_j c_i = 0, \quad i = j + 1, \ldots, n .$$

Thus $g_j = 0$ and x_j is an optimal solution.

We have developed the algorithm thus far under the ordering assumption, implicit in (3.29), that the first j columns of D^{-1} are normalized conjugate directions. Choosing the next conjugate direction according to (3.45) will violate this assumption. We must therefore allow for the possibility that any of the columns of D^{-1} are normalized conjugate directions. At a general iteration j, let

$$D'_j = [\, d_{1j}, \ldots, d_{nj} \,] \quad \text{and}$$
$$D^{-1}_j = [\, c_{1j}, \ldots, c_{nj} \,]$$

be the analogs, respectively, of D' and D^{-1} in (3.29). We identify the conjugate direction columns of D^{-1}_j using the ordered index set

$$J_j = \{\, \alpha_{1j}, \ldots, \alpha_{nj} \,\},$$

where

$$\alpha_{ij} = \begin{cases} -1 \,, & \text{if } d_{ij} = Cc_{ij} \,, \\ 0 \,, & \text{otherwise.} \end{cases}$$

The conjugate direction columns of D^{-1}_j are all those c_{ij} for which $\alpha_{ij} = -1$. All c_{ij} for which $\alpha_{ij} = 0$ are candidates for the next search direction. In terms of the index set, (3.30) and (3.31) become

$$c'_{ij}Cc_{kj} = 0 \,, \quad \text{for all } i \,, \; k \text{ with } i \neq k \,, \; \alpha_{ij} = \alpha_{kj} = -1 \text{ and}$$
$$c'_{ij}Cc_{ij} = 1 \,, \quad \text{for all } i \text{ with } \alpha_{ij} = -1 \,,$$

respectively. Equation (3.33) becomes

$$g'_j c_{ij} = 0 \,, \quad \text{for all } i \text{ with } \alpha_{ij} = -1 \,.$$

The search direction selection rule (3.45) becomes

$$|\, g'_j c_{kj} \,| = \max \{\, |\, g'_j c_{ij} \,| \mid \text{all } i \text{ with } \alpha_{ij} = 0 \,\} \,.$$

Having chosen k, D'_{j+1} is obtained from D'_j by replacing column k with $(c'_{k+j}Cc_{kj})^{-1/2}Cc_{kj}$. Procedure Φ could be used to compute D^{-1}_{j+1};

$$D^{-1}_{j+1} = \Phi(D^{-1}_j \,, \; (c'_{kj}Cc_{kj})^{-1/2}Cc_{kj} \,, \; k) \,.$$

However, as we have seen in obtaining (3.40) to (3.42) the conjugate direction columns of D^{-1}_j are unchanged by the updating. The following procedure accounts for this and results in significant computational savings.

Procedure Φ_1 .

Let $D' = [\, d_1, \ldots, d_n \,]$, $D^{-1} = [\, c_1, \ldots, c_n \,]$, and $J = \{\, \alpha_1, \ldots, \alpha_n \,\}$. Let d and k be a given n-vector and an index, respectively, with $d'c_k \neq 0$ and $d'c_i = 0$ for all i with $\alpha_i = -1$. Let

$$\hat{D}' = [\, d_1, \ldots, d_{k-1} \,, \; d \,, \; d_{k+1}, \ldots, d_n \,]$$

and $\hat{D}^{-1} = [\hat{c}_1, \ldots, \hat{c}_n]$. We use the notation $\Phi_1(D^{-1}, d, k, J)$ to denote $\hat{D}^{-1}$. An explicit way to compute the columns of $\hat{D}^{-1}$ is:

$$\hat{c}_i = c_i - \frac{d'c_i}{d'c_k} c_k, \quad \text{for all } i \text{ with } i \neq k \text{ and } \alpha_i \neq -1,$$

$$\hat{c}_k = \frac{1}{d'c_k} c_k, \quad \text{and}$$

$$\hat{c}_i = c_i, \quad \text{for all } i \text{ with } \alpha_i = -1.$$

Note that $\hat{D}^{-1} = \Phi(D^{-1}, d, k)$, so $\hat{D}^{-1}$ could be obtained by using Procedure Φ. However, doing so would require the explicit and unnecessary computation of $d'c_i$ for all i with $\alpha_i = -1$. Procedure Φ_1 differs from Procedure Φ only in that it suppresses the updating of columns that are known not to change.

Next we give a detailed formulation of the algorithm.

ALGORITHM   1

Model Problem:

$$\text{minimize:} \quad c'x + \tfrac{1}{2}x'Cx.$$

Initialization:
Start with any x_0, set $J_0 = \{0, \ldots, 0\}$ and $D_0^{-1} = I$. Compute $f(x_0) = c'x_0 + \tfrac{1}{2}x_0'Cx_0$, $g_0 = c + Cx_0$, and set $j = 0$.

Step 1:   Computation of Search Direction s_j.
Let $D_j^{-1} = [c_{1j}, \ldots, c_{nj}]$ and $J_j = \{\alpha_{1j}, \ldots, \alpha_{nj}\}$. If $\alpha_{ij} \neq 0$ for $i = 1, \ldots, n$, stop with optimal solution x_j. Otherwise, compute the smallest index k such that

$$|g_j'c_{kj}| = \max\{|g_j'c_{ij}| \,|\, \text{all } i \text{ with } \alpha_{ij} = 0\}.$$

If $g_j'c_{kj} = 0$, stop with optimal solution x_j. Otherwise, set $s_j = c_{kj}$ if $g_j'c_{kj} > 0$ and $s_j = -c_{kj}$ if $g_j'c_{kj} < 0$. Go to Step 2.

Step 2:   Computation of Step Size σ_j.
Compute $s_j'Cs_j$. If $s_j'Cs_j = 0$, print x_j, s_j, and the message "the objective function is unbounded from below" and stop. Otherwise, set

$$\sigma_j = \frac{g_j's_j}{s_j'Cs_j}$$

and go to Step 3.

Step 3:   **Update.**

Set $x_{j+1} = x_j - \sigma_j s_j$, $g_{j+1} = c + C x_{j+1}$, and $f(x_{j+1}) = c' x_{j+1} + \frac{1}{2} x'_{j+1} C x_{j+1}$. Set $d_j = (s'_j C s_j)^{-1/2} C s_j$, $D_{j+1}^{-1} = \Phi_1(D_j^{-1}, d_j, k, J_j)$, and $J_{j+1} = \{ \alpha_{1,j+1}, \ldots, \alpha_{n,j+1} \}$, where

$$\alpha_{i,j+1} = \alpha_{ij}, \quad i = 1, \ldots, n, \quad i \neq k,$$
$$\alpha_{k,j+1} = -1.$$

Replace j with $j + 1$ and go to Step 1.

We illustrate Algorithm 1 by applying it to the following example.

Example 3.3

$$\text{minimize:} \quad -7x_1 - 8x_2 - 9x_3 + x_1^2 + x_2^2 + x_3^2$$
$$+ \ x_1 x_2 + x_1 x_3 + x_2 x_3.$$

Here

$$c = \begin{bmatrix} -7 \\ -8 \\ -9 \end{bmatrix} \quad \text{and} \quad C = \begin{bmatrix} 2 & 1 & 1 \\ 1 & 2 & 1 \\ 1 & 1 & 2 \end{bmatrix}.$$

Initialization:

$$x_0 = \begin{bmatrix} 0 \\ 0 \\ 0 \end{bmatrix}, \quad J_0 = \{ 0, 0, 0 \}, \quad D_0^{-1} = \begin{bmatrix} 1 & 0 & 0 \\ 0 & 1 & 0 \\ 0 & 0 & 1 \end{bmatrix}, \quad f(x_0) = 0,$$

$$g_0 = \begin{bmatrix} -7 \\ -8 \\ -9 \end{bmatrix}, \quad j = 0.$$

Iteration 0

Step 1: $\ |g'_0 c_{30}| = \max \{ |-7|, |-8|, |-9| \} = 9, \ k = 3,$

$$s_0 = \begin{bmatrix} 0 \\ 0 \\ -1 \end{bmatrix}.$$

Step 2: $\ s'_0 C s_0 = 2, \ \sigma_0 = \dfrac{9}{2}.$

Step 3: $\ x_1 = \begin{bmatrix} 0 \\ 0 \\ 0 \end{bmatrix} - \dfrac{9}{2} \begin{bmatrix} 0 \\ 0 \\ -1 \end{bmatrix} = \begin{bmatrix} 0 \\ 0 \\ 9/2 \end{bmatrix}, \ g_1 = \begin{bmatrix} -5/2 \\ -7/2 \\ 0 \end{bmatrix},$

$$f(x_1) = \frac{-81}{4},$$

$$d_0 = \begin{bmatrix} -1/\sqrt{2} \\ -1/\sqrt{2} \\ -2/\sqrt{2} \end{bmatrix}, \quad D_1^{-1} = \begin{bmatrix} 1 & 0 & 0 \\ 0 & 1 & 0 \\ -1/2 & -1/2 & -1/\sqrt{2} \end{bmatrix},$$

$$J_1 = \{0, 0, -1\}, \quad j = 1.$$

Iteration 1

Step 1: $\quad |g_1'c_{21}| = \max\left\{ |\frac{-5}{2}|, |\frac{-7}{2}|, - \right\} = \frac{7}{2}, \quad k = 2,$

$$s_1 = \begin{bmatrix} 0 \\ -1 \\ 1/2 \end{bmatrix}.$$

Step 2: $\quad s_1'Cs_1 = \frac{3}{2}, \quad \sigma_1 = \frac{7/2}{3/2} = \frac{7}{3}.$

Step 3: $\quad x_2 = \begin{bmatrix} 0 \\ 0 \\ 9/2 \end{bmatrix} - \frac{7}{3}\begin{bmatrix} 0 \\ -1 \\ 1/2 \end{bmatrix} = \begin{bmatrix} 0 \\ 7/3 \\ 10/3 \end{bmatrix}, \quad g_2 = \begin{bmatrix} -4/3 \\ 0 \\ 0 \end{bmatrix},$

$$f(x_2) = \frac{-73}{3},$$

$$d_1 = \begin{bmatrix} -\sqrt{2}/2\sqrt{3} \\ -3\sqrt{2}/2\sqrt{3} \\ 0 \end{bmatrix}, \quad D_2^{-1} = \begin{bmatrix} 1 & 0 & 0 \\ -1/3 & -\sqrt{2}/\sqrt{3} & 0 \\ -1/3 & 1/\sqrt{6} & -1/\sqrt{2} \end{bmatrix},$$

$$J_2 = \{0, -1, -1\}, \quad j = 2.$$

Iteration 2

Step 1: $\quad |g_2'c_{12}| = \max\left\{ |\frac{-4}{3}|, -, - \right\} = \frac{4}{3}, \quad k = 1,$

$$s_2 = \begin{bmatrix} -1 \\ 1/3 \\ 1/3 \end{bmatrix}.$$

Step 2: $\quad s_2'Cs_2 = \frac{4}{3}, \quad \sigma_2 = \frac{4/3}{4/3} = 1.$

Step 3: $x_3 = \begin{bmatrix} 0 \\ 7/3 \\ 10/3 \end{bmatrix} - \begin{bmatrix} -1 \\ 1/3 \\ 1/3 \end{bmatrix} = \begin{bmatrix} 1 \\ 2 \\ 3 \end{bmatrix}$, $g_3 = \begin{bmatrix} 0 \\ 0 \\ 0 \end{bmatrix}$,

$f(x_3) = -25$,

$d_2 = \begin{bmatrix} -2\sqrt{3}/3 \\ 0 \\ 0 \end{bmatrix}$, $D_3^{-1} = \begin{bmatrix} -\sqrt{3}/2 & 0 & 0 \\ 1/2\sqrt{3} & -\sqrt{2}/\sqrt{3} & 0 \\ 1/2\sqrt{3} & 1/\sqrt{6} & -1/\sqrt{2} \end{bmatrix}$,

$J_3 = \{-1, -1, -1\}$, $j = 3$.

Iteration 3

Step 1: $\alpha_{i3} \neq 0$ for $i = 1, 2, 3$; stop with optimal solution $x_3 = (1, 2, 3)'$.   ◇

A Matlab program, Alg1.m, which implements Algorithm 1 is given in Figure 3.6, Section 3.6. A second Matlab program which sets up the data for Example 3.3 and the results of applying Alg1.m to it are shown in Figures 3.8 and 3.9, respectively, Section 3.6.

We further illustrate Algorithm 1 by applying it to a problem that is unbounded from below.

Example 3.4

$$\text{minimize}: \ x_1 - x_2 + x_1^2 + x_2^2 + 2x_1x_2.$$

Here

$$c = \begin{bmatrix} 1 \\ -1 \end{bmatrix} \text{ and } C = \begin{bmatrix} 2 & 2 \\ 2 & 2 \end{bmatrix}.$$

Initialization:

$x_0 = \begin{bmatrix} 0 \\ 0 \end{bmatrix}$, $J_0 = \{0, 0\}$, $D_0^{-1} = \begin{bmatrix} 1 & 0 \\ 0 & 1 \end{bmatrix}$, $f(x_0) = 0$,

$g_0 = \begin{bmatrix} 1 \\ -1 \end{bmatrix}$, $j = 0$.

Iteration 0

Step 1: $|g_0'c_{10}| = \max\{|1|, |-1|\} = 1$, $k = 1$,

$$s_0 = \begin{bmatrix} 1 \\ 0 \end{bmatrix}.$$

Step 2: $s_0' C s_0 = 2$, $\sigma_0 = \dfrac{1}{2}$.

Step 3: $x_1 = \begin{bmatrix} 0 \\ 0 \end{bmatrix} - \dfrac{1}{2} \begin{bmatrix} 1 \\ 0 \end{bmatrix} = \begin{bmatrix} -1/2 \\ 0 \end{bmatrix}$, $g_1 = \begin{bmatrix} 0 \\ -2 \end{bmatrix}$,

$$f(x_1) = \dfrac{-1}{4},$$

$$d_0 = \begin{bmatrix} 2/\sqrt{2} \\ 2/\sqrt{2} \end{bmatrix}, \quad D_1^{-1} = \begin{bmatrix} 1/\sqrt{2} & -1 \\ 0 & 1 \end{bmatrix},$$

$$J_1 = \{ -1, 0 \}, \quad j = 1.$$

Iteration 1

Step 1: $| g_1' c_{21} | = \max \{ -, | -2 | \} = 2$, $k = 2$,

$$s_1 = \begin{bmatrix} 1 \\ -1 \end{bmatrix}.$$

Step 2: $s_1' C s_1 = 0$, $x_1 = (-1/2, 0)'$, $s_1 = (1, -1)'$, and stop, the problem is unbounded from below.

This completes Example 3.4.   ◇

A Matlab program which sets up the data for Example 3.4 and a second Matlab program which shows the output from solving it using Algorithm 1 are shown in Figures 3.10 and 3.11, respectively, in Section 3.6.

The amount of computer time required to solve a particular problem using Algorithm 1 is proportional to the total number of arithmetic operations performed (additions/subtractions and multiplications/divisions). The reader will be asked to show (Exercise 3.17) that a total of approximately $\frac{11}{2}n^3$ arithmetic operations are required when Algorithm 1 continues for n iterations. This operation count is approximate in the sense that it ignores terms of degree 2 or smaller. However, these terms will be small in comparison to the cubic term when n is large. The operation count can be reduced by over 50% by performing the calculations in a more efficient, if somewhat indirect, manner (Exercise 3.18).

Two critical properties of Algorithm 1 are given in (3.29) and (3.33). Assuming that they are satisfied at some iteration j, (3.43) and (3.39), respectively, show that they are also satisfied at iteration $j + 1$. Since they are trivially satisfied for $j = 0$, it follows from mathematical induction that they are satisfied for all j. In terms of the index set, these two results are formulated as follows.

Lemma 3.3 Let x_j, g_j, $D_j^{-1} = [\, c_{1j}, \ldots, c_{nj} \,]$, and $J_j = \{\, \alpha_{1j}, \ldots, \alpha_{nj} \,\}$ be obtained from the jth iteration of Algorithm 1. Let $D_j' = [\, d_{1j}, \ldots, d_{nj} \,]$. Then

(a) $g_j' c_{ij} = 0$, for all i with $\alpha_{ij} = -1$.

(b) $d_{ij} = C c_{ij}$, for all i with $\alpha_{ij} = -1$.

There are two points in Algorithm 1 at which termination can occur with the message "Optimal solution obtained x_j." The optimality of x_j in these two cases is justified in the proof of the following theorem.

Theorem 3.9 Let Algorithm 1 be applied to $\min \{\, c'x + \frac{1}{2} x'Cx \,\}$. Then termination occurs after $j \leq n$ iterations with either an optimal solution x_j or the information that the problem is unbounded from below. In the latter case, Algorithm 1 terminates with x_j and s_j such that $f(x_j - \sigma s_j) \to -\infty$ as $\sigma \to +\infty$.

Proof: Exactly j of the α_{ij}'s have value -1 at iteration j. If termination has not occurred prior to iteration n, $\alpha_{1n} = \cdots = \alpha_{nn} = -1$ and Lemma 3.3(a) asserts that

$$g_n' c_{1n} = \cdots = g_n' c_{nn} = 0 \, ;$$

that is,

$$g_n' D^{-1} = 0 \, .$$

This implies that $g_n = 0$ and, by Theorem 3.3, x_n is indeed optimal.

Next suppose that termination occurs with an optimal solution at some iteration $j < n$. From Step 1, this means that $g_j' c_{kj} = 0$. The definition of k implies that

$$g_j' c_{ij} = 0 \, , \quad \text{for all } i \text{ with } \alpha_{ij} = 0 \, . \tag{3.46}$$

Letting $D_j' = [\, d_{1j}, \ldots, d_{nj} \,]$, Lemma 3.1 implies that

$$g_j = \sum_{\alpha_{ij} = -1} (g_j' c_{ij}) d_{ij} + \sum_{\alpha_{ij} = 0} (g_j' c_{ij}) d_{ij} \, ,$$

where

$$\sum_{\alpha_{ij} = -1} \quad \text{and} \quad \sum_{\alpha_{ij} = 0}$$

mean the sum over all i with $\alpha_{ij} = -1$ and $\alpha_{ij} = 0$, respectively. From Lemma 3.3(a), all coefficients in the first sum vanish and from (3.46) so do all of those in the second sum.

Finally, if termination occurs with the message that the problem is unbounded from below, then from Step 2, $s_j'Cs_j = 0$. From Taylor's series,

$$f(x_j - \sigma s_j) = f(x_j) - \sigma g_j's_j,$$

and from Step 1, $g_j's_j > 0$. This implies that $f(x_j - \sigma s_j) \to -\infty$ as $\sigma \to +\infty$ and the problem is indeed unbounded from below. $\qquad \square$

On completion of Algorithm 1, there may be many α_{ij}'s with value zero. This would be the case, for example, if x_0 just happened to be an optimal solution. Since $g(x_0) = 0$, Algorithm 1 would terminate in Step 1 with $\alpha_{10} = \cdots = \alpha_{n0} = 0$. For some types of post-optimality analysis, it is useful to have as many conjugate directions as possible in the final data and we next describe how this can be done.

Let Algorithm 1 terminate at iteration j with optimal solution x_j and let $D_j^{-1} = [c_{1j}, \ldots, c_{nj}]$ and $J_j = \{\alpha_{1j}, \ldots, \alpha_{nj}\}$ be the associated data. If there is an index k with $\alpha_{kj} = 0$ and $c_{kj}'Cc_{kj} > 0$, then we can set

$$\left.\begin{aligned}
d_j &= (c_{kj}Cc_{kj})^{-1/2}Cc_{kj}, \\
D_{j+1}^{-1} &= \Phi_1(D_j^{-1}, d_j, k, J_j), \\
J_{j+1} &= \{\alpha_{1,j+1}, \ldots, \alpha_{n,j+1}\}, \quad \text{where} \\
\alpha_{i,j+1} &= \alpha_{ij}, \quad i = 1, \ldots, n, \quad i \neq k, \\
\alpha_{k,j+1} &= -1, \quad \text{and} \\
x_{j+1} &= x_j.
\end{aligned}\right\} \tag{3.47}$$

Doing this increases the number of conjugate directions by exactly one. The updating may be continued until at some subsequent iteration ν either

$$\alpha_{i\nu} \neq 0, \quad \text{for all } i = 1, \ldots, n, \quad \text{or} \tag{3.48}$$

$$c_{i\nu}'Cc_{i\nu} = 0, \text{ for all } i \text{ with } \alpha_{i\nu} = 0. \tag{3.49}$$

When (3.48) or (3.49) is satisfied, we say that $\{c_{i\nu} \mid \text{all } i \text{ with } \alpha_{i\nu} = -1\}$ is a <u>complete set of normalized conjugate directions</u> for C. The reader will be asked to show (Exercise 3.13) that the number of conjugate directions in a complete set is rank C. Note that from Exercise 3.7(a), (3.49) is equivalent to

$$Cc_{i\nu} = 0 , \quad \text{for all } i \text{ with } \alpha_{i\nu} = 0 .$$

3.4 QP With Linear Equality Constraints: Theory

In this section, we consider the quadratic minimization problem subject to linear equality constraints

$$\min \{ \, c'x + \tfrac{1}{2}x'Cx \mid a_i'x = b_i , \quad i = 1,\ldots,r \, \} , \tag{3.50}$$

where the objective function $f(x) = c'x + \tfrac{1}{2}x'Cx$ is as in the previous two sections, $a_1,\ldots,a_r$ are n-vectors, and $b_1,\ldots,b_r$ are scalars. The <u>gradient</u> of the ith constraint is a_i. The model problem is thus to minimize a quadratic function of n variables subject to r linear equality constraints. An equivalent formulation of (3.50) is

$$\min \{ \, c'x + \tfrac{1}{2}x'Cx \mid Ax = b \, \} , \tag{3.51}$$

where

$$A' = [\, a_1,\ldots,a_r \,] \quad \text{and} \quad b = (\, b_1,\ldots,b_r \,)' .$$

The <u>feasible</u> <u>region</u> for (3.51) is

$$R = \{ \, x \mid Ax = b \, \} .$$

A point x_0 is <u>feasible</u> for (3.51) if $x_0 \in R$ and <u>infeasible</u> otherwise. x_0 is a <u>global</u> <u>minimizer</u> or an <u>optimal</u> <u>solution</u> (or simply <u>optimal</u>) for (3.51) if $x_0 \in R$ and $f(x_0) \leq f(x)$ for all $x \in R$. The objective function for (3.51) is <u>unbounded</u> <u>from</u> <u>below</u> over its feasible region [or more briefly, problem (3.51) is unbounded from below] if there are n-vectors x_0 and s_0 such that $Ax_0 = b$, $As_0 = 0$, and $f(x_0 - \sigma s_0) \to -\infty$ as $\sigma \to +\infty$.

Let x_0 be an optimal solution. We first obtain a necessary condition for optimality. From Theorem 3.1, (Orthogonal Decomposition Theorem), $g(x_0) = c + Cx_0$ may be written as

$$g(x_0) = h_1 + h_2 ,$$

where $h_1 \in$ span $\{\, a_1, \ldots, a_r \,\}$ and $Ah_2 = 0$. We may write h_1 as $h_1 = -A'u$ for some r-vector u.[5] Then

$$g'(x_0)h_2 = h'_1 h_2 + h'_2 h_2 = -u'Ah_2 + h'_2 h_2,$$

so that

$$g'(x_0)h_2 = h'_2 h_2. \tag{3.52}$$

Here, it is helpful to think of h_2 as a search direction. Note that for any scalar σ

$$A(x_0 - \sigma h_2) = Ax_0 = b\,,$$

so $x_0 - \sigma h_2 \in R$ for all σ. From Taylor's series,

$$f(x_0 - \sigma h_2) = f(x_0) - \sigma g'(x_0)h_2 + \tfrac{1}{2}\sigma^2 h'_2 Ch_2$$

which gives $f(x_0 - \sigma h_2)$ as an explicit quadratic function of σ. Since x_0 is optimal, $f(x_0 - \sigma h_2)$ must be minimized when $\sigma = 0$. It follows from elementary calculus that its derivative must vanish when $\sigma = 0$; that is,

$$g'(x_0)h_2 = 0\,.$$

But from (3.52), $g'(x_0)h_2$ is the sum of the squares of the components of h_2. Thus $h_2 = 0$ and

$$-g(x_0) = A'u. \tag{3.53}$$

Letting $u = (\, u_1, \ldots, u_r \,)'$, (3.53) may be written as

$$-g(x_0) = u_1 a_1 + \cdots + u_r a_r\,.$$

This shows that the negative gradient of the objective at an optimal solution can be written as a linear combination of the gradients of the equality constraints. Each u_i is called the <u>multiplier</u> <u>associated</u> <u>with</u> <u>constraint</u> i, $i = 1, \ldots, r$.

We have proven

Theorem 3.10 *Let x_0 be an optimal solution for* min $\{\, c'x + \tfrac{1}{2}x'Cx \mid Ax = b \,\}$. *Then $Ax_0 = b$, and there exists an r-vector u with $-g(x_0) = A'u$.*

Theorem 3.10 asserts that $-g(x_0) = A'u$ is a necessary condition for optimality. Provided that $f(x)$ is convex, this condition is also sufficient for optimality, as we show in

---

[5] We could equally as well write $h_1 = A'u$. However, we introduce the minus sign here to make the final characterization more compatible with that for inequality constraints developed in the next chapter.

Theorem 3.11 *Let $f(x) = c'x + \frac{1}{2}x'Cx$ be convex. Then x_0 is an optimal solution for* min $\{ c'x + \frac{1}{2}x'Cx \mid Ax = b \}$ *if and only if $Ax_0 = b$ and there is an r-vector u with $-g(x_0) = A'u$.*

Proof: The "only if" part is a restatement of Theorem 3.10. To prove the "if" part, assume that x_0 satisfies $Ax_0 = b$ and that there is an r-vector u satisfying $-g(x_0) = A'u$. We must show that x_0 is optimal. Let x be any point in R. Since $Ax_0 = b$ and $Ax = b$, subtraction gives

$$A(x - x_0) = 0 . \tag{3.54}$$

From Taylor's series,

$$f(x) = f(x_0) + g'(x_0)(x - x_0) + \tfrac{1}{2}(x - x_0)'C(x - x_0) . \tag{3.55}$$

Because $-g(x_0) = A'u$, it follows from (3.54) that

$$g'(x_0)(x - x_0) = -u'A(x - x_0) = 0 .$$

Substitution of this into (3.55) gives

$$f(x) - f(x_0) = \tfrac{1}{2}(x - x_0)'C(x - x_0) .$$

Because f is convex, $f(x) \geq f(x_0)$. Since x is an arbitrary point in R, it follows that

$$f(x_0) \leq f(x) , \quad \text{for all } x \in R ,$$

and x_0 is indeed optimal. □

The optimality conditions of Theorem 3.11 are illustrated in

Example 3.5

$$\begin{aligned}
\text{minimize:} \quad & -4x_1 - 4x_2 + x_1^2 + x_2^2 \\
\text{subject to:} \quad & x_1 + x_2 = 2. \quad (1)
\end{aligned}$$

This problem is illustrated in Figure 3.4. The feasible region for this problem is the line $x_1 + x_2 = 2$ from which it is clear that the optimal solution is the point at which the level set for the objective function is tangent to the line $x_1 + x_2 = 2$. This point is $x_0 = (1 , 1)'$. Because $g(x_0) = (-2 , -2)'$, the optimality condition $-g(x_0) = A'u$ becomes

$$-g(x_0) = \begin{bmatrix} 2 \\ 2 \end{bmatrix} = u_1 \begin{bmatrix} 1 \\ 1 \end{bmatrix}$$

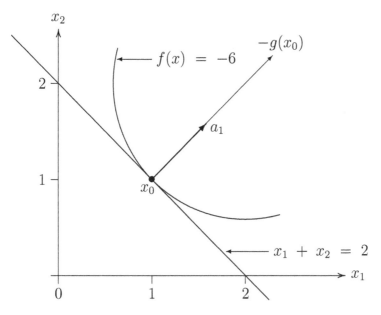

Figure 3.4.   Geometry of Example 3.5.

and is satisfied with $u_1 = 2$.   ◊

Theorems 3.10 and 3.11 are generalizations of Theorems 3.2 and 3.3, respectively, to the case of quadratic minimization subject to linear equality constraints. When $r = 0$, the condition $-g(x_0) = A'u$ reduces to $g(x_0) = 0$.

Theorem 3.11 may be used to show that

$$\min \{ c'x + \tfrac{1}{2}x'Cx \mid Ax = b \} \tag{3.56}$$

may be solved by solving certain linear equations, provided that the objective function is convex. An optimal solution x_0 must satisfy

$$Ax_0 = b \text{ and } -g(x_0) = A'u .$$

Replacing $g(x_0)$ with $c + Cx_0$ and writing the results in partitioned matrix form gives

$$\begin{bmatrix} C & A' \\ A & 0 \end{bmatrix} \begin{bmatrix} x_0 \\ u \end{bmatrix} = \begin{bmatrix} -c \\ b \end{bmatrix} . \tag{3.57}$$

These are $n + r$ linear equations in the $n + r$ unknowns represented by the components of x_0 and u. They may have either no solution, a unique solution, or an infinite number of solutions. The first possibility corresponds to either R being null or (3.56) being unbounded from below. The second possibility corresponds to a unique optimal solution, and the third corresponds to either alternate optimal solutions, a multiplicity of choices for u, or both. If the coefficient matrix of the equations (3.57) is nonsingular, then of course the equations possess a unique solution and consequently (3.56) has a unique optimal solution. The reader will be asked to verify (Exercise 3.21) that the coefficient matrix of (3.57) is nonsingular if and only if a certain condition is satisfied. A special case of this result is that the coefficient matrix will be nonsingular provided that A has full row rank and that C is positive definite.

Analogous to Theorem 3.4, we prove

Theorem 3.12 *Let x_0 be an optimal solution for* $\min \{ c'x + \frac{1}{2}x'Cx \mid Ax = b \}$. *If $f(x) = c'x + \frac{1}{2}x'Cx$ is strictly convex, then x_0 is the unique optimal solution.*

Proof: Suppose to the contrary that x_1 is also an optimal solution and that $x_1 \neq x_0$. Since $Ax_1 = b$ and $Ax_0 = b$, subtracting gives $A(x_1 - x_0) = 0$. Since x_0 is optimal, Theorem 3.10 asserts that $-g(x_0) = A'u$. With $A(x_1 - x_0) = 0$, this gives

$$g'(x_0)(x_1 - x_0) = -u'A(x_1 - x_0) = 0 .$$

With Taylor's series, this last gives

$$f(x_1) = f(x_0) + \tfrac{1}{2}(x_1 - x_0)'C(x_1 - x_0) .$$

Both x_1 and x_0 are optimal, so $f(x_1) = f(x_0)$. Thus

$$(x_1 - x_0)'C(x_1 - x_0) = 0 .$$

The definition of strict convexity implies $(x_1 - x_0) = 0$, or, that $x_1 = x_0$. The assumption that there is a second optimal solution leads to a contradiction and is therefore false.   □

Thus far, we have assumed the existence of an optimal solution for (3.51). Analogous to Theorem 3.7, we next demonstrate the existence of an optimal solution provided that R is nonnull and the objective function is bounded from below over R.

Theorem 3.13 *Suppose that there is a γ with $f(x) = c'x + \frac{1}{2}x'Cx \geq \gamma$ for all $x \in R = \{ x \mid Ax = b \}$ and $R \neq \emptyset$. Then there exists an optimal solution for $\min \{ c'x + \frac{1}{2}x'Cx \mid Ax = b \}$.*

Proof: Let x_0 be any point in R. Let

$$H = \begin{bmatrix} C & A' \\ A & 0 \end{bmatrix} \quad \text{and} \quad h = \begin{bmatrix} -c \\ b \end{bmatrix}.$$

Using the Orthogonal Decomposition Theorem (Theorem 3.1), we can write h as $h = h_1 + h_2$, where

$$h_1 = Hw, \quad H'h_2 = 0,$$

and w is an $(n + r)$-vector. Partition h_2 as $h_2 = (s', v')'$, where s and v are n- and r-vectors, respectively. Writing out the partitions of $H'h_2 = 0$ gives

$$Cs + A'v = 0, \tag{3.58}$$

$$As = 0. \tag{3.59}$$

Premultiplying (3.58) by s' and using (3.59) gives

$$s'Cs = 0. \tag{3.60}$$

Premultiplying (3.58) by x_0' gives $x_0'Cs = -(Ax_0)'v$. Because $g(x_0) = c + Cx_0$ and $Ax_0 = b$, we have

$$g'(x_0)s = c's + x_0'Cs = c's - (Ax_0)'v$$
$$= c's - b'v.$$

From the definition of h and h_2, $h'h_2 = -(c's - b'v)$. Furthermore, the Orthogonal Decomposition Theorem (Theorem 3.1) asserts that $h'h_2 = h_2'h_2$. Therefore

$$g'(x_0)s = -h_2'h_2. \tag{3.61}$$

From (3.59), $A(x_0 + \sigma s) = Ax_0 = b$ for all scalars σ. Furthermore, from Taylor's series, (3.60), and (3.61),

$$f(x_0 + \sigma s) = f(x_0) - \sigma h_2'h_2. \tag{3.62}$$

Now if $h_2 \neq 0$, $h_2'h_2$ being the sum of the squares of the components of h_2 must be strictly positive. Then (3.62) implies that $f(x_0 + \sigma s) \rightarrow -\infty$ as $\sigma \rightarrow +\infty$. But this would be in contradiction to the hypothesis that f is bounded from below over R. Therefore $h_2 = 0$ and $h = h_1$.

Partitioning w as $w = (x_1', u')'$, the partitions defining h may be written

$$Cx_1 + A'u = -c, \tag{3.63}$$

$$Ax_1 = b.$$

We have now shown the existence of an x_1 satisfying $-g(x_1) = A'u$ and $Ax_1 = b$. It remains to show that x_1 is optimal. Note that Theorem 3.11 cannot be used to make this conclusion because we have not assumed that f is convex.

Let x be any point in R. Since $Ax = b$ and $Ax_1 = b$, subtraction gives $A(x - x_1) = 0$. Also, (3.63) is equivalent to $-g(x_1) = A'u$. Thus $g'(x_1)(x - x_1) = -u'A(x - x_1) = 0$ and, from Taylor's series,

$$
\begin{aligned}
f(x) &= f(x_1) + g'(x_1)(x - x_1) + \tfrac{1}{2}(x - x_1)'C(x - x_1) \\
&= f(x_1) + \tfrac{1}{2}(x - x_1)'C(x - x_1).
\end{aligned}
\tag{3.64}
$$

Suppose that $(x - x_1)'C(x - x_1) < 0$. Then $A[x_1 + \sigma(x - x_1)] = Ax_1 = b$ so $x_1 + \sigma(x - x_1) \in R$ for all scalars σ. It follows from (3.64) that

$$
f(x_1 + \sigma(x - x_1)) = f(x_1) + \tfrac{1}{2}\sigma^2(x - x_1)'C(x - x_1)
$$

and thus $f(x_1 + \sigma(x - x_1)) \to -\infty$ as $\sigma \to +\infty$, in contradiction to the hypothesis of the theorem. The assumption that $(x - x_1)'C(x - x_1) < 0$ leads to a contradiction and is therefore false. Thus $(x - x_1)'C(x - x_1) \geq 0$ for all $x \in R$ and from (3.64),

$$
f(x_1) \leq f(x), \quad \text{for all } x \in R.
$$

By definition, x_1 is optimal. □

3.5 QP With Linear Equality Constraints: Algorithm 2

In this section, we present an algorithm for the solution of the quadratic minimization problem subject to linear equality constraints

$$
\min \{ c'x + \tfrac{1}{2}x'Cx \mid a_i'x = b_i, \ i = 1, \ldots, r \}.
\tag{3.65}
$$

We assume that the objective function is convex. We also assume that $a_1, \ldots, a_r$ are linearly independent and that a feasible point for (3.65) is known. These latter two assumptions will be removed in Chapter 5. The algorithm for the solution of (3.65) is a simple modification of Algorithm 1, requiring only a change in the initial data. The algorithm will terminate in at most $n - r$ steps with either an optimal solution, or the information that (3.65) is unbounded from below.

Let x_0 be any feasible point for (3.65). Similar to Algorithm 1, we would like to construct a finite sequence of points x_1, x_2, ... according to

$$x_{j+1} = x_j - \sigma s_j .$$

Suppose that $x_j \in R$. Then for any constraint i, $a_i'x_j = b_i$ and constraint i will also be satisfied by x_{j+1}; that is, $a_i'x_{j+1} = b_i$, provided that $a_i's_j = 0$. Therefore each of x_1, ... will be feasible provided that

$$a_i's_j = 0 , \quad i = 1,\ldots,r , \tag{3.66}$$

for $j = 0$, 1, $\ldots$.

Now let D_0' be any nonsingular matrix for which the first r columns are $a_1,\ldots,a_r$, respectively. Let the remaining columns be denoted by $d_{r+1},\ldots,$ d_n, and let $c_1,\ldots,c_n$ denote the columns of D_0^{-1}. Then

$$D_0' = [\, a_1,\ldots,a_r , \ d_{r+1},\ldots,d_n \,] \quad \text{and}$$
$$D_0^{-1} = [\, c_1,\ldots,c_r , \ c_{r+1},\ldots,c_n \,] .$$

If s_0 is chosen parallel to any of $c_{r+1},\ldots,c_n$, then by definition of the inverse matrix, (3.66) will be satisfied for $j = 0$. This suggests that an algorithm for the solution of (3.65) can be obtained from Algorithm 1 by first, beginning with an x_0 and D_0^{-1} as just described and second, removing the first r columns of D_0^{-1} as candidates for the search direction. Suppose that this is done and, for simplicity, suppose that s_0 is chosen parallel to column $r + 1$ of D_0^{-1}, s_1 is chosen parallel to column $r + 2$ of D_1^{-1}, ... etc. At iteration j we have

$$D_j' = [a_1,\ldots,a_r , \ Cc_{r+1},\ldots,Cc_{r+j} , \ d_{r+j+1},\ldots,d_n] \tag{3.67}$$

and

$$D_j^{-1} = [\, c_1,\ldots,c_r , \quad c_{r+1},\ldots, \quad c_{r+j} , \ c_{r+j+1},\ldots,c_n \,] ,$$

where we have suppressed the iteration index j on the columns of D_j' and D_j^{-1}. If s_j is chosen parallel to any of $c_{r+j+1},\ldots,c_n$, then by definition of the inverse matrix, (3.66) will be satisfied and consequently x_{j+1} will be feasible.

Lemma 3.3 asserts that

$$g_j'c_{r+1} = g_j'c_{r+2} = \cdots = g_j'c_{r+j} = 0 .$$

Thus from Lemma 3.1 ,

$$-g_j = (-g_j'c_1)a_1 + \cdots + (-g_j'c_r)a_r$$
$$+ (-g_j'c_{r+j+1})d_{r+j+1} \quad + \cdots + (-g_j'c_n)d_n. \tag{3.68}$$

This shows that $-g_j$ is the sum of two vectors. The first is in span $\{\, a_1,\ldots,a_r \,\}$ and the second is in span $\{\, d_{r+j+1},\ldots,d_n \,\}$. The vectors $d_{r+j+1},\ldots,d_n$ are

$n - r - j$ in number and this number decreases by one at each iteration. If the algorithm continues for $n - r$ iterations, then no such vectors remain and (3.68) becomes

$$-g_{n-r} = (-g'_{n-r}c_1)a_1 + \cdots + (-g'_{n-r}c_r)a_r . \tag{3.69}$$

With $u_1 = -g'_{n-r}c_1, \ldots, u_r = -g'_{n-r}c_r$, it follows from Theorem 3.11 that x_{n-r} is optimal. If termination occurs with optimal solution x_j for some $j < n - r$ then from Step 1,

$$g'_j c_{r+j+1} = \cdots = g'_j c_n = 0 ,$$

(3.68) reduces to (3.69), and again x_j is optimal. Finally, if termination occurs at iteration j with s_j and the message that (3.65) is unbounded from below then s_j satisfies (3.66) so that $x_j - \sigma s_j \in R$ for all scalars σ. Furthermore, $g'_j s_j > 0$ by construction and $s'_j C s_j = 0$ because of the manner in which termination occurred. Taylor's series then reduces to

$$f(x_j - \sigma s_j) = f(x_j) - \sigma g'_j s_j$$

and $f(x_j - \sigma s_j) \to -\infty$ and $\sigma \to +\infty$. Thus, (3.65) is indeed unbounded from below.

We have developed the algorithm under the simplifying order assumption implicit in (3.67). The ordering assumption may be removed by using the ordered index set $J_j = \{ \alpha_{1j}, \ldots, \alpha_{nj} \}$ of Algorithm 1, appropriately modified to account for the equality constraints. A simple way to do this is to set $\alpha_{ij} = k$ if column i of D'_j is a_k. For example, with this convention the index set corresponding to (3.67) is

$$J_j = \{1, 2, \ldots, r, -1, -1, \ldots, -1, 0, 0, \ldots, 0\}.$$

Because Step 1 of Algorithm 1 chooses s_j parallel to some c_{kj} with $\alpha_{kj} = 0$ and all columns of D'_j which are gradients of constraints having strictly positive indices, it follows that (3.66) will be satisfied at each iteration.

In order to initiate the algorithm, we require any $x_0 \in R$, a nonsingular $D'_0 = [d_1, \ldots, d_n]$, D_0^{-1}, and an index set $J_0 = \{ \alpha_{10}, \ldots, \alpha_{n0} \}$ satisfying $0 \le \alpha_{i0} \le r, i = 1, \ldots, n, d_i = a_{\alpha_{i0}}$ for all i with $1 \le a_{\alpha_{i0}} \le r$ and each of $1, 2, \ldots, r$ is in J_0. This means that each $a_1, \ldots, a_r$ must be a column of D'_0 and the associated index is the corresponding index of J_0. The remaining indices of J_0 have value 0. As with Algorithm 1, the matrix D_0 is required only for explanation and need not be stored by the algorithm.

We now give a detailed formulation of the algorithm.

ALGORITHM 2

Model Problem:

$$\text{minimize:} \quad c'x + \tfrac{1}{2}x'Cx$$
$$\text{subject to:} \quad a_i'x = b_i, \quad i = 1,\ldots,r.$$

Initialization:
Start with any feasible x_0, set $J_0 = \{\alpha_{10},\ldots,\alpha_{n0}\}$, and D_0^{-1}, where $D_0' = [\, d_1,\ldots,d_n \,]$ is nonsingular, $0 \le \alpha_{i0} \le r$ for $i = 1,\ldots,n$, $d_i = \alpha_{\alpha_{i0}}$ for all i with $1 \le \alpha_{i0} \le r$, and, each of $1,\ldots,r$ is in J_0. Compute $f(x_0) = c'x_0 + \tfrac{1}{2}x_0'Cx_0$, $g_0 = c + Cx_0$, and set $j = 0$.

Step 1:   **Computation of Search Direction s_j.**
Same as Step 1 of Algorithm 1.

Step 2:   **Computation of Step Size σ_j.**
Same as Step 2 of Algorithm 1.

Step 3:   **Update.**
Same as Step 3 of Algorithm 1.

We illustrate Algorithm 2 by applying it to the following example.

Example 3.6

$$\text{minimize:} \quad -12x_1 - 13x_2 - 14x_3 + x_1^2 + x_2^2 + x_3^2$$
$$+ x_1x_2 + x_1x_3 + x_2x_3$$
$$\text{subject to:} \quad x_1 + x_2 + x_3 = 9. \quad (1)$$

Here

$$c = \begin{bmatrix} -12 \\ -13 \\ -14 \end{bmatrix} \quad \text{and} \quad C = \begin{bmatrix} 2 & 1 & 1 \\ 1 & 2 & 1 \\ 1 & 1 & 2 \end{bmatrix}.$$

Initialization:

$$x_0 = \begin{bmatrix} 9 \\ 0 \\ 0 \end{bmatrix}, \quad J_0 = \{1, 0, 0\}, \quad D_0^{-1} = \begin{bmatrix} 1 & -1 & -1 \\ 0 & 1 & 0 \\ 0 & 0 & 1 \end{bmatrix},$$

$$f(x_0) = -27, \quad g_0 = \begin{bmatrix} 6 \\ -4 \\ -5 \end{bmatrix}, \quad j = 0.$$

Iteration 0

Step 1: $|g_0' c_{30}| = \max\{-, |-10|, |-11|\} = 11, \quad k = 3,$

$$s_0 = \begin{bmatrix} 1 \\ 0 \\ -1 \end{bmatrix}.$$

Step 2: $s_0' C s_0 = 2, \quad \sigma_0 = \dfrac{11}{2}.$

Step 3: $x_1 = \begin{bmatrix} 9 \\ 0 \\ 0 \end{bmatrix} - \dfrac{11}{2} \begin{bmatrix} 1 \\ 0 \\ -1 \end{bmatrix} = \begin{bmatrix} 7/2 \\ 0 \\ 11/2 \end{bmatrix}, \quad g_1 = \begin{bmatrix} 1/2 \\ -4 \\ 1/2 \end{bmatrix},$

$$f(x_1) = \dfrac{-229}{4},$$

$$d_0 = \begin{bmatrix} 1/\sqrt{2} \\ 0 \\ -1/\sqrt{2} \end{bmatrix}, \quad D_1^{-1} = \begin{bmatrix} 1/2 & -1/2 & 1/\sqrt{2} \\ 0 & 1 & 0 \\ 1/2 & -1/2 & -1/\sqrt{2} \end{bmatrix},$$

$$J_1 = \{1, 0, -1\}, \quad j = 1.$$

Iteration 1

Step 1: $|g_1' c_{21}| = \max\left\{-, \left|\dfrac{-9}{2}\right|, -\right\} = \dfrac{9}{2}, \quad k = 2,$

$$s_1 = \begin{bmatrix} 1/2 \\ -1 \\ 1/2 \end{bmatrix}.$$

Step 2: $s_1' C s_1 = \dfrac{3}{2}, \quad \sigma_1 = \dfrac{9/2}{3/2} = 3.$

Step 3: $x_2 = \begin{bmatrix} 7/2 \\ 0 \\ 11/2 \end{bmatrix} - 3 \begin{bmatrix} 1/2 \\ -1 \\ 1/2 \end{bmatrix} = \begin{bmatrix} 2 \\ 3 \\ 4 \end{bmatrix}, \quad g_2 = \begin{bmatrix} -1 \\ -1 \\ -1 \end{bmatrix},$

$$f(x_2) = -64,$$

$$d_1 = \begin{bmatrix} \sqrt{2}/2\sqrt{3} \\ -\sqrt{2}/\sqrt{3} \\ \sqrt{2}/2\sqrt{3} \end{bmatrix}, \quad D_2^{-1} = \begin{bmatrix} 1/3 & 1/\sqrt{6} & 1/\sqrt{2} \\ 1/3 & -2/\sqrt{3} & 0 \\ 1/3 & 1/\sqrt{6} & -1/\sqrt{2} \end{bmatrix},$$

$$J_2 = \{1, -1, -1\}, \quad j = 2.$$

Iteration 2

Step 1: $\alpha_{i2} \neq 0$ for $i = 1, 2, 3$; stop with optimal solution $x_2 = (2, 3, 4)'$.   ◇

A Matlab program which sets up the data for Example 3.6 and a second which shows the results are shown in Figures 3.12 and 3.13, respectively in Section 3.6.

We further illustrate Algorithm 2 by applying it to a problem that is unbounded from below.

Example 3.7

$$
\begin{aligned}
\text{minimize:} \quad & -2x_1 - 2x_2 + x_3 \\
& + \tfrac{1}{2}(x_1^2 + 4x_2^2 + x_3^2 + 4x_1x_2 - 2x_1x_3 - 4x_2x_3) \\
\text{subject to:} \quad & x_1 - x_2 + x_3 = 1. \quad (1)
\end{aligned}
$$

Here

$$
c = \begin{bmatrix} -2 \\ -2 \\ 1 \end{bmatrix} \quad \text{and} \quad C = \begin{bmatrix} 1 & 2 & -1 \\ 2 & 4 & -2 \\ -1 & -2 & 1 \end{bmatrix}.
$$

Initialization:

$$
x_0 = \begin{bmatrix} 0 \\ 0 \\ 1 \end{bmatrix}, \quad J_0 = \{0, 0, 1\}, \quad D_0^{-1} = \begin{bmatrix} 1 & 0 & 0 \\ 0 & 0 & 0 \\ -1 & 1 & 1 \end{bmatrix},
$$

$$
f(x_0) = \frac{3}{2}, \quad g_0 = \begin{bmatrix} -3 \\ -4 \\ 2 \end{bmatrix}, \quad j = 0.
$$

Iteration 0

Step 1: $|g_0'c_{10}| = \max\{|-5|, |-2|, -\} = 5, \quad k = 1,$

$$
s_0 = \begin{bmatrix} -1 \\ 0 \\ 1 \end{bmatrix}.
$$

Step 2: $s_0'Cs_0 = 4$, $\sigma_0 = \dfrac{5}{4}$.

Step 3: $x_1 = \begin{bmatrix} 0 \\ 0 \\ 1 \end{bmatrix} - \dfrac{5}{4} \begin{bmatrix} -1 \\ 0 \\ 1 \end{bmatrix} = \begin{bmatrix} 5/4 \\ 0 \\ -1/4 \end{bmatrix}$, $g_1 = \begin{bmatrix} -1/2 \\ 1 \\ -1/2 \end{bmatrix}$,

$$f(x_1) = \dfrac{-13}{8} ,$$

$$d_0 = \begin{bmatrix} -1 \\ -2 \\ 1 \end{bmatrix} , \quad D_1^{-1} = \begin{bmatrix} -1/2 & -1/2 & 1/2 \\ 0 & 1 & 0 \\ 1/4 & 3/2 & 1/2 \end{bmatrix} ,$$

$$J_1 = \{ -1 , 0 , 1 \} , \quad j = 1 .$$

Iteration 1

Step 1: $|g_1'c_{21}| = \max \left\{ - , \left| \dfrac{1}{2} \right| , - \right\} = \dfrac{1}{2}$, $k = 2$,

$$s_1 = \begin{bmatrix} -1/2 \\ 1 \\ 3/2 \end{bmatrix} .$$

Step 2: $s_1'Cs_1 = 0$, $x_1 = (5/4 , 0 , -1/4)'$, $s_1 = (-1/2 , 1 , 3/2)'$, and stop, the objective function is unbounded from below. $\Diamond$

A Matlab program which formulates the data for Example 3.7 and its solution by Algorithm 1 are given in Figures 3.14 and 3.15, Section 3.6. Algorithm 2 differs from Algorithm 1 only in the initial data requirements. Consequently, Lemma 3.3 applies to Algorithm 2 as well as to Algorithm 1. A Matlab program which implements Algorithm 2 is given in Figure 3.16. It is used to establish the properties of Algorithm 2 in

Theorem 3.14 *Let Algorithm 2 be applied to* min $\{ c'x + \frac{1}{2}x'Cx \mid Ax = b \}$. *Then termination occurs after* $j \leq n - r$ *iterations with either an optimal solution* x_j *or the information that the problem is unbounded from below. In the former case, the multipliers for the constraints are given by* $u_{\alpha_{ij}} = -g_j'c_{ij}$ *for all i with* $1 \leq \alpha_{ij} \leq r$, *where* $D_j^{-1} = [c_{1j}, \ldots, c_{nj}]$ *and* $J_j = \{ \alpha_{1j}, \ldots, \alpha_{nj} \}$. *In the latter case, Algorithm 2 terminates with* x_j *and* s_j *such that* $A(x_j - \sigma s_j) = b$ *for all* σ *and* $f(x_j - \sigma s_j) \to -\infty$ *as* $\sigma \to +\infty$.

Proof: Since $\alpha_{kj} = 0$, it follows from the definition of the inverse matrix that

$$a_i's_j = 0 , \quad i = 1, \ldots, r ,$$

for $j = 0, 1, \ldots$. Consequently, each of $x_1, x_2, \ldots$ is feasible.

At iteration j, there are $n - r$ elements of J_j having value zero. At each subsequent iteration, exactly one of these is changed to -1. Therefore termination must occur in at most $n - r$ iterations.

For any iteration $j \geq 0$, let $D'_j = [d_{1j}, \ldots, d_{nj}]$. Then Lemma 3.2 implies that

$$-g_j = \sum_{\alpha_{ij} = -1} (-g'_j c_{ij}) d_{ij} + \sum_{\alpha_{ij} = 0} (-g'_j c_{ij}) d_{ij} + \sum_{1 \leq \alpha_{ij} \leq r} (-g'_j c_{ij}) d_{ij}.$$

From Lemma 3.3, every coefficient in the first term has value zero. When Algorithm 2 terminates with "optimal solution x_j," either there are no α_{ij}'s with value zero or, from Step 1,

$$g'_j c_{ij} = 0, \quad \text{for all } i \text{ with } \alpha_{ij} = 0.$$

In either case the second summation vanishes and

$$-g_j = \sum_{1 \leq \alpha_{ij} \leq r} (-g'_j c_{ij}) d_{ij}. \tag{3.70}$$

For all i with $1 \leq \alpha_{ij} \leq r$ define

$$u_{\alpha_{ij}} = -g'_j c_{ij}$$

and note that $d_{ij} = a_{\alpha_{ij}}$. Thus (3.70) becomes

$$-g_j = u_1 a_1 + \cdots + u_r a_r$$

and the optimality of x_j follows from Theorem 3.10.

Finally, if termination occurs with the message that "the problem is unbounded from below," then from Step 2 $s'_j C s_j = 0$ and from Step 1, $g'_j s_j > 0$. From Taylor's series,

$$f(x_j - \sigma s_j) = f(x_j) - \sigma g'_j s_j$$

so that $f(x_j - \sigma s_j) \to -\infty$ as $\sigma \to +\infty$ and the problem is indeed unbounded from below. $\qquad \square$

Theorem 3.14 gives an explicit formula for obtaining the multipliers from the final data of Algorithm 2. These multipliers will play an important role in Chapter 5 in solving quadratic programming problems with linear inequality constraints.

The problem of Example 3.6 was solved using Algorithm 2. The algorithm terminated with an optimal solution at iteration 2. From the final data and Theorem 3.14, we obtain the single multiplier

$$u_1 = -g'_2 c_{12} = 1.$$

3.6   Computer Programs

In this section, we develop Matlab programs to implement each of the algorithms formulated in Chapter 3. We show the results of applying each algorithm to the text examples used to illustrate the algorithm. When required, we shall also include Matlab programs which facilitate the presentation of these results.

```
1  %  Phi1.m
2  function [DhatInv] = Phi1(Dinv,d,k,J,n)
3  %  Computes DhatInv from Dinv, d, k using Procdure Phi1.
4  %  Suppresses update of conjugate direction columns.
5  check = d' * Dinv(:,k);
6  if abs(check) < 1.e-6
7      errmsg = 'new matrix is singular'
8      soln = 'Press Ctrl+C to continue'
9      pause
10     return
11 end
12 check = 1. / check;
13 DhatInv(:,k) = Dinv(:,k) * check;
14 for i = 1:n
15     if  i≠k
16         if J(i) ≠ −1
17             dci = d' * Dinv(:,i);
18             DhatInv(:,i) = Dinv(:,i) − dci*DhatInv(:,k);
19         end
20         if J(i) == −1;
21             DhatInv(:,i) = Dinv(:,i);
22         end
23     end
24 end
25 end
```

Figure 3.5.   Procedure Phi1: Phi1.m.

Discussion of Phi1.m Figure 3.5 shows the Matlab code for Procedure Φ_1. The code begins by computing $d'c_k$ (using the notation that defines the procedure) in line 5. If this is too small, lines 7 and 8 print error messages to the console and wait for user intervention. Otherwise, the new k-th column is computed and the nonconjugate direction columns are updated using lines 15 to 19. In lines 20 to 22, the conjugate direction columns are copied unchanged into the new matrix.

```
1   %  Alg1.m
2   function[x,J,fx,Dinv,msg,alg,j] = Alg1(c,C,n,tol,x,J,Dinv)
3   %  Alg1 implements QP Algorithm 1 to solve an
4   %  unconstrained quadratic minimization problem.
5
6   %  Initialization
7   alg = 'Algorithm 1';
8   fx = c'*x + x'*C*x/2.;
9   g = c + C *x;
10  for j = 0:n
11      %  Step 1: Computation of Search Direction s_j
12      %  nzero = # components of J having value 0
13      nzerolist = find(J == 0);
14      nzero = length(nzerolist);
15      if nzero == 0
16          msg = 'Optimal solution obtained';
17      return
18      end
19      if nzero ≥ 1
20          dotprod = abs(g'* Dinv(:,nzerolist) );
21          big = max(dotprod);
22          k = nzerolist(find(big==dotprod,1));
23          gjckj = g'* Dinv(:,k);
24          % get the correct sign for gjckj
25          if gjckj > 0.;
26              s = Dinv(:,k);
27          elseif gjckj ≤ 0.
28              s = - Dinv(:,k);
29          end
30          if abs(big) < tol
31              msg = 'Optimal solution obtained';
32          return
33          end
34      end
35
36      %  Step 2 of Algorithm 1
37      %  optimal step size (sigtil)
38
39      stCs = s'*C*s;
40      if abs(stCs) < tol
41          msg = 'Unbounded from below';
42          return
43      end
44      sigtil = big/stCs;
45      sigma = sigtil;
46
47      %  Step 3 of Algorithm 1
48
49      x = x - sigma*s;
50      g = c + C*x;
```

```
51      fx = c'*x +x'*C*x/2.;
52      d =  sqrt(stCs)*C*s;
53      [Dinvnew] = Phi1(Dinv,d,k,J,n);
54      Dinv = Dinvnew;
55      J(k) = -1;
56   end
57
58   end
```

Figure 3.6.   Algorithm 1: Alg1.m.

Discussion of Alg1.m Figure 3.6 shows the Matlab code for Alg1.m. In order to show the logic of the code more clearly, we show the match up between the Matlab code symbols and the algebraic quantities used in the formal statement of Algorithm 1 as follows: $x = x_j$, $fx = f(x_j)$, $g = g(x_j)$, $J = J_j = \{\ \alpha_{1j}, \ldots, \alpha_{nj}\ \}$, $Dinv = D_j^{-1} = [\ c_{1j}, \ldots, c_{nj}\ ]$, $Dinv(:,k) = c_{kj}$, $s = s_j$, $gjckj = g_j'c_{kj}$, $stCs = s_j'Cs_j$ and $sigma = \sigma$.

After initialization, Step 1 begins at line 11 by computing *nzerolist* which is an ordered list of indices of J_j for which $\alpha_{ij} = 0$. *nzero* is computed as the number of elements in this list. If this number is zero, then all α_{ij}'s must be -1 and the current x is optimal (lines 15-18). If there is at least one i with $\alpha_{ij} = 0$, Alg1.m proceeds to compute the vector of the absolute value of the inner product of the gradient with every column of *Dinv* which is in *nzerolist*. *big* is computed as the largest such value and k is it's position. s is then set to be the *k-th* column of *Dinv* (or its negative). If *big* is smaller than tolerance, then x is optimal.

Next, Alg1 proceeds to calculate the optimal step size. If *stCs* is smaller than tolerance, $f(x - \sigma s)$ is behaving in a linear manner and can be reduced to $-\infty$; i.e., the problem is unbounded from below. Otherwise, the optimal step size, *sigtil*, is computed.

In Step 3, x, g, fx are all updated and $d = (s_j'Cs_j)^{1/2}Cs_j$ is used to update the *k-th* column of *Dinv* using Procedure Φ_1.

```
1    %   checkopt1.m
2    function checkopt1(x,n,c,C,fx,j,msg,alg)
3    %   given a proposed unconstrained optimal
4    %   solution, checkopt1 computes and prints the norm
5    %   of the gradient plus other results from Algorithm 1.
6    normgrad = norm(c + C*x);
7    iter = j;
8    fprintf('Optimization Summary\n');
9    fprintf('%s\n',msg);
10   fprintf('Algorithm: %s\n',alg);
11   if strcmp(msg,'Optimal solution obtained')
12       fprintf('Optimal objective value   %.8f\n',fx);
```

```
13   fprintf('Iterations    %d\n',iter);
14   fprintf('Norm of Gradient              %.8f\n',normgrad);
15   fprintf('\n\n   Optimal Solution\n')
16   for i=1:n
17       fprintf('   %d %12.8f\n',i,x(i))
18   end
19 end
```

Figure 3.7. Check optimality: checkopt1.m.

Discussion of checkopt1.m. This routine prints a summary of the optimization results. The string "msg" contains either "Optimal solution obtained" or "Unbounded from below" and is printed in line 9. Line 10 prints the algorithm number. In this case, it is always "1," but in subsequent material it could refer to one of several algorithms. If "Optimal solution obtained" is the case, then more information is printed including the norm of the gradient (which should be close to zero), and, the optimal solution.

```
1 %  eg3p3.m  Example 3.3
2 %   Illustrates Algorithm 1
3 c = [−7 −8 −9]'; C = [ 2 1 1; 1 2 1; 1 1 2]; n = 3;
4 x = [0 0 0]'; tol = 1.e−6; Dinv = eye(n); J = zeros(1,n);
5 checkdata(C)
6 [x,J,fx,Dinv,msg,alg,j] = Alg1(c,C,n,tol,x,J,Dinv);
7 checkopt1(x,n,c,C,fx,j,msg,alg);
```

Figure 3.8. Example 3.3: eg3p3.m.

```
1 eg3p3
2 Optimization Summary
3 Optimal solution obtained
4 Algorithm: Algorithm 1
5 Optimal objective value   −25.00000000
6 Iterations   3
7 Norm of Gradient              0.00000000
8
9
10   Optimal Solution
11   1   1.00000000
12   2   2.00000000
13   3   3.00000000
```

Figure 3.9. Example 3.3: output.

Discussion of eg3p3.m and its output. Figure 3.8 shows the Matlab code setting up the data for Example 3.3. Figure 3.9 shows the output from applying this to the data of Example 3.3. This output was formulated and printed with checkopt1.m Note that the norm of the gradient of the objective function is very small (zero, actually). Also, note that the optimal solution obtained by Alg1 is in agreement with that obtained in Example 3.3.

```
1  %   eg3p4.m   Example 3.4
2  %   Illustrates Algorithm 1
3  c = [1 -1]'; C = [ 2 2; 2 2 ]; n = 2;
4  x = [0 0]'; tol = 1.e-6; Dinv = eye(n); J = zeros(1,n);
5  checkdata(C)
6  [x,J,fx,Dinv,msg,alg,j] = Alg1(c,C,n,tol,x,J,Dinv);
7  checkopt1(x,n,c,C,fx,j,msg,alg);
```

Figure 3.10.   Example 3.4: eg3p4.m.

```
1  eg3p4
2  Optimization Summary
3  Unbounded from below
4  Algorithm: Algorithm 1
```

Figure 3.11.   Example 3.4: output.

Discussion of eg3p4.m and its output. Figure 3.10 shows the Matlab code setting up the data for Example 3.4. Figure 3.11 shows the output from applying this to the data of Example 3.4 and that this problem is unbounded from below in agreement with the results of Example 3.4.

```
1  % eg3p6.m   Example 3.6
2  % Solves a problem with linear equality constraints by using
3  % Algorithm 1 with appropriate initial data
4  n  = 3; tol = 1.e-6; c = [-12 -13 -14 ]';
5  C = [2 1 1; 1 2 1; 1 1 2]; x = [9 0 0]'; J = [1 0 0]';
6  Dinv = [1 -1 -1; 0 1 0; 0 0 1];
7  checkdata(C);
8  [x,J,fx,Dinv,msg,alg,j] = Alg1(c,C,n,tol,x,J,Dinv);
9  checkopt1(x,n,c,C,fx,j,msg,alg);
```

Figure 3.12.   Example 3.6: eg3p6.m.

```
 1  eg3p6
 2  Optimization Summary
 3  Optimal solution obtained
 4  Algorithm: Algorithm 1
 5  Optimal objective value    -64.00000000
 6  Iterations    2
 7  Norm of Gradient              1.73205081
 8
 9
10      Optimal Solution
11      1    2.00000000
12      2    3.00000000
13      3    4.00000000
```

Figure 3.13.   Example 3.6: output.

Discussion of eg3p6.m and its output. Figure 3.12 shows the Matlab code setting up the data for Example 3.6. Figure 3.13 shows the output from applying this to the data of Example 3.6. Note that the norm of the gradient of the objective function is not zero. This is because the initial data was formulated for a problem with a linear equality constraint.

```
 1  %eg3p7.m   Example 3.7
 2  % Solves a problem with linear equality constraints by using
 3  % Algorithm 1 with appropriate initial data
 4  n   = 3; tol = 1.e-6; c = [-2 -2 1 ]'; x = [0 0 1]';
 5  C = [1 2 -1; 2 4 -2; -1 -2 1]; J = [0 0 1];
 6  Dinv = [1 -1 -1; 0 1 0; 0 0 1];
 7  [x,J,fx,Dinv,msg,alg,j] = Alg1(c,C,n,tol,x,J,Dinv);
 8  checkopt1(x,n,c,C,fx,j,msg,alg);
```

Figure 3.14.   Example 3.7: eg3p7.m.

```
 1  eg3p7
 2  Optimization Summary
 3  Unbounded from below
 4  Algorithm: Algorithm 1
```

Figure 3.15.   Example 3.7: output.

Discussion of eg3p7.m and its output. Figure 3.14 shows the Matlab code setting up the data for Example 3.7. Figure 3.15 shows the output from applying this to the data of Example 3.7. This correctly shows that the problem Example 3.7 is unbounded from below.

```
1  %  Alg2.m
2  function[x,J,fx,Dinv,msg,alg,j] = Alg2(c,C,n,tol,x,J,Dinv)
3  %  Alg2 implements QP Algorithm 2 to solve a
4  %  linear equality constrained quadratic minimization
5  %  problem.
6  [x,J,fx,Dinv,msg,alg,j] = Alg1(c,C,n,tol,x,J,Dinv)
7  alg = 'Algorithm 2';
8  end
```

Figure 3.16.   Algorithm 2.

Discussion of Algorithm2. Algorithm2 is identical to Algorithm 1 except for a different initial data requirement.

3.7   Exercises

3.1 Verify (3.3) and (3.4)

3.2 Verify (3.6) by substituting for $g(x_0)$ and simplifying.

3.3 Use Algorithm 1 to solve each of the following problems. In each case, use $x_0 = (0 , 0)'$.

   (a) minimize: $- 19x_1 - 15x_2 + \frac{1}{2}(5x_1^2 + 3x_2^2 + 6x_1x_2)$
   (b) minimize: $3x_1 + 6x_2 + \frac{1}{2}(2x_1^2 + 8x_2^2 + 8x_1x_2)$
   (c) minimize: $3x_1 + 5x_2 + \frac{1}{2}(2x_1^2 + 8x_2^2 + 8x_1x_2)$
   (d) minimize: $- 3x_1 - x_2 + \frac{1}{2}(2x_1^2 + x_2^2 + 2x_1x_2)$

3.4 Use Algorithm 1 to solve min $\{ c'x + \frac{1}{2}x'Cx \}$ for each of the following pairs of c and C. In each case, use $x_0 = (0 , 0 , 0)'$.

$$\text{(a)} \quad c = \begin{bmatrix} -11 \\ -12 \\ -13 \end{bmatrix}, \quad C = \begin{bmatrix} 2 & 1 & 1 \\ 1 & 2 & 1 \\ 1 & 1 & 2 \end{bmatrix}$$

$$\text{(b)} \quad c = \begin{bmatrix} -21 \\ -15 \\ -21 \end{bmatrix}, \quad C = \begin{bmatrix} 5 & 2 & 4 \\ 2 & 5 & 1 \\ 4 & 1 & 5 \end{bmatrix}$$

$$\text{(c)} \quad c = \begin{bmatrix} -2 \\ -4 \\ -6 \end{bmatrix}, \quad C = \begin{bmatrix} 2 & 4 & 6 \\ 4 & 8 & 12 \\ 6 & 12 & 18 \end{bmatrix}$$

$$\text{(d)} \quad c = \begin{bmatrix} -1 \\ -4 \\ -6 \end{bmatrix}, \quad C = \begin{bmatrix} 2 & 4 & 6 \\ 4 & 8 & 12 \\ 6 & 12 & 18 \end{bmatrix}$$

3.5 Use Algorithm 2 to solve each of the following problems using the indicated initial data.

(a)

$$\text{minimize}: \quad -10x_1 \quad - \quad 8x_2 \quad + \quad x_1^2 + x_2^2$$
$$\text{subject to}: \qquad x_1 \quad + \quad x_2 \quad = \quad 1. \qquad (1)$$

Initial Data: $x_0 = (1,0)'$, $J_0 = \{1,0\}$, $D_0^{-1} = \begin{bmatrix} 1 & -1 \\ 0 & 1 \end{bmatrix}$.

(b)

$$\text{minimize}: \quad 3x_1 \quad - \quad 3x_2 \quad + \quad \tfrac{1}{3}(x_1^2 + 2x_2^2)$$
$$\text{subject to}: \qquad x_1 \quad + \quad x_2 \quad = \quad 1. \qquad (1)$$

Initial Data: $x_0 = (1/2, 1/2)'$, $J_0 = \{1,0\}$, $D_0^{-1} = \begin{bmatrix} 1 & -1 \\ 0 & 1 \end{bmatrix}$.

3.6 Use Algorithm 2 to solve each of the following problems using the indicated initial data.

(a)

$$\text{minimize}: \quad -9x_1 \quad + \quad 8x_2 \quad - \quad 13x_3 \quad + \quad \tfrac{1}{2}(x_1^2 + 2x_2^2 + 3x_3^2)$$
$$\text{subject to}: \qquad 2x_1 \quad - \quad 3x_2 \quad + \qquad x_3 \quad = \quad -1. \qquad (1)$$

Initial Data: $x_0 = (0,0,-1)'$ $J_0 = \{1,0,0\}$,
$$D_0^{-1} = \begin{bmatrix} 1/2 & 3/2 & -1/2 \\ 0 & 1 & 0 \\ 0 & 0 & 1 \end{bmatrix}.$$

(b)

$$\text{minimize}: \quad x_1^2 \quad + \quad x_2^2 \quad + \quad x_3^2 \quad + \quad x_4^2$$
$$\text{subject to}: \qquad x_1 \quad + \quad x_2 \quad + \quad x_3 \quad + \quad x_4 \quad = \quad 1, \quad (1)$$
$$\qquad\qquad\qquad x_1 \quad - \quad x_2 \quad - \quad x_3 \quad - \quad x_4 \quad = \quad 0. \quad (2)$$

Initial Data: $x_0 = (1/2, 1/2, 0, 0)'$, $J_0 = \{1, 2, 0, 0\}$,
$$D_0^{-1} = \begin{bmatrix} 1/2 & 1/2 & 0 & 0 \\ 1/2 & -1/2 & -1 & -1 \\ 0 & 0 & 1 & 0 \\ 0 & 0 & 0 & 1 \end{bmatrix}.$$

3.7 (a) Show that if C is positive semidefinite and $s'Cs = 0$, then $Cs = 0$.
Hint: Use Theorem 3.3.

(b) A symmetric matrix C is indefinite if there are s_1 and s_2 such that $s_1'Cs_1 > 0$ and $s_2'Cs_2 < 0$. It is well-known from linear algebra that C is indefinite if and only if C has at least one positive and at least one negative eigenvalue. Show by example that part (a) does not hold for indefinite C.

3.8 Suppose that $f(x) = c'x + \frac{1}{2}x'Cx$ is convex and that T is an (n, n) matrix.

(a) Show that $h(x) \equiv f(Tx)$ is also convex.

(b) If f is strictly convex, under what conditions is h strictly convex?

3.9 Let C be positive definite. Show

(a) C is nonsingular.

(b) C^{-1} is positive definite.

3.10 Show that the general two variable quadratic function

$$f(x) = c_1 x_1 + c_2 x_2 + \frac{1}{2}(\gamma_{11} x_1^2 + 2\gamma_{12} x_1 x_2 + \gamma_{22} x_2^2)$$

is convex if and only if $\gamma_{11} \geq 0$, $\gamma_{22} \geq 0$, and $\gamma_{11}\gamma_{22} \geq \gamma_{12}^2$.

3.11 Let γ_{ij} denote element (i, j) of C. For C positive semidefinite, show that $\gamma_{ii} \geq 0$ for $i = 1, 2, \ldots, n$ and that $\gamma_{ii}\gamma_{jj} \geq \gamma_{ij}^2$ for $i, j = 1, 2, \ldots, n$.

3.12 Let C be positive definite and let $s_1, s_2, \ldots, s_n$ be a set of normalized conjugate directions. Let

$$M_k = s_1 s_1' + s_2 s_2' + \cdots + s_k s_k',$$

and note that M_k is the sum of k rank one matrices. Let $S_k = \text{span}$ $\{ s_1, s_2, \ldots, s_k \}$.

(a) Show that M_k is a left inverse of C on S_k; i.e., show that $M_k Cx = x$ for all $x \in S_k$.

(b) Is M_k a right inverse of C on S_k?

(c) Show that $M_n = C^{-1}$.

(d) Suppose that x_0 is any given point and that x_1 is defined by

$$x_1 = x_0 - ((g_0's_1)s_1 + (g_0's_2)s_2 + \cdots + (g_0's_n)s_n).$$

Show that x_1 is the optimal solution for min $\{ c'x + \frac{1}{2}x'Cx \}$, where $g_0 = g(x_0)$.

3.13 Let C be an (n, n) symmetric positive semidefinite matrix. Show that the number of conjugate directions in a complete set of conjugate directions for C is equal to rank C.

3.14 Let C be positive semidefinite. Show that any two eigenvectors of C having distinct, nonzero eigenvalues, are conjugate directions.

3.15 Show that at iteration j of Algorithm 1 at least j components of g_j will have value zero. *Hint:* Use Lemma 3.1, Lemma 3.3, and the fact that $D_0 = I$.

3.16 Suppose that Algorithm 1 is applied to

minimize: $nx_1 + (n - 1)x_2 + \cdots + x_n + \frac{1}{2}(x_1^2 + x_2^2 + \cdots + x_n^2)$

beginning with $x_0 = 0$, $D_0^{-1} = I$, and $J_0 = \{0, \ldots, 0\}$. Determine explicit expressions for x_j, D_j^{-1}, and J_j for each iteration j.

3.17 The purpose of this exercise is to establish the total number of arithmetic operations required by Algorithm 1 to solve a problem having n variables.

(a) At iteration j, show that Steps 1, 2, and 3 require approximately $n(n - j)$, n^2, and $3n^2 + 2n(n - j)$ arithmetic operations, respectively.

(b) Assuming that Algorithm 1 requires n iterations, show that a total of approximately $\frac{11}{2}n^3$ arithmetic operations are required.

3.18 The purpose of this exercise is to show that the operation count for Algorithm 1 may be reduced by performing calculations involving C in a more efficient manner. Suppose that we compute $w_j = Cs_j$ in Step 2. Then $s_j'Cs_j$ can be computed as $s_j'w_j$. In Step 3, g_{j+1} can be computed from Taylor's series as $g_{j+1} = g_j - \sigma_j w_j$. Furthermore, $f(x_{j+1})$ can also be computed from Taylor's series as

$$f(x_{j+1}) = f(x_j) - \sigma_j g_j' s_j + \frac{1}{2}\sigma_j^2 s_j' w_j .$$

Finally, d_j may be computed as $(s_j'w_j)^{-1/2}w_j$. Show that performing the calculations in this way reduces the operations count to approximately $\frac{5}{2}n^3$, a reduction of about 54.54% (see Exercise 3.17).

3.19 Suppose that Algorithm 1 is changed as follows. In Step 1, instead of choosing k to be the smallest index such that

$$|g_j'c_{kj}| = \max\{|g_j'c_{ij}| \,|\, \text{all } i \text{ with } \alpha_{ij} = 0\}$$

k is now chosen to be the smallest index such that

$$f\left(x_j - \frac{g'_j c_{kj}}{c'_{kj} C c_{kj}} c_{kj}\right) = \min\left\{ f\left(x_j - \frac{g'_j c_{ij}}{c'_{ij} C c_{ij}} c_{ij}\right) \mid \text{all } i \text{ with } \alpha_{ij} = 0 \right\}$$

and then s_j is calculated from c_{kj} as before. That is, s_j is chosen to be the candidate column of D_j^{-1} which gives the greatest reduction in f. How many additional arithmetic operations are required to choose k in this way? Compare this with the operation counts obtained in Exercises 3.17 and 3.18.

3.20 (a) Suppose that Algorithm 2 requires the maximum number of iterations, namely, $j = n - r$. This implies that $\alpha_{ij} \neq 0$ for $i = 1, \ldots, n$. Show that

$$s'Cs > 0 \quad \text{for all } s \neq 0 \text{ with } a'_i s = 0, \quad i = 1, \ldots, r.$$

(b) Suppose that Algorithm 1 requires n iterations. Show that C is positive definite. *Hint:* Use part (a).

3.21 Let C be an (n, n) positive semidefinite symmetric matrix. Let A be a (k, n) matrix having linearly independent rows. Let

$$H = \begin{bmatrix} C & A' \\ A & 0 \end{bmatrix}.$$

Prove that H is nonsingular if and only if $s'Cs > 0$ for all s with $As = 0$ and $s \neq 0$.

3.22 (a) Consider the problem

$$\min\left\{ c'x + \tfrac{1}{2}x'Cx \mid Ax = b \right\}.$$

Assume that A is (r, n) and that rank $A = r$. Suppose that the problem has been solved by Algorithm 2 and that $j = n - r$ iterations were required. Let $D_j^{-1} = [c_{1j}, \ldots, c_{nj}]$, $J_j = \{\alpha_{1j}, \ldots, \alpha_{nj}\}$, and let x_0 be any point satisfying $Ax_0 = b$. Although x_j has been determined as the optimal solution, it is the purpose of this exercise is to show that the point x_1 given by

$$x_1 = x_0 - \sum_{\alpha_{ij}=-1} (g'_0 c_{ij}) c_{ij}$$

is also optimal for the given problem. Verify that x_1 is indeed optimal. Note the slight abuse of notation. The x_1 given above is different than the x_1 determined by Algorithm 2.

(b) Consider the problem

$$\min \{ (c + tq)'x + \tfrac{1}{2}x'Cx \mid Ax = b \},$$

where q is a given n-vector and t is a scalar parameter which is to vary from $-\infty$ to $+\infty$. The solution of this problem is required for all such t. This problem is a special case of a parametric quadratic programming problem (see Chapter 7). Assume that A is as in part (a), and suppose that the problem has been solved for $t = 0$ by Algorithm 2 and that $j = n - r$ iterations were required. Let D_j^{-1} and J_j be as in part (a) and let x_0 be any point satisfying $Ax_0 = b$. Show that

$$x^*(t) = h_1 + th_2,$$

is optimal for all t with $-\infty \le t \le +\infty$, where

$$h_1 = x_0 - \sum_{\alpha_{ij} = -1} [(c + Cx_0)'c_{ij}]c_{ij} \text{ and } h_2 = - \sum_{\alpha_{ij} = -1} (q'c_{ij})c_{ij}.$$

(c) In addition, show that h_1 may also be calculated from the following (which is independent of x_0):

$$h_1 = \sum_{1 \le \alpha_{ij} \le r} b_{\alpha_{ij}} c_{ij} - \sum_{\alpha_{ij} = -1} (c'c_{ij})c_{ij}.$$

Chapter 4

Quadratic Programming Theory

A quadratic programming problem is that of minimizing a quadratic function subject to linear inequality and/or equality constraints. The majority of the results of this chapter require the assumption of convexity of the objective function. Therefore we assume throughout this chapter, unless explicitly stated to the contrary, that the objective function for each quadratic minimization problem is convex. By definition, this is equivalent to the assumption that the Hessian matrix of the objective function is positive semidefinite. Those results which do not require convexity assumptions will be discussed in Chapter 9.

Generally speaking, we will use a model problem having "$\leq$" constraints only. Having obtained results for this model problem, it is a simple process to reformulate the result for a model problem having both "$\leq$" and "$=$" constraints. Problems with "$\geq$" constraints may be replaced by an equivalent problem having "$\leq$" constraints simply by multiplying both sides of the "$\geq$" constraint by -1. We could begin with a model problem having all three types of constraints ("$\leq$," "$=$," "$\geq$") made explicit, but formulating results for it would be quite cumbersome and very little would be gained. Thus, we primarily work with "$\leq$" constraints. Also, we do not generally give explicit consideration to maximization problems. This is without loss of generality since maximizing a function is equivalent to minimizing the negative of the function.

In Section 4.1, we establish the existence of an optimal solution for a quadratic programming problem provided that the feasible region is nonnull and that the objective function is bounded from below over the feasible region. Conditions which are both necessary and sufficient for optimality are also given. Duality is the subject of Section 4.2. Associated with any quadratic programming problem (called the primal problem) is a second quadratic programming problem called the dual problem. Weak, strong, and converse duality theorems are demonstrated which show the relationship between the primal and dual problems. This may be regarded as a generalization of the duality theory for linear programming problems. Alternate optimal solutions are discussed in Section 4.3 and a condition is given which is sufficient for the uniqueness of an optimal solution. In Section 4.4, we examine the effect of small changes in some of the problem data on the optimal solution. In particular, we will show that provided a technical assumption is satisfied, both the optimal solutions for the primal and dual are locally linear functions of the linear part of the objective function and the right-hand side vector of the constraints.

4.1   QP Optimality Conditions

In this section, we establish the existence of an optimal solution for a quadratic programming problem provided that the objective function is bounded from below over the feasible region. In addition, we will demonstrate conditions which are both necessary and sufficient for optimality for a particular model problem. These results are then used to derive analogous optimality conditions for a variety of other model problems.

We consider the model problem

$$\min \{\ c'x \ + \ \tfrac{1}{2}x'Cx \mid a_i'x \ \leq \ b_i\ , \ \ i \ = \ 1,\ldots,m\ \}, \qquad (4.1)$$

where $a_1,\ldots,a_m$ are n-vectors, $b_1,\ldots,b_m$ are scalars, and $f(x) = c'x + \tfrac{1}{2}x'Cx$ denotes the objective function. A somewhat more compact formulation of (4.1) is

$$\min \{\ c'x \ + \ \tfrac{1}{2}x'Cx \mid Ax \ \leq \ b\ \}, \qquad (4.2)$$

where $A' = [\ a_1,\ldots,a_m\ ]$ and $b = (\ b_1,\ldots,b_m\ )'$. Our model problem thus has n variables and m inequality constraints. The <u>feasible</u> <u>region</u> for (4.1) and (4.2) is

$$R = \{\ x \mid Ax \ \leq \ b\ \}.$$

A point x_0 is <u>feasible</u> for (4.2) if $x_0 \in R$ and <u>infeasible</u> otherwise. Constraint i is <u>inactive</u>, <u>active</u> or <u>violated</u> at x_0 according to whether $a_i'x_0 < b_i, a_i'x_0 = b_i$ or $a_i'x_0 > b_i$, respectively. x_0 is an <u>optimal</u> <u>solution</u> (or simply <u>optimal</u>) if $x_0 \in R$ and $f(x_0) \leq f(x)$ for all $x \in R$. The objective function for (4.2) is <u>unbounded</u> <u>from</u> <u>below</u> over the points in its feasible region [or more briefly, problem (4.2) is unbounded from below] if there are n-vectors x_0 and s_0 such that $x_0 - \sigma s_0 \in R$ for all $\sigma \geq 0$ and $f(x_0 - \sigma s_0) \to -\infty$ as $\sigma \to +\infty$.

Numerical methods for linear programming are based on the fact that the feasible region possesses only a finite number of extreme points and at least one extreme point is optimal. The analogous quantity to an extreme point for quadratic programming is a quasistationary point which is defined as follows. For any $x_0 \in R$, define

$$I(x_0) = \{\, i \mid a_i' x_0 = b_i \,, \; 1 \leq i \leq m \,\};$$

that is, $I(x_0)$ is the set of indices of those constraints active at x_0. x_0 is a <u>quasistationary point</u> for (4.1) if $x_0 \in R$ and x_0 is an optimal solution for

$$\min \{\, c'x + \tfrac{1}{2} x' C x \mid a_i' x = b_i \,, \; \text{for all } i \in I(x_0) \,\}. \tag{4.3}$$

Any extreme point of R is a quasistationary point for (4.1). This is because the feasible region for (4.3) consists of precisely the point x_0, which is then necessarily optimal. An optimal solution for (4.1) is also a quasistationary point.

In general, (4.1) may possess many quasistationary points. Pick any subset of $\{\, 1\,, \, 2, \ldots, m \,\}$. Let I_0 denote this subset. If

$$\min \{\, c'x + \tfrac{1}{2} x' C x \mid a_i' x = b_i \,, \; \text{all } i \in I_0 \,\} \tag{4.4}$$

possesses an optimal solution, let it be denoted by x_0. Either x_0 is in R or it is not. If it is, then it is a quasistationary point. One could theoretically enumerate all possible quasistationary points by enumerating all subsets of $\{\, 1\,, \, 2, \ldots, m \,\}$ and solving the corresponding (4.4).[1] There are, however, 2^m such subsets so a rather large amount of work would be required.

We give a geometric illustration of quasistationary points in

Example 4.1

$$
\begin{aligned}
\text{minimize:} \quad & -10x_1 - 4x_2 + x_1^2 + x_2^2 \\
\text{subject to:} \quad & x_1 && \leq \;\; 4, && (1) \\
& \quad\;\; -x_2 && \leq -1, && (2) \\
& -x_1 && \leq -1, && (3) \\
& \quad\;\; x_2 && \leq \;\; 3. && (4)
\end{aligned}
$$

The feasible region is shown in Figure 4.1. The level sets for the objective function are concentric circles centered at $(5\,,\, 2)'$. Each of the extreme points $x_1, \ldots, x_4$ is a quasistationary point. With $I_0 = \{\, 3 \,\}$, the optimal solution for (4.4) is $x_5 = (1\,,\, 2)'$. Since $x_5 \in R$, x_5 is also a quasistationary point. Similarly, so too is x_6. With $I_0 = \{\, 4 \,\}$, the optimal solution for (4.4) is x_7. Since $x_7 \notin R$, it is not a quasistationary point. Similarly, neither is x_8. $\quad\diamond$

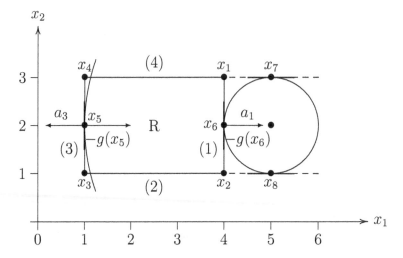

Figure 4.1.   Quasi-stationary points for Example 4.1.

It is useful to differentiate among two types of quasistationary points. A quasistationary point x_0 is <u>nondegenerate</u> if the gradients of those constraints active at x_0 are linearly independent. A quasistationary point x_0 is <u>degenerate</u> if the gradients of those constraints active at x_0 are linearly dependent. Inspection of Figure 4.1 shows that all quasistationary points for Example 4.1 are nondegenerate. Degenerate quasistationary points are illustrated in

Example 4.2

$$
\begin{array}{lrcrcl}
\text{minimize:} & -10x_1 & - & 4x_2 & + & x_1^2 + x_2^2 \\
\text{subject to:} & x_1 & & & \leq & 4, \quad (1) \\
& & - & x_2 & \leq & -1, \quad (2) \\
& -x_1 & & & \leq & -1, \quad (3) \\
& & & x_2 & \leq & 3, \quad (4) \\
& -2x_1 & & & \leq & -2, \quad (5) \\
& -x_1 & - & x_2 & \leq & -2. \quad (6)
\end{array}
$$

This problem differs from Example 4.1 by the addition of constraints (5) and (6). The feasible region is shown in Figure 4.2 and is identical to that for Example 4.1. Each quasistationary point of Example 4.1 is also a quasistationary point for the present example. Constraints (3) and (5) are both active at x_5

---

[1]There is a difficulty with this. (4.4) may not have a unique optimal solution. One optimal solution may be in R while another is not. If we found one not in R, we might incorrectly conclude that there is no quasistationary point associated with I_0 .

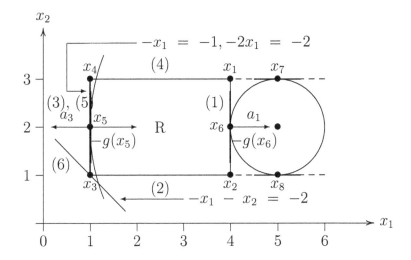

Figure 4.2 Degenerate and nondegenerate quasistationary points for Example 4.2

and their gradients are linearly dependent ($a_5 = 2a_3$). Constraints (2), (3), (5) and (6) are active at x_3. Since 4 vectors in a space of dimension 2 must be linearly dependent, x_3 is a degenerate quasistationary point. Similarly because the three constraints (3), (4) and (5) are active at x_4, x_4 is also a degenerate quasistationary point. ◇

We will next demonstrate a lemma which shows that if $f(x)$ is bounded from below for $x \in R$ and if x_0 is any point in R, then there exists a quasistationary point x_1 such that $f(x_1) \leq f(x_0)$. In doing so, it is convenient to use the notion of a maximum feasible step size which the reader may have previously encountered in linear programming. Let x_0 be any point in R and let s_0 be any n-vector which we think of as a search direction. The largest value of σ for which $x_0 - \sigma s_0 \in R$ is called the <u>maximum feasible step size</u> and is denoted by σ_0. The requirement that $x_0 - \sigma s_0$ be in R is

$$a_i'(x_0 - \sigma s_0) \leq b_i , \quad i = 1, \ldots, m .$$

Simplifying these inequalities gives

$$\sigma_0 = \min \left\{ \frac{a_i' x_0 - b_i}{a_i' s_0} \mid \text{all } i = 1, \ldots, m \text{ with } a_i' s_0 < 0 \right\} .$$

If $a_i' s_0 \geq 0$ for $i = 1, \ldots, m$, then $x_0 - \sigma s_0 \in R$ for all σ and we set $\sigma_0 = +\infty$. If $\sigma_0 < +\infty$, let l be the smallest index such that

$$\sigma_0 = \frac{a_l' x_0 - b_l}{a_l' s_0} .$$

Then constraint l becomes active at $x_0 - \sigma_0 s_0$.

Lemma 4.1 *Suppose that f is convex and that there is a γ such that $f(x) \geq \gamma$ for all $x \in R \equiv \{ x \mid a'_i x \leq b_i, \; i = 1, \ldots, m \}$. Let x_0 be any point in R. Then there is a quasistationary point $\tilde{x}$ with $f(\tilde{x}) \leq f(x_0)$.*

Proof: Consider the problem

$$\min \{ c'x + \tfrac{1}{2}x'Cx \mid a'_i x = b_i, \; \text{all } i \in I(x_0) \} . \tag{4.5}$$

This is a problem of quadratic minimization subject to linear equality constraints. If $\{ a_i \mid i \in I(x_0) \}$ are linearly dependent, then we can replace $I(x_0)$ with a subset of indices of active constraints having maximal size while leaving the feasible region for (4.5) unchanged. Thus we may assume without loss of generality that the gradients of the constraints of (4.5) are linearly independent. Algorithm 2 may then be applied to (4.5) with initial point x_0. According to Theorem 3.14, termination will occur in a finite number of steps with either an optimal solution $\hat{x}$ (Case 1) or a point $\hat{x}$ and a search direction s_0 such that $f(\hat{x} - \sigma s_0) \to -\infty$ as $\sigma \to +\infty$ (Case 2). We pursue these two possibilities separately.

Case 1:   optimal solution at $\hat{x}$

Let $s_0 = x_0 - \hat{x}$ and let σ_0 be the maximum feasible step size for x_0 and s_0. Because $\hat{x}$ is optimal for (4.5), $f(x_0 - \sigma s_0)$ is nonincreasing for $0 \leq \sigma \leq \sigma_0$. If $\sigma_0 \geq 1$, then $x_0 - 1 \cdot s_0 \equiv \hat{x}$ is in R and $\hat{x}$ is a quasistationary point with $f(\hat{x}) \leq f(x_0)$. In this case the lemma is proven with $\tilde{x} = \hat{x}$. If $\sigma_0 < 1$, let l be the index of the restricting constraint; that is,

$$\sigma_0 = \frac{a'_l x_0 - b_l}{a'_l s_0} .$$

Define $x_1 = x_0 - \sigma_0 s_0$. Then $f(x_1) \leq f(x_0)$. Furthermore, $a'_i x_0 = b_i$ and $a'_i \hat{x} = b_i$ for all $i \in I(x_0)$. Since $a'_l s_0 < 0$, this implies that $\{ a_i \mid i \in I(x_0) \}$ together with a_l are linearly independent. Therefore the number of constraints active at x_1 which have linearly independent gradients is at least one greater than at x_0.

Case 2:   (4.5) is unbounded from below

Let σ_0 denote the maximum feasible step size for $\hat{x}$ and s_0. If $\sigma_0 = +\infty$, then $x_0 - \sigma s_0$ is feasible for (4.1) for all $\sigma \geq 0$ and $f(x_0 - \sigma s_0) \to -\infty$ as $\sigma \to +\infty$. But this is in contradiction to the hypothesis of the lemma and therefore cannot occur. Therefore $\sigma_0 < +\infty$. Let l be the index of the restricting constraint and define $x_1 = \hat{x} - \sigma_0 s_0$. Then $f(x_1) \leq f(\hat{x}) \leq f(x_0)$. As in Case 1, the number of constraints active at x_1 which have linearly independent gradients, is at least one greater than x_0.

If Case 1 occurs with quasistationary point x_1 with $f(x_1) \leq f(x_0)$, then the lemma is proven with $\tilde{x} = x_1$. Otherwise, both Case 1 and Case 2 conclude with $x_1 \in R$, $f(x_1) \leq f(x_0)$, and for which the number of constraints active at x_1 having linearly independent gradients is at least one greater than at x_0. In this case, we can replace $I(x_0)$ with $I(x_1)$ in (4.5) and again argue that Case 1 or Case 2 must occur. If Case 1 occurs with quasistationary point x_2 and $f(x_2) \leq f(x_1)$ then the lemma is proven with $\tilde{x} = x_2$. Otherwise, both Case 1 and Case 2 conclude with $x_2 \in R$, $f(x_2) \leq f(x_1)$, and for which the number of constraints active at x_2 having linearly independent gradients is at least one greater that at x_1.

This argument may be repeated several times, if necessary, until either from Case 1 a quasistationary point x_k with $f(x_k) \leq f(x_{k-1})$ is obtained and the lemma is proven with $\tilde{x} = x_k$, or, an x_k with $f(x_k) \leq f(x_{k-1})$ is obtained and the number of constraints active at x_k having linearly independent gradients is exactly n. In this latter case, x_k is an extreme point and therefore a quasistationary point. The proof of the lemma is then complete with $\tilde{x} = x_k$. $\qquad\square$

Let x_0 be a quasistationary point for (4.1). Associated with x_0 is the quasistationary set $S(x_0)$ defined by

$$S(x_0) = \{\, x \mid a_i'x = b_i \,, \text{ for all } i \in I(x_0) \,, \, f(x) = f(x_0) \,\} \,.$$

In other words, $S(x_0)$ is just the set of alternate optimal solutions for

$$\min \{\, c'x + \tfrac{1}{2}x'Cx \mid a_i'x = b_i \,, \text{ for all } i \in I(x_0) \,\} \,.$$

Because of alternate optimal solutions, (4.1) may possess infinitely many quasistationary points. However, because each quasistationary set is associated with a subset of $\{1, 2, \ldots, m\}$ there are at most 2^m quasistationary sets.

We are now ready to demonstrate the existence of an optimal solution for a quadratic programming problem.

Theorem 4.1 *Let $f(x)$ be bounded from below for all $x \in R = \{\, x \mid Ax \leq b \,\}$. Then there exists an optimal solution $\hat{x}$ for*

$$\min \{\, c'x + \tfrac{1}{2}x'Cx \mid Ax \leq b \,\}$$

and $\hat{x}$ is a quasistationary point.

Proof: There are finitely many quasistationary sets. Suppose that they are p in number. Let x_i be a quasistationary point from the ith quasistationary set. Define k such that

$$f(x_k) = \min \{\, f(x_i) \mid i = 1, \ldots, p \,\} \,.$$

Now let x be any point in R. From Lemma 4.1, there is a quasistationary point $\tilde{x}$ with $f(\tilde{x}) \leq f(x)$. Now $\tilde{x} \in S(x_i)$ for some i with $1 \leq i \leq p$ so that $f(x_i) = f(\tilde{x})$. But then by definition of k,

$$f(x_k) \leq f(x_i) = f(\tilde{x}) \leq f(x) .$$

Therefore $f(x_k) \leq f(x)$ for all $x \in R$ and $\tilde{x} \equiv x_k$ is an optimal solution for the given problem. $\square$

Because a quasistationary point is an optimal solution to a quadratic minimization problem subject to linear equality constraints, Theorem 3.11 can be applied to characterize such a point. By definition, x_0 is a quasistationary point if and only if $x_0 \in R$ and x_0 is an optimal solution for

$$\min \{ c'x + \tfrac{1}{2}x'Cx \mid a_i'x = b_i , \text{ all } i \subset I(x_0) \} . \tag{4.6}$$

Theorem 3.11 asserts that x_0 is optimal for (4.6) if and only if there exist numbers u_i, $i \in I(x_0)$, with

$$-g(x_0) = \sum_{i \in I(x_0)} u_i a_i . \tag{4.7}$$

This notation is somewhat awkward because the constraints of (4.6) are not indexed sequentially. However by defining

$$u_i = 0 , \text{ for } i \notin I(x_0), \tag{4.8}$$

we obtain an m-vector $u = (u_1, \ldots, u_m)'$ and (4.7) becomes

$$-g(x_0) = u_1 a_1 + \cdots + u_m a_m .$$

Furthermore, (4.8) is equivalent to the condition

$$u_i(a_i'x_0 - b_i) = 0, \quad i = 1, \ldots, m . \tag{4.9}$$

This is because for $i \notin I(x_0)$, $a_i'x_0 - b_i < 0$ so that (4.9) implies that $u_i = 0$. Furthermore, for $i \in I(x_0)$ $a_i'x_0 - b_i = 0$ so that (4.9) is satisfied with no restriction imposed on u_i.

Summarizing we have proven

Theorem 4.2 x_0 *is a quasistationary point for*

$$\min \{ c'x + \tfrac{1}{2}x'Cx \mid a_i'x \leq b_i , \quad i = 1, \ldots, m \}$$

if and only if

(a) $a_i'x_0 \leq b_i$, $i = 1, \ldots, m$,

(b) $-g(x_0) = u_1 a_1 + \cdots + u_m a_m$,

(c) $u_i(a_i'x_0 - b_i) = 0$, $i = 1, \ldots, m$.

We now turn to the problem of obtaining necessary and sufficient conditions for optimality for the quadratic programming problem

$$\min \{ c'x + \tfrac{1}{2}x'Cx \mid a_i'x \leq b_i , \ i = 1, \ldots, m \}. \tag{4.10}$$

We saw from geometrical examples in Chapter 1 that an optimal solution x_0 is characterized by the conditions that x_0 is feasible and that $-g(x_0)$ lies in the cone spanned by the gradients of those constraints active at x_0. These geometric conditions may be formulated algebraically for the model problem (4.10) as follows:

$$a_i'x_0 \leq b_i , \ i = 1, \ldots, m , \tag{4.11}$$

$$-g(x_0) = u_1 a_1 + \cdots + u_m a_m \ ; \ u_i \geq 0, \ i = 1, \ldots, m, \tag{4.12}$$

$$u_i(a_i'x_0 - b_i) = 0 , \ i = 1, \ldots, m . \tag{4.13}$$

(4.11) is just a restatement of feasibility. (4.12) associates a nonnegative multiplier u_i with each constraint i. (4.13) forces the multiplier associated with each inactive constraint to have value zero. Then (4.12) expresses $-g(x_0)$ as a nonnegative linear combination of the gradients of active constraints.

Notice that (4.11) to (4.13) differ from the necessary and sufficient conditions for a quasistationary point (Theorem 4.2) only in that the multipliers for the former are required to be nonnegative.

With $C = 0$, (4.10) reduces to the linear programming problem

$$\min \{ c'x \mid a_i'x \leq b_i , \ i = 1, \ldots, m \} , \tag{4.14}$$

and, with $g(x_0) = c + Cx_0 = c$, (4.11) to (4.13) reduce to

$$a_i'x_0 \leq b_i , \ i = 1, \ldots, m ,$$
$$-c = u_1 a_1 + \cdots + u_m a_m \ ; \ u_i \geq 0 , \ i = 1, \ldots, m ,$$
$$u_i(a_i'x_0 - b_i) = 0 , \ i = 1, \ldots, m ,$$

which are both necessary and sufficient for x_0 to be an optimal solution for (4.14) [Theorem 3.1, Best and Ritter, 1985]. Therefore (4.11) to (4.13) reduce to the appropriate optimality conditions for a linear programming problem.

They may also be regarded as a specialization of the nonlinear programming problem to the quadratic case.

We next prove that (4.11) to (4.13) are both necessary and sufficient for optimality for (4.10). This is one of the most fundamental results in quadratic programming and we sometimes call (4.11) to (4.13) the <u>optimality</u> <u>conditions</u> for (4.10).

Theorem 4.3 x_0 is an optimal solution for

$$\min \{ c'x + \tfrac{1}{2}x'Cx \mid a_i'x \leq b_i , \quad i = 1,\ldots,m \}$$

if and only if

(a) $a_i'x_0 \leq b_i , \quad i = 1,\ldots,m,$

(b) $-g(x_0) = u_1 a_1 + \cdots + u_m a_m ; \quad u_i \geq 0 , \quad i = 1,\ldots,m,$

(c) $u_i(a_i'x_0 - b_i) = 0 , \quad i = 1,\ldots,m.$

Proof: Suppose first that x_0 is optimal. Let s be any n-vector such that

$$a_i's \geq 0 , \quad \text{for all } i \in I(x_0) .$$

Consider the point $x_0 - \sigma s$, where σ is a nonnegative scalar. For $i \in I(x_0)$, $a_i'(x_0 - \sigma s) = b_i - \sigma a_i's \leq b_i$. Therefore $x_0 - \sigma s$ satisfies all constraints $i \in I(x_0)$ for all $\sigma \geq 0$. Since the remaining constraints are inactive at x_0, there is a $\hat{\sigma} > 0$ such that $x_0 - \sigma s \in R$ for all σ with $0 \leq \sigma \leq \hat{\sigma}$. Since x_0 is optimal, $f(x_0 - \sigma s)$ must be nondecreasing in σ for small and positive σ. From elementary calculus, this means that the derivative of $f(x_0 - \sigma s)$ with respect to σ, evaluated at $\sigma = 0$ must be nonnegative. From Taylor's series,
$$f(x_0 - \sigma s) = f(x_0) - \sigma g'(x_0)s + \tfrac{1}{2}\sigma^2 s'Cs$$
so that the derivative in question is $-g'(x_0)s$. Therefore $-g'(x_0)s \geq 0$ and we have shown that

$$a_i's \geq 0 \text{ for all } i \in I(x_0) \text{ implies that } -g'(x_0)s \geq 0 . \tag{4.15}$$

Now consider the linear programming problem

$$\min \{ -g'(x_0)s \mid -a_i's \leq 0 , \text{ for all } i \in I(x_0) \} . \tag{4.16}$$

It follows from (4.15) that the objective function value for (4.16) can be no smaller than zero. Furthermore, this bound is achieved for $s = 0$. Consequently, $s = 0$ is an optimal solution for (4.16). From the Strong Duality

Theorem of linear programming [Theorem 3.8, Best and Ritter, 1985], there exists an optimal solution for the dual problem

$$\max \left\{ 0 \mid - \sum_{i \, \in \, I(x_0)} u_i a_i = g(x_0) , \ u_i \geq 0 , \text{ all } i \in I(x_0) \right\} .$$

Defining $u_i = 0$ for all $i \notin I(x_0)$ now shows that all three conditions in the statement of the theorem are satisfied.

Conversely, suppose x_0 satisfies the three stated conditions. Let x be any point in R. Taking the inner product of both sides of the second condition with $x - x_0$ gives

$$g'(x_0)(x - x_0) = - \sum_{i \, \in \, I(x_0)} (u_i a_i' x - u_i a_i' x_0) .$$

The third condition states that $u_i a_i' x_0 = u_i b_i$ for all $i \in I(x_0)$. Therefore

$$g'(x_0)(x - x_0) = \sum_{i \, \in \, I(x_0)} u_i(b_i - a_i' x) .$$

Because $x \in R$ and each $u_i \geq 0$, each term in this sum is nonnegative. Thus the sum itself is nonnegative and

$$g'(x_0)(x - x_0) \geq 0 . \tag{4.17}$$

From Taylor's series,

$$f(x) = f(x_0) + g'(x_0)(x - x_0) + \tfrac{1}{2}(x - x_0)'C(x - x_0).$$

Because f is convex, the last term is nonnegative. With (4.17), this implies that

$$f(x) \geq f(x_0) .$$

Since this last is true for all $x \in R$, we have shown that x_0 is indeed an optimal solution. □

Theorem 4.3 may be reformulated in a more compact way by relying more heavily on matrix notation. Suppose that our model problem is

$$\min \{ c'x + \tfrac{1}{2}x'Cx \mid Ax \leq b \}. \tag{4.18}$$

We claim that the optimality conditions for this are equivalent to

$$Ax_0 \leq b, \tag{4.19}$$

$$-g(x_0) = A'u , \ u \geq 0, \tag{4.20}$$

$$u'(Ax_0 - b) = 0, \tag{4.21}$$

where u is an m-vector. (4.19) and (4.20) are restatements of their counterparts in Theorem 4.3. With $u = (u_1, \ldots, u_m)'$, expansion of (4.21) gives

$$\sum_{i=1}^{m} u_i(a_i'x_0 - b_i) = 0 .$$

From (4.19) and (4.20), each term in this sum is nonpositive. Since the terms sum to zero, each term itself must have value zero; that is,

$$u_i(a_i'x_0 - b_i) = 0 , \quad i = 1, \ldots, m .$$

Summarizing, we have

Theorem 4.4 x_0 *is an optimal solution for* $\min \{ c'x + \frac{1}{2}x'Cx \mid Ax \leq b \}$ *if and only if*

(a) $Ax_0 \leq b$,

(b) $-g(x_0) = A'u , \quad u \geq 0$,

(c) $u'(Ax_0 - b) = 0$.

We have obtained the optimality conditions for the model problem of Theorem 4.4. It is useful to derive optimality conditions for other model problems. Consider the model problem

$$\min \{ c'x + \tfrac{1}{2}x'Cx \mid A_1x \leq b_1 , \quad A_2x = b_2 \}, \tag{4.22}$$

which has explicit inequality and equality constraints. We deduce the optimality conditions for (4.22) by first rewriting it in the form of (4.18), stating the optimality conditions using Theorem 4.4, and then simplifying the conditions. (4.22) differs from (4.18) only in that the former includes equality constraints. Since the equality constraints $A_2x = b_2$ are equivalent to the set of inequalities $A_2x \leq b_2$ and $A_2x \geq b_2$, (4.22) may be written as

$$\min \{ c'x + \tfrac{1}{2}x'Cx \mid A_1x \leq b_1 , \quad A_2x \leq b_2 , \quad -A_2x \leq -b_2 \},$$

which has the same form as (4.18) with

$$A = \begin{bmatrix} A_1 \\ A_2 \\ -A_2 \end{bmatrix} \quad \text{and} \quad b = \begin{bmatrix} b_1 \\ b_2 \\ -b_2 \end{bmatrix} .$$

Letting

$$u = \begin{bmatrix} u_1 \\ u_2 \\ u_3 \end{bmatrix} ,$$

the three optimality conditions of Theorem 4.4 for (4.22) are

$$A_1 x_0 \le b_1 \;, \quad A_2 x_0 = b_2 \;,$$
$$-g(x_0) = A_1' u_1 + A_2'(u_2 - u_3) \;, \quad u_1 \ge 0 \;, \quad u_2 \ge 0 \;, \quad u_3 \ge 0 \;,$$
$$u_1'(A_1 x_0 - b_1) + (u_2 - u_3)'(A_2 x_0 - b_2) = 0 \;.$$

These conditions depend only on $(u_2 - u_3)$ and while both u_2 and u_3 are required to be nonnegative, each component of $(u_2 - u_3)$ may be either positive or negative. Replacing $(u_2 - u_3)$ with u_2 which is then unrestricted in sign, and using $A_2 x_0 = b_2$ to simplify the third condition, the optimality conditions for (4.22) are thus

$$A_1 x_0 \le b_1 \;, \quad A_2 x_0 = b_2 \;,$$
$$-g(x_0) = A_1' u_1 + A_2' u_2 \;, \quad u_1 \ge 0 \;,$$
$$u_1'(A_1 x_0 - b_1) = 0 \;.$$

Theorem 4.4 the model problem (4.22) becomes

Theorem 4.5 x_0 *is an optimal solution for* $\min \{ c'x + \frac{1}{2}x'Cx \mid A_1 x \le b_1 \;, \quad A_2 x = b_2 \}$ *if and only if*

(a) $A_1 x_0 \le b_1 \;, \quad A_2 x_0 = b_2,$

(b) $-g(x_0) = A_1' u_1 + A_2' u_2 \;, \quad u_1 \ge 0,$

(c) $u_1'(A_1 x_0 - b_1) = 0.$

As a further example, we next derive the optimality conditions for the model problem

$$\min \{ c'x + \tfrac{1}{2}x'Cx \mid Ax = b \;, \quad x \ge 0 \}. \tag{4.23}$$

Rewriting the nonnegativity constraints in the equivalent form $-Ix \le 0$, (4.23) becomes

$$\min \{ c'x + \tfrac{1}{2}x'Cx \mid -Ix \le 0 \;, \quad Ax = b \} \;,$$

which is the same form as (4.22) with A_1, b_1, A_2, and b_2 replaced with $-I$, 0, A, and b, respectively. Theorem 4.5 asserts that the optimality conditions for (4.23) are

$$Ax_0 = b \;, \quad x_0 \ge 0,$$
$$-g(x_0) = A' u_1 - u_2 \;, \quad u_2 \ge 0, \tag{4.24}$$
$$u_2' x_0 = 0,$$

and the appropriate version of Theorem 4.4 is

Theorem 4.6 x_0 *is an optimal solution for* min $\{ c'x + \frac{1}{2}x'Cx \mid Ax = b, x \geq 0 \}$ *if and only if*

(a) $Ax_0 = b, x_0 \geq 0,$

(b) $-g(x_0) = A'u_1 - u_2, u_2 \geq 0,$

(c) $u_2'x_0 = 0.$

Because of the explicit presence of nonnegativity constraints in (4.23), the optimality conditions for (4.23) may be expressed in a more compact, but equivalent, manner by eliminating u_2 and replacing u_1 with u. Doing so in (4.24) gives

$$Ax_0 = b, x_0 \geq 0,$$
$$-g(x_0) \leq A'u,$$
$$(g(x_0) + A'u)'x_0 = 0,$$

and Theorem 4.6 becomes

Theorem 4.7 x_0 *is an optimal solution for* min $\{ c'x + \frac{1}{2}x'Cx \mid Ax = b, x \geq 0 \}$ *if and only if*

(a) $Ax_0 = b, x_0 \geq 0,$

(b) $-g(x_0) \leq A'u,$

(c) $(g(x_0) + A'u)'x_0 = 0.$

Tables 4.1(a) and 4.1(b) summarize the optimality conditions for a variety of model problems. Theorem 4.4 applies to each model problem with the indicated optimality conditions for the remaining entries of Table 4.1(a) and 4.1(b). The first four entries of Table 4.1(a) have been verified in the preceding text. The reader is asked to verify the last entry of Table 4.1(a) plus both entries of Table 4.1(b) in Exercise 4.6.

model problem: optimality conditions:	$\min \{ c'x + \frac{1}{2}x'Cx \mid Ax \leq b \}$ $Ax \leq b$ $-c - Cx = A'u, \ u \geq 0$ $u'(Ax - b) = 0$
model problem: optimality conditions:	$\min \{ c'x + \frac{1}{2}x'Cx \mid A_1x \leq b_1, \ A_2x = b_2 \}$ $A_1x \leq b_1, \ A_2x = b_2$ $-c - Cx = A_1'u_1 + A_2'u_2, \ u_1 \geq 0$ $u_1'(A_1x - b_1) = 0$
model problem: optimality conditions:	$\min \{ c'x + \frac{1}{2}x'Cx \mid Ax = b, \ x \geq 0 \}$ $Ax = b, \ x \geq 0$ $-c - Cx = A'u_1 - u_2, \ u_2 \geq 0$ $u_2'x = 0$ $\qquad$ OR $Ax = b, \ x \geq 0$ $-c - Cx \leq A'u$ $(c + Cx + A'u)'x = 0$
model problem: optimality conditions:	$\min \{ c'x + \frac{1}{2}x'Cx \mid Ax \geq b \}$ $Ax \geq b$ $-c - Cx = A'u, \ u \leq 0$ $u'(Ax - b) = 0$
model problem: optimality conditions:	$\min \{ c'x + \frac{1}{2}x'Cx \mid Ax \leq b, \ x \geq 0 \}$ $Ax \leq b, \ x \geq 0$ $-c - Cx = A'u_1 - u_2, \ u_1 \geq 0, \ u_2 \geq 0$ $u_1'(Ax - b) = 0, \ u_2'x = 0$ $\qquad$ OR $Ax \leq b, \ x \geq 0$ $-c - Cx \leq A'u, \ u \geq 0$ $u'(Ax - b) = 0, \ (c + Cx + A'u)'x = 0$

Table 4.1(a) Optimality Conditions for Various Model Problems.

mod prob:	$\min \{ c'x + \frac{1}{2}x'Cx \mid A_1x \leq b_1, \ A_2x = b_2, \ x \leq b_3, \ x \geq b_4 \}$
opt con's:	$A_1x \leq b_1, \ A_2x = b_2, \ x \leq b_3, \ x \geq b_4$
	$-c - Cx = A_1'u_1 + A_2'u_2 + u_3 + u_4, \ u_1 \geq 0, \ u_3 \geq 0, \ u_4 \leq 0$
	$u_1'(A_1x - b_1) = 0, \ u_3'(x - b_3) = 0, \ u_4'(x - b_4) = 0$
mod prob:	$\min \{ c_1'x_1 + c_2'x_2 + \frac{1}{2}(x_1'C_{11}x_1 + x_2'C_{22}x_2 + x_1'C_{12}x_2 + x_2'C_{21}x_1) \mid$
	$\quad A_1x_1 + A_2x_2 = b, \ x_2 \geq 0 \}$
opt con's:	$A_1x_1 + A_2x_2 = b, \ x_2 \geq 0$
	$-c_1 - C_{11}x_1 - C_{12}x_2 = A_1'u, \ -c_2 - C_{21}x_1 - C_{22}x_2 \leq A_2'u$
	$(c_2 + C_{21}x_1 + C_{22}x_2 + A_2'u)'x_2 = 0$

Table 4.1(b)   Optimality Conditions for Various Model Problems, continued.

4.2   QP Duality

For any quadratic programming problem (QP), there is a second QP, formulated in terms of the data of the given one, called the dual problem. Solution of the given problem automatically provides an optimal solution for the dual problem. This and other properties analogous to those of the duality theory of linear programming are developed in this section.

We motivate the definition of the dual problem as follows. For our usual model QP

$$\min \{ c'x + \tfrac{1}{2}x'Cx \mid Ax \leq b \}$$

let $f(x) = c'x + \frac{1}{2}x'Cx$. Define

$$L(x, u) = f(x) + u'(Ax - b).$$

We first consider the problem of fixing x, maximizing $L(x, u)$ over all $u \geq 0$, and then minimizing the result over all x; that is,

$$\min_{x} \max_{u \geq 0} L(x, u). \tag{4.25}$$

Let x be fixed. If at least one component of $Ax - b$ is strictly positive, then the maximum of L is $+\infty$ and is obtained by letting the associated component of u tend to $+\infty$. Since we are interested in minimizing L over x, any point satisfying $Ax \leq b$ will give a better objective function value than one for which one or more components of $Ax - b$ are strictly positive. The minimization of L may thus be restricted to those x for which $Ax \leq b$. For any such x, the maximum value of L over $u \geq 0$ is simply $f(x)$ and is attained for $u = 0$. Therefore (4.25) reduces to

$$\min \{ c'x + \tfrac{1}{2}x'Cx \mid Ax \leq b \}, \tag{4.26}$$

which is just our usual model QP.

Next consider the problem of fixing u, minimizing $L(x, u)$ over all x, and then maximizing the result over all $u \geq 0$; that is,

$$\max_{u \geq 0} \min_{x} L(x, u). \tag{4.27}$$

Note that (4.27) differs from (4.25) only in that the order of minimization and maximization have been reversed. Let $u \geq 0$ be fixed. Rearrangement of the terms of L gives

$$L(x, u) = -b'u + (c + A'u)'x + \tfrac{1}{2}x'Cx .$$

Since u is fixed, $L(x, u)$ is a quadratic function of x. Since C is positive semidefinite, Theorem 3.3 asserts that a necessary and sufficient condition that x minimize $L(x, u)$ is that the gradient of $L(x, u)$ with respect to x vanish; that is

$$c + A'u + Cx = 0 .$$

This last condition forces x to minimize $L(x, u)$ over x for fixed u. Consequently, (4.27) may be written as

$$\max \{ c'x + \tfrac{1}{2}x'Cx + u'(Ax - b) \mid Cx + A'u = -c, u \geq 0 \} \tag{4.28}$$

which is a quadratic programming problem.

Because of the special relationship between (4.25) and (4.27), we introduce the following definition. For the <u>primal</u> <u>problem</u>

$$\min \{ c'x + \tfrac{1}{2}x'Cx \mid Ax \leq b \}$$

the <u>dual</u> <u>problem</u> is

$$\max \{ c'x + \tfrac{1}{2}x'Cx + u'(Ax - b) \mid Cx + A'u = -c, u \geq 0 \}.$$

Example 4.3
Formulate the dual of

$$
\begin{array}{rllll}
\text{minimize:} & -4x_1 & - & 10x_2 & + & x_1^2 + x_2^2 \\
\text{subject to:} & -\ x_1 & + & x_2 & \leq & 2, & (1) \\
& x_1 & + & x_2 & \leq & 6, & (2) \\
& x_1 & & & \leq & 5, & (3) \\
& & - & x_2 & \leq & 0, & (4) \\
& -\ x_1 & & & \leq & -1. & (5)
\end{array}
$$

(This is the problem of Example 1.1.) By definition, the dual is

maximize: $\quad -4x_1 - 10x_2 + x_1^2 + x_2^2 + u_1(-x_1 + x_2 - 2)$
$$\qquad\qquad + u_2(x_1 + x_2 - 6) + u_3(x_1 - 5) + u_4(-x_2)$$
$$\qquad\qquad + u_5(-x_1 + 1)$$

subject to: $\quad 2x_1 \qquad\quad - u_1 + u_2 + u_3 \qquad - u_5 = 4, \qquad (1)$
$$\qquad\qquad\quad 2x_2 + u_1 + u_2 \qquad - u_4 \qquad = 10, \qquad (2)$$
$$\qquad\qquad\qquad u_1, \ldots, u_5 \geq 0.$$

$\diamond$

The dual QP is apparently different than the dual LP in that the former includes "x" variables while the latter does not. Nevertheless, LP duality may be considered a special case of QP duality as we now show. Setting $C = 0$ in the QP primal problem gives the LP

$$\min \{ c'x \mid Ax \leq b \}. \qquad (4.29)$$

Setting $C = 0$ in the dual QP gives

$$\max \{ c'x + u'(Ax - b) \mid A'u = -c, \ u \geq 0 \}.$$

For any dual feasible u, $A'u = -c$ so that $c'x = -u'Ax$ and the dual objective function reduces to $-b'u$. Thus the dual QP becomes

$$\max \{ -b'u \mid A'u = -c, \ u \geq 0 \}. \qquad (4.30)$$

(4.30) is the dual LP associated with the primal LP (4.29) [pg. 80, Best and Ritter, 1985]. Therefore our definition of a QP primal-dual pair is a generalization of an LP primal-dual pair.

We next present several results showing the relationship between primal and dual quadratic programming problems. These are appropriate generalizations of the corresponding duality theorems for linear programming. In the following, it is convenient to let

$$L(x, u) = c'x + \tfrac{1}{2}x'Cx + u'(Ax - b)$$

denote the objective function for the dual QP.

Theorem 4.8 *(Weak Duality Theorem) Let x be feasible for the primal problem and (y, u) be feasible for the dual problem. Then $f(x) \geq L(y, u)$.*

Proof: Because f is convex, Theorem 3.5(a) gives

$$f(x) - f(y) \geq g'(y)(x - y) ,$$

and since (y , u) is feasible for the dual

$$g(y) = c + Cy = -A'u .$$

Comparing the primal and dual objective function values and using the above two relations gives

$$\begin{aligned}
f(x) - L(y, u) &= f(x) - f(y) - u'(Ay - b) \\
&\geq g'(y)(x - y) - u'(Ay - b) \\
&\geq -u'A(x - y) - u'(Ay - b) \\
&\geq u'(b - Ax).
\end{aligned}$$

Since $Ax \leq b$ and $u \geq 0$, it follows that $f(x) \geq L(y, u)$ which completes the proof of the theorem. $\square$

Suppose that x_0 and (y_0 , u_0) are feasible for the primal and dual problems, respectively, and that $f(x_0) = L(y_0 , u_0)$. If x is any other feasible point for the primal then the Weak Duality Theorem (Theorem 4.8) states that $f(x) \geq L(y_0 , u_0) = f(x_0)$ and consequently x_0 is an optimal solution for the primal. Similarly, if (y , u) is any feasible point for the dual then $L(y, u) \leq f(x_0) = L(y_0 , u_0)$, which shows that (y_0 , u_0) is an optimal solution for the dual. This proves

Corollary 4.1 *If x_0 is feasible for the primal problem and (y_0 , u_0) is feasible for the dual problem and $f(x_0) = L(y_0 , u_0)$, then x_0 is optimal for the primal and (y_0 , u_0) is optimal for the dual.*

If the feasible region for the primal problem has at least one point x_0, then the Weak Duality Theorem (Theorem 4.8) states that $L(y, u) \leq f(x_0)$ for any (y , u) feasible for the dual; that is, $f(x_0)$ is an upper bound for $L(y, u)$. Therefore if $L(y, u)$ is unbounded from above over the feasible region for the dual, it follows by contradiction that the feasible region for the primal is empty. Similarly, if $f(x)$ is unbounded from below over R then there is no feasible point for the dual. This proves

Corollary 4.2 *If $L(y, u)$ is unbounded from above over the dual feasible region, then there is no feasible point for the primal problem. If $f(x)$ is unbounded from below over the feasible region for the primal, then there is no feasible point for the dual problem.*

The Weak Duality Theorem (Theorem 4.8) states that $f(x)$ can be no less than $L(y, u)$. The Strong Duality Theorem (Theorem 4.9) states that at optimality the two objective function values coincide.

Theorem 4.9 *(Strong Duality Theorem) Let x_0 be an optimal solution for the primal problem. Then there exists a u_0 such that (x_0, u_0) is an optimal solution for the dual problem and $f(x_0) = L(x_0, u_0)$. Furthermore, if u_0 is* **any** *vector which with x_0 satisfies the optimality conditions for the primal problem, then (x_0, u_0) is an optimal solution for the dual problem and $f(x_0) = L(x_0, u_0)$.*

Proof: With x_0 being optimal for the primal problem, Theorem 4.4 asserts the existence of u_0 which with x_0 satisfies the optimality conditions

$$Ax_0 \leq b,$$
$$Cx_0 + A'u_0 = -c, \quad u_0 \geq 0,$$
$$u_0'(Ax_0 - b) = 0.$$

The second optimality condition implies that (x_0, u_0) is feasible for the dual. The third condition implies that $f(x_0) = L(x_0, u_0)$. Corollary 4.1 now implies that (x_0, u_0) is optimal for the dual.   □

Let x_0 be optimal for the primal. The proof of Theorem 4.9 uses the result from Theorem 4.4 that x_0 satisfies the optimality conditions and proceeds by demonstrating that x_0 together with the vector u_0 of the optimality conditions constitute an optimal solution for the dual. For these reasons, each of the three optimality conditions is frequently referred to by the parenthetical names following:

$$
\begin{aligned}
Ax &\leq b, & \text{(primal feasibility)} \\
-c - Cx &= A'u, \quad u \geq 0, & \text{(dual feasibility)} \\
u'(Ax - b) &= 0. & \text{(complementary slackness)}
\end{aligned}
$$

We have referred to u_i as the multiplier associated with constraint i. It is also referred to as the dual variable associated with constraint i.

An algorithm for the solution of the primal problem typically terminates after constructing a vector u_0 which with the candidate optimal solution x_0 satisfies the optimality conditions of Theorem 4.4. It follows from the proof of the Strong Duality Theorem (Theorem 4.9) that the pair (x_0, u_0) is optimal for the dual problem. Thus obtaining an optimal solution for the dual problem is an automatic consequence of solving the primal problem.

Consider the problem

$$\min \{ \, c'x + \tfrac{1}{2}x'Cx \mid Ax \le b \, \}$$

and the dual,

$$\max \{ \, c'x + \tfrac{1}{2}x'Cx + u'(Ax - b) \mid Cx + A'u = -c, \, u \ge 0 \, \},$$

further. The dual problem has $n + m$ variables. The Hessian matrix of the dual objective function is

$$\begin{bmatrix} C & A' \\ A & 0 \end{bmatrix},$$

which is not, in general, negative semidefinite[2]. However, the dual is equivalent to a minimization problem for which the Hessian is indeed positive semidefinite as we now show. For any dual feasible $(x \, , \, u)$,

$$Cx + A'u = -c \, .$$

Multiplying on the left by x' gives

$$x'Cx + u'Ax = -c'x \, .$$

Using this, the dual objective function simplifies to

$$-b'u - \tfrac{1}{2}x'Cx \, .$$

After rewriting as an equivalent minimization problem, the dual becomes

$$\min \{ \, b'u + \tfrac{1}{2}x'Cx \mid Cx + A'u = -c \, , \; u \ge 0 \, \}. \qquad (4.31)$$

The Hessian matrix for this reformulated problem is

$$\begin{bmatrix} C & 0 \\ 0 & 0 \end{bmatrix},$$

which is indeed positive semidefinite.

Because the objective function for the reformulated dual is convex, the optimality conditions for (4.31) are both necessary and sufficient for optimality. From the second entry of Table 4.1(b), these optimality conditions are

$$Cx_0 + A'u_0 = -c \, , \;\; u_0 \ge 0 \, ,$$
$$C(x_0 - x_1) = 0 \, , \;\; Ax_1 \le b \, ,$$
$$u_0'(b - Ax_1) = 0 \, ,$$

---

[2]We are interested in **negative** semidefinite because the dual is a **maximization** problem.

where we have used $-x_1$ to denote the multipliers associated with the equality constraints of (4.31). These show that in general, there is no reason to expect that the x_0 portion of an optimal solution for the dual to be feasible for the primal. However, x_1 is feasible for the primal and from the second optimality condition, $Cx_0 = Cx_1$. Substituting this into the first gives $Cx_1 + A'u_0 = -c$, $u_0 \geq 0$. Thus x_1 satisfies all of the optimality conditions for the primal and from Theorem 4.4 is therefore optimal. We have shown

Theorem 4.10 *(Converse Duality Theorem) Let (x_0 , u_0) be an optimal solution for the dual problem. Then there is an x_1 with $C(x_0 - x_1) = 0$ such that x_1 is an optimal solution for the primal problem and $f(x_1) = L(x_0 , u_0)$. Furthermore, (x_1 , u_0) is an optimal solution for the dual problem.*

The previous analysis shows that there may be many optimal solutions for the dual problem. The "x" portion of some of these may not be feasible (and therefore not optimal) for the primal, but at least one will be primal optimal. This is further illustrated in

Example 4.4

$$
\begin{aligned}
\text{minimize:} \quad & -2x_1 - 2x_2 + \tfrac{1}{2}(x_1^2 + x_2^2 + 2x_1x_2) \\
\text{subject to:} \quad & x_1 + x_2 \leq 1, \quad (1) \\
& -x_1 \qquad\quad \leq 0, \quad (2) \\
& \qquad -x_2 \leq 0. \quad (3)
\end{aligned}
$$

Observe that the objective function for the primal problem may be written as $f(x) = \tfrac{1}{2}(x_1 + x_2 - 2)^2 - 2$, so that the level sets for f are pairs of parallel lines symmetric about the line $x_1 + x_2 = 2$. From Figure 4.3, it is apparent that every point on the line segment joining $(1 , 0)'$ and $(0 , 1)'$ is an optimal solution. This can be verified using the optimality conditions as follows. From the third optimality condition, we set $u_2 = u_3 = 0$. The remaining conditions require that

$$
\begin{bmatrix} -x_1 - x_2 + 2 \\ -x_1 - x_2 + 2 \end{bmatrix} = u_1 \begin{bmatrix} 1 \\ 1 \end{bmatrix} \quad \text{and} \quad x_1 + x_2 = 1 ,
$$

which are satisfied with $u_1 = 1$ and $x_1 + x_2 = 1$. Therefore all points on the line segment joining $(1 , 0)'$ and $(0 , 1)'$ are indeed optimal and the optimal primal objective function value is $-3/2$.

The dual problem is

$$
\text{maximize:} \quad -2x_1 - 2x_2 + \tfrac{1}{2}(x_1^2 + x_2^2 + 2x_1x_2)
$$

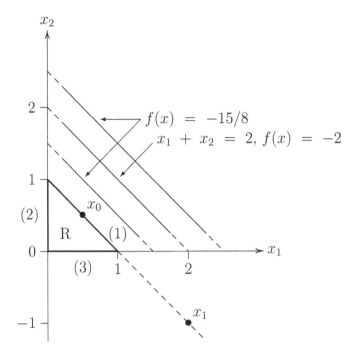

Figure 4.3. Geometry of Example 4.4.

$$+ \quad u_1(x_1 + x_2 - 1) + u_2(-x_1) + u_3(-x_2)$$

subject to:
$$x_1 + x_2 + u_1 - u_2 \quad = 2, \quad (1)$$
$$x_1 + x_2 + u_1 \quad - u_3 = 2, \quad (2)$$
$$u_1, u_2, u_3 \geq 0.$$

Consider the pair (x_0, u_0), where $x_0 = (2, -1)'$ and $u_0 = (1, 0, 0)'$. Note that (x_0, u_0) is feasible for the dual and that $L(x_0, u_0) = -3/2$. From Corollary 4.1, (x_0, u_0) is optimal for the dual. However, x_0 is not feasible for the primal. A similar argument shows that $(x_0 - \sigma s, u_0)$, with $s = (1, -1)'$, is dual optimal for all scalars σ. As σ varies from $-\infty$ to $+\infty$, $x_0 - \sigma s$ is the set of points on the line $x_1 + x_2 = 1$. Therefore any point on that line together with u_0 is dual optimal. However, only the portion of the line from $\sigma = 1$ to $\sigma = 2$ is feasible (and optimal) for the primal. $\Diamond$

In Example 4.4, C is positive semidefinite but not positive definite, thus accounting for the alternate optimal solutions. As an immediate consequence of the Converse Duality Theorem (Theorem 4.10) and the fact that C being positive definite implies that $Cx = 0$ if and only if $x = 0$, we have

Theorem 4.11 *(Strict Converse Duality Theorem) Let* $f(x)$ *be strictly con-*

vex. If (x_0 , u_0) is an optimal solution for the dual problem, then x_0 is an optimal solution for the primal problem and $f(x_0) = L(x_0 , u_0)$.

We next show that if both primal and dual problems have feasible solutions, then they both have optimal solutions.

Theorem 4.12 *If the primal problem has a feasible solution x_0 and the dual problem has a feasible solution (x_1 , u_1) then both the primal and dual problems have optimal solutions.*

Proof: From the Weak Duality Theorem (Theorem 4.8), the primal objective function is bounded from below by the number $L(x_1 , u_1)$ and the dual objective function is bounded from above by the number $f(x_0)$. Since each problem has at least one feasible point, it follows from Theorem 4.1 that each has an optimal solution. □

We have seen that the optimality conditions can be formulated for a variety of model quadratic programming problems. Using similar procedures, the dual corresponding to any model primal problem can be constructed.

$$\min \{ c'x + \tfrac{1}{2}x'Cx \mid A_1x \leq b_1 , \quad A_2x = b_2 \}, \tag{4.32}$$

then the dual problem is formulated by writing the equality constraints in the equivalent form $A_2x \leq b_2$ and $-A_2x \leq -b_2$. Letting $u_1, u_2,$ and u_3 be the multiplier vectors for the constraints $A_1x \leq b_1, A_2x \leq b_2,$ and $-A_2x \leq -b_2$, respectively, and applying the definition of the dual gives the dual problem

maximize: $c'x + \tfrac{1}{2}x'Cx + u_1'(A_1x - b_1) + (u_2 - u_3)'(A_2x - b_2)$

subject to: $Cx + A_1'u_1 + A_2'(u_2 - u_3) = -c,$

$u_1 \geq 0 , u_2 \geq 0 , u_3 \geq 0.$

Since the above depends on $(u_2 - u_3)$ and $(u_2 - u_3)$ is unrestricted in sign, it can be replaced by another vector, say u_2. The dual problem for (4.32) is thus

maximize: $c'x + \tfrac{1}{2}x'Cx + u_1'(A_1x - b_1) + u_2'(A_2x - b_2)$

subject to: $Cx + A_1'u_1 + A_2'u_2 = -c, \quad u_1 \geq 0.$ $\left. \right\} \tag{4.33}$

As a second example, we next formulate the dual of the primal problem

$$\min \{ c'x + \tfrac{1}{2}x'Cx \mid Ax = b , \quad x \geq 0 \} . \tag{4.34}$$

Rewriting the nonnegativity constraints in the equivalent form $-Ix \leq 0$, (4.33) becomes

$$\min \{ c'x + \tfrac{1}{2}x'Cx \mid -Ix \leq 0, \; Ax = b \},$$

which is of the same form as (4.32) with A_1, b_1, A_2, and b_2 replaced with $-I$, 0, A, and b, respectively. The dual of (4.34) is then obtained from (4.33) as

$$\left. \begin{aligned} \text{maximize:} \quad & c'x + \tfrac{1}{2}x'Cx + u_1'(Ax - b) - u_2'x \\ \text{subject to:} \quad & Cx + A'u_1 - u_2 = -c, \; u_2 \geq 0. \end{aligned} \right\} \quad (4.35)$$

Because of the explicit presence of nonnegativity constraints in (4.34) the dual problem (4.35) may be expressed in a more compact, but equivalent, manner. Replacing u_1 with u and eliminating u_2 gives the dual of (4.34) as

$$\max \{ -b'u - \tfrac{1}{2}x'Cx \mid Cx + A'u \geq -c \}.$$

primal:	$\min \{ c'x + \tfrac{1}{2}x'Cx \mid Ax \leq b \}$
dual:	$\max \{ c'x + \tfrac{1}{2}x'Cx + u'(Ax - b) \mid Cx + A'u = -c, u \geq 0 \}$
	OR
	$\max \{ -b'u - \tfrac{1}{2}x'Cx \mid Cx + A'u = -c, \; u \geq 0 \}$
primal:	$\min \{ c'x + \tfrac{1}{2}x'Cx \mid A_1x \leq b_1, \; A_2x = b_2 \}$
dual:	$\max \{ c'x + \tfrac{1}{2}x'Cx + u_1'(A_1x - b_1) + u_2'(A_2x - b_2) \mid$
	$\quad Cx + A_1'u_1 + A_2'u_2 = -c, \; u_1 \geq 0 \}$
primal:	$\min \{ c'x + \tfrac{1}{2}x'Cx \mid Ax = b, \; x \geq 0 \}$
dual:	$\max \{ c'x + \tfrac{1}{2}x'Cx + u_1'(Ax - b) - u_2'x \mid Cx + A'u_1 - u_2 = -c,$
	$\quad u_2 \geq 0 \}$
	OR
	$\max \{ -b'u - \tfrac{1}{2}x'Cx \mid Cx + A'u \geq -c \}$
primal:	$\min \{ c'x + \tfrac{1}{2}x'Cx \mid Ax \geq b \}$
dual:	$\max \{ c'x + \tfrac{1}{2}x'Cx + u'(Ax - b) \mid Cx + A'u = -c, \; u \leq 0 \}$

Table 4.2(a) Dual Problems for Various Model Primal Problems.

Using similar procedures, the dual problems may be formulated for any model primal quadratic programming problem. Analogous to Tables 4.1(a) and (b), Tables 4.2(a) and (b) give the dual problem for several possible model primal problems. The verification of each is straightforward and is left as an exercise (Exercise 4.7). Of course, each theorem in this section holds for any properly formulated primal-dual pair.

primal:	$\min \{ c'x + \frac{1}{2}x'Cx \mid Ax \leq b , \ x \geq 0 \}$
dual:	$\max \{ c'x + \frac{1}{2}x'Cx + u_1'(Ax - b) - u_2'x \mid Cx + A'u_1 - u_2 = -c ,$
	$u_1 \geq 0 , \ u_2 \geq 0 \}$
	OR
	$\max \{ -b'u - \frac{1}{2}x'Cx \mid Cx + A'u \geq -c , \ u \geq 0 \}$
primal:	$\min \{ c'x + \frac{1}{2}x'Cx \mid A_1x \leq b_1 , \ A_2x = b_2 , \ x \leq b_3 , \ x \geq b_4 \}$
dual:	$\max \{ c'x + \frac{1}{2}x'Cx + u_1'(A_1x - b_1) + u_2'(A_2x - b_2) + u_3'(x - b_3) +$
	$u_4'(x - b_4) \mid Cx + A_1'u_1 + A_2'u_2 + u_3 + u_4 = -c , \ u_1 \geq 0 ,$
	$u_3 \geq 0 , \ u_4 \leq 0 \}$
primal:	$\min \{ c_1'x_1 + c_2'x_2 + \frac{1}{2}(x_1'C_{11}x_1 + x_2'C_{22}x_2 + x_1'C_{12}x_2 + x_2'C_{21}x_1) \mid$
	$A_1x_1 + A_2x_2 = b , \ x_2 \geq 0 \}$
dual:	$\max \{ c_1'x_1 + c_2'x_2 + \frac{1}{2}(x_1'C_{11}x_1 + x_2'C_{22}x_2 + x_1'C_{12}x_2 + x_2'C_{21}x_1) +$
	$u_2'(-x_2) + u_2'(A_1x_1 + A_2x_2 - b) \mid C_{11}x_1 + C_{12}x_2 + A_1'u_2 = -c_1 ,$
	$C_{21}x_1 + C_{22}x_2 - u_1 + A_2'u_2 = -c_2 , \ u_1 \geq 0 \}$
primal:	$\min \{ c'x + \frac{1}{2}x'Cx \mid Ax \leq b \}$ where C is positive definite
dual:	$\max \{ -\frac{1}{2}c'C^{-1}c - (AC^{-1}c + b)'u - \frac{1}{2}u'AC^{-1}A'u \mid u \geq 0 \}$

Table 4.2(b)   Dual Problems for Various Model Primal Problems.

The reader familiar with the duality theory of linear programming may have expected this section to include a result stating that the dual of the dual QP is the same as the primal QP. Unfortunately, this is not true in general.

$$\min \{ c'x + \frac{1}{2}x'Cx \mid Ax \leq b \}. \tag{4.36}$$

We have seen that the dual problem is equivalent to

$$\min \{ b'u + \frac{1}{2}x'Cx \mid Cx + A'u = -c , \ u \geq 0 \}.$$

This has the same form as the third primal entry of Table 4.2(b). The dual of the dual problem is thus

$$\max \{ \frac{1}{2}x'Cx + v'(Cx + c) \mid Cx + Cv = 0 , \ Av \geq b \} . \tag{4.37}$$

Because $v'C = -x'C$ for any feasible (x , v), we have $v'Cx = -x'Cx$. Using this to simplify the objective function, replacing v with $-v$, and writing (4.37) as an equivalent minimization problem gives the dual of the dual problem for (4.36) as

$$\min \{ c'v + \frac{1}{2}x'Cx \mid C(x - v) = 0, \ Av \leq b \} . \tag{4.38}$$

Before proceeding further, it is useful to consider an example.

Example 4.5

Formulate and solve the dual of the dual of the primal problem given in Example 4.4.

According to (4.38), the dual of the dual is

$$
\begin{aligned}
\text{minimize}: \quad & -2v_1 - 2v_2 + \tfrac{1}{2}(x_1^2 + x_2^2 + 2x_1x_2) \\
\text{subject to}: \quad & x_1 + x_2 - v_1 - v_2 = 0, \quad (1) \\
& x_1 + x_2 - v_1 - v_2 = 0, \quad (2) \\
& v_1 + v_2 \leq 1, \quad (3) \\
& -v_1 \leq 0, \quad (4) \\
& -v_2 \leq 0, \quad (5)
\end{aligned}
$$

The objective function value for $x_0 = (1/2, 1/2)'$ and $v_0 = (1/2, 1/2)'$ is $-3/2$. Since this is identical to the optimal primal and dual objective function values, it follows from Corollary 4.1 that (x_0, v_0) is optimal for the problem at hand. Furthermore, from Example 4.4, x_0 is optimal for the primal problem. However, the same arguments show that $x_1 = (2, -1)'$ and $v_1 = (1/2, 1/2)'$ is also optimal although x_1 is not feasible for the primal problem. $\diamond$

From Example 4.5 we conclude that the dual of the dual is in general different from that of the primal. However, one instance in which the two problems are identical is when C is positive definite. In this case, C is invertible and the constraints of (4.38) imply that $x = v$, and (4.38) becomes

$$
\min \{ c'x + \tfrac{1}{2}x'Cx \mid Ax \leq b \},
$$

which is just the primal problem. We have demonstrated

Theorem 4.13 *If the objective function for the primal problem is strictly convex, then the dual of the dual quadratic programming problem is equivalent to the primal.*

Strict convexity is not the sole requirement for which the conclusion of Theorem 4.13 remains valid. Note that if we take $C = 0$ (which is positive semidefinite) then (4.38) becomes

$$
\min \{ c'v \mid Av \leq b \},
$$

which is precisely the primal problem.

The dual QP is formulated in terms of both primal and dual variables whereas the dual LP requires only dual variables. If C is positive definite, the dual QP may be expressed solely in terms of u and, in addition, the only constraints are nonnegativity constraints. To see this, consider our usual primal problem

$$\min \{ c'x + \tfrac{1}{2}x'Cx \mid Ax \leq b \}$$

and its dual

$$\max \{ c'x + \tfrac{1}{2}x'Cx + u'(Ax - b) \mid Cx + A'u = -c, \ u \geq 0 \}. \quad (4.39)$$

For any dual feasible pair (x, u), $Cx = -A'u - c$. Because C is positive definite, C is invertible so that

$$x = -C^{-1}A'u - C^{-1}c.$$

Substitution of this expression into (4.39) gives, upon simplification, that (4.39) is equivalent to

$$\max \{ -\tfrac{1}{2}c'C^{-1}c - (AC^{-1}c + b)'u - \tfrac{1}{2}u'AC^{-1}A'u \mid u \geq 0 \}.$$

Expressed in this form, the dual problem depends only on u. If u_0 is an optimal solution, the corresponding optimal value of x is $x_0 = -C^{-1}A'u_0 - C^{-1}c$. From the Strict Converse Duality Theorem (Theorem 4.11), x_0 is the optimal solution for the primal.

4.3   Unique and Alternate Optimal Solutions

This section is concerned with conditions under which an optimal solution of a quadratic programming problem is uniquely determined. We derive such conditions by first characterizing the set of all optimal solutions and then requiring this set to be a single point.

The set of all optimal solutions for a quadratic programming problem is sometimes called the <u>entire</u> <u>family</u> <u>of</u> <u>optimal</u> <u>solutions</u>.

We give a geometrical illustration of some of the possibilities in

Example 4.6

$$
\begin{aligned}
\text{minimize:} \quad & -6x_1 - 6x_2 + x_1^2 + x_2^2 + 2x_1x_2 \\
\text{subject to:} \quad & x_1 + x_2 \leq 2, \quad (1) \\
& -x_1 \leq 0, \quad (2) \\
& -x_2 \leq 0. \quad (3)
\end{aligned}
$$

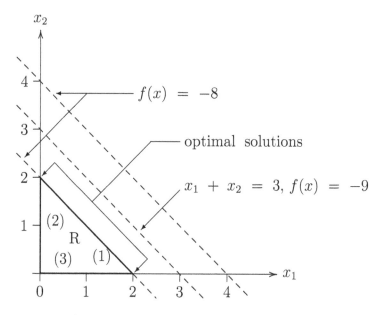

Figure 4.4. Geometry of Example 4.6.

The feasible region and level sets for the objective function are shown in Figure 4.4. Because the objective function can be written as $f(x) = (x_1 + x_2 - 3)^2 - 9$, the level sets are pairs of lines parallel to the line $x_1 + x_2 = 3$. Since this is parallel to the first constraint, all points on the line segment joining $(2, 0)'$ and $(0, 2)'$ are optimal solutions.

In particular, $x_0 = (1, 1)'$ is optimal. Since $g(x_0) = (-2, -2)'$, the optimality conditions are satisfied with $u_1 = 2$, $u_2 = u_3 = 0$. Let s be parallel to $(1, -1)'$. Then $s = \sigma(1, -1)'$ for some scalar σ and $x_0 - \sigma s$ is feasible for all σ with $-1 \le \sigma \le 1$. For σ in this interval, $x_0 - \sigma s$ is precisely the line segment joining $(2, 0)'$ and $(0, 2)'$; i.e., the set of all optimal solutions for this problem. Note also that $g'(x_0)s = 0$ and $s'Cs = 0$. $\Diamond$

The following lemma and theorem generalizes the results of Example 4.6.

Lemma 4.2 *Let x_0 and x_1 be optimal solutions for*

$$\min \{ c'x + \tfrac{1}{2}x'Cx \mid A_1x \le b_1, \quad A_2x = b_2 \}.$$

Then all points on the line segment joining x_0 and x_1 are also optimal, $g'(x_0)(x_1 - x_0) = 0$, and $(x_1 - x_0)'C(x_1 - x_0) = 0$.

Proof: Let σ be any scalar with $0 \le \sigma \le 1$. Because $f(x)$ is convex, Theorem

3.6(a) asserts that

$$f(\sigma x_1 + (1 - \sigma)x_0) \leq \sigma f(x_1) + (1 - \sigma)f(x_0).$$

Since both x_0 and x_1 are optimal, $f(x_0) = f(x_1)$ and thus

$$f(\sigma x_1 + (1 - \sigma)x_0) \leq f(x_0). \tag{4.40}$$

Since x_1 and x_0 are both feasible, so too is $\sigma x_1 + (1 - \sigma)x_0$ for all σ with $0 \leq \sigma \leq 1$ (Exercise 4.8). But then (4.40) implies that $\sigma x_1 + (1 - \sigma)x_0$ is also an optimal solution. Furthermore, since $\sigma x_1 + (1 - \sigma)x_0 = x_0 + \sigma(x_1 - x_0)$, Taylor's series gives

$$f(\sigma x_1 + (1 - \sigma)x_0) = f(x_0) + \sigma g'(x_0)(x_1 - x_0) + \tfrac{1}{2}\sigma^2(x_1 - x_0)'C(x_1 - x_0).$$

Since this must hold for all σ with $0 \leq \sigma \leq 1$, it follows that $g'(x_0)(x_1 - x_0) = 0$ and $(x_1 - x_0)'C(x_1 - x_0) = 0$. $\qquad\square$

The next result characterizes the entire family of optimal solutions for a quadratic programming problem.

Theorem 4.14 *Let x_0 be an optimal solution for the model problem*

$$\left.\begin{array}{rll}
\text{minimize}: & c'x + \tfrac{1}{2}x'Cx & \\
\text{subject to}: & a_i'x \leq b_i, & i = 1, \ldots, m, \\
& a_i'x = b_i, & i = m+1, \ldots, m+r,
\end{array}\right\} \tag{4.41}$$

and let $u = (u_1, \ldots, u_{m+r})'$ be the associated vector of multipliers. Define the set S to consist of all vectors $x_0 - s$ where s satisfies $x_0 - s \in R$,

$$a_i's = 0, \quad \text{for all } i \text{ with } 1 \leq i \leq m, \quad a_i'x_0 = b_i, \quad \text{and } u_i > 0,$$
$$a_i's \geq 0, \quad \text{for all } i \text{ with } 1 \leq i \leq m, \quad a_i'x_0 = b_i, \quad \text{and } u_i = 0,$$
$$a_i's = 0, \quad \text{for all } i \text{ with } m+1 \leq i \leq m+r,$$

and $s'Cs = 0$. Then S is the entire family of optimal solutions for (4.41).

Proof: Let T denote the entire family of optimal solutions for (4.41). We show that $S = T$ by showing $T \subseteq S$ and $S \subseteq T$.

Suppose first that $x_0 - s$ is an optimal solution. Then $x_0 - s \in R$ and from Lemma 4.2, $g'(x_0)s = 0$ and $s'Cs = 0$. Because x_0 and $x_0 - s$ are feasible, $a'_{m+1}s = \cdots = a'_{m+r}s = 0$ so that from the optimality conditions for x_0,

$$-g'(x_0)s = \sum_{i=1}^{m} u_I a'_i s + \sum_{i=1}^{r} u_{m+i} a'_{m+i} s$$

$$= \sum_{i=1}^{m} u_I a'_i s = 0. \tag{4.42}$$

Let i be such that $1 \le i \le m$ and $a'_i x_0 = b_i$. Since $x_0 - s$ is feasible, $a'_i(x_0 - s) \le b_i$ which implies that $a'_i s \ge 0$. Therefore

$$a'_i s \ge 0 , \quad \text{for all } i \text{ with } 1 \le i \le m \text{ and } a'_i x_0 = b_i . \tag{4.43}$$

From the complementary slackness part of the optimality conditions for x_0, all terms in (4.42) associated with constraints inactive at x_0, must vanish. Since $u_i \ge 0$ for all i with $1 \le i \le m$ and $a'_i x_0 = b_i$, (4.43) implies that the remaining terms in (4.42) are all nonnegative. Since they sum to zero, each term itself must vanish. Therefore $a'_i s = 0$ for all i with $1 \le i \le m$, $a'_i x_0 = b_i$, and $u_i > 0$. This implies that $x_0 - s \in S$ and thus $T \subseteq S$.

Conversely, suppose that $x = x_0 - s \in S$. Since $-g(x_0) = u_1 a_1 + \cdots + u_{m+r} a_{m+r}$, it follows by definition of S that $g'(x_0)s = 0$. Since $s'Cs = 0$, it follows from Taylor's series that

$$f(x_0 - s) = f(x_0) - g'(x_0)s + \tfrac{1}{2}s'Cs$$
$$= f(x_0) .$$

Thus $x_0 - s$ is a feasible point which gives the same objective function value as an optimal solution. Therefore $x_0 - s$ is also optimal and $S \subseteq T$.

We have shown that $T \subseteq S$ and $S \subseteq T$. This implies that $S = T$, which completes the proof of the theorem. □

Theorem 4.14 can be used to establish necessary and sufficient conditions for a unique optimal solution.

Theorem 4.15 *Let x_0 be an optimal solution for*

$$\begin{aligned} \text{minimize}: \quad & c'x + \tfrac{1}{2}x'Cx \\ \text{subject to}: \quad & a'_i x \le b_i, \quad i = 1, \dots, m, \\ & a'_i x = b_i, \quad i = m+1, \dots, m+r, \end{aligned}$$

and let $u = (u_1, \ldots, u_{m+r})'$ denote the vector of multipliers associated with it. Then x_0 is the unique optimal solution if and only if $s'Cs > 0$ for all $s \neq 0$ satisfying

$$a_i's = 0 , \quad \text{for all } i \text{ with } 1 \leq i \leq m , \quad a_i'x_0 = b_i , \quad \text{and } u_i > 0 ,$$
$$a_i's \geq 0 , \quad \text{for all } i \text{ with } 1 \leq i \leq m , \quad a_i'x_0 = b_i , \quad \text{and } u_i = 0 ,$$
$$a_i's = 0 , \quad \text{for } i = m + 1, \ldots, m + r .$$

Proof: First observe that if $x_0 - s \in R$, then

$$a_i's = 0 , \quad \text{for all } i \text{ with } 1 \leq i \leq m \text{ and } a_i'x_0 = b_i ,$$
$$a_i's = 0 , \quad \text{for } i = m + 1, \ldots, m + r .$$

Conversely, if s satisfies these conditions then $x_0 - \sigma s \in R$ for all positive σ sufficiently small. Because x_0 is the unique optimal solution if and only if the entire family of optimal solutions is x_0, the theorem now follows from Theorem 4.14.   □

For the special case of unconstrained quadratic minimization, Theorem 4.15 asserts that an optimal solution is the unique optimal solution for min $\{ c'x + \frac{1}{2}x'Cx \}$ if and only if $s'Cs > 0$ for all $s \neq 0$. That is, x_0 is unique if and only if C is positive definite. This establishes the converse of Theorem 3.4.

For the special case of quadratic minimization subject to linear equality constraints, Theorem 4.15 asserts that an optimal solution is the unique optimal solution for min $\{ c'x + \frac{1}{2}x'Cx \mid Ax = b \}$ if and only if $s'Cs > 0$ for all $s \neq 0$ with $As = 0$. If C were positive definite, then $s'Cs > 0$ for all $s \neq 0$ and in particular $s'Cs > 0$ for all $s \neq 0$ for which $As = 0$. Therefore Theorem 4.15 implies Theorem 3.12.

Let x_0 be an optimal solution for (4.41) and let $u = (u_1, \ldots, u_{m+r})'$ denote the associated vector of multipliers. The complementary slackness condition asserts that

$$u_i(a_i'x_0 - b_i) = 0 , \quad \text{for } i = 1, \ldots, m .$$

This condition forces either u_i, $a_i'x_0 - b_i$, or both to vanish. This last possibility can result in some technical difficulties. In order to exclude such cases, we say that x_0 satisfies the <u>strict</u> <u>complementary</u> <u>slackness</u> condition for (4.41) if

$$a_i'x_0 = b_i \text{ implies } u_i > 0 , \quad i = 1, \ldots, m .$$

Note that no requirement is imposed on the multipliers for the equality constraints.

When the strict complementary slackness assumption is satisfied, Theorem 4.15 can be used to give a simpler sufficient condition for uniqueness. It is also a simpler condition to recognize computationally.

Theorem 4.16 *Let x_0 be an optimal solution for*

$$\begin{aligned} \text{minimize}: \quad & c'x + \tfrac{1}{2}x'Cx \\ \text{subject to}: \quad & a_i'x \leq b_i, \quad & i = 1, \ldots, m, \\ & a_i'x = b_i, \quad & i = m+1, \ldots, m+r. \end{aligned}$$

Suppose that x_0 satisfies the strict complementary slackness condition and that $s'Cs > 0$ for all $s \neq 0$ satisfying

$$\begin{aligned} a_i's &= 0, \quad \text{for all } i \text{ with } 1 \leq i \leq m \text{ and } a_i'x_0 = b_i, \\ a_i's &= 0, \quad \text{for } i = m+1, \ldots, m+r. \end{aligned}$$

Then x_0 is the unique optimal solution. In addition, if the gradients of those constraints active at x_0 are linearly independent, then the associated vector of multipliers is uniquely determined.

Proof: The uniqueness of x_0 is a direct consequence of Theorem 4.15 and the strict complementary slackness condition.

Suppose that there are two distinct vectors of multipliers u_0 and u_1. Let I_0 denote the set of indices of those constraints active at x_0, including the equality constraints. Since x_0 is the unique optimal solution,

$$-g(x_0) = \sum_{i \in I_0} (u_0)_i a_i$$

and

$$-g(x_0) = \sum_{i \in I_0} (u_1)_i a_i.$$

Subtraction gives

$$\sum_{i \in I_0} ((u_0)_i - (u_1)_i) a_i = 0.$$

Since a_i, all $i \in I_0$, are linearly independent, this implies that $u_0 = u_1$, and indeed, the multipliers are uniquely determined. □

4.4   Sensitivity Analysis

In many applications, the data used to define a quadratic programming problem is not precisely determined. This may be a consequence of rounding errors in measurements or it may be that precise data is prohibitively expensive to obtain. It is therefore important to analyze the behavior of the optimal solution and the optimal objective function value resulting from changes in the problem data.

In this section, we restrict our attention to the effects arising from small changes in the right-hand side of the constraint functions and/or the linear part of the objective function. As in linear programming, this is often called sensitivity or post-optimality analysis. An analogous study of certain types of large changes in this part of the problem data is made in Chapter 8 under the heading of parametric quadratic programming.

We introduce the most important ideas and results of this section by means of a small example.

Example 4.7

$$
\begin{aligned}
\text{minimize:} \quad & -2x_1 - 3x_2 + \tfrac{1}{2}(x_1^2 + x_2^2) \\
\text{subject to:} \quad & x_1 + x_2 \le 2, \quad (1) \\
& -x_1 \quad\quad\ \le 0, \quad (2) \\
& \quad\ -x_2 \le 0. \quad (3)
\end{aligned}
$$

The problem is illustrated in Figure 4.5. The objective function is strictly convex, its level sets are concentric circles centered at $(2 , 3)'$, and only the first constraint is active at the optimal solution. From complementary slackness we can set $u_2 = u_3 = 0$. The remaining part of the optimality conditions is

$$
\begin{aligned}
x_1 \quad\quad + u_1 &= 2 , \\
x_2 + u_1 &= 3 , \\
x_1 + x_2 \quad\quad &= 2 .
\end{aligned}
$$

These are readily solved to give the optimal solution $x_1 = 1/2$, $x_2 = 3/2$, and $u_1 = 3/2$. $\Diamond$

Next we consider the similar problem

Example 4.8

$$
\text{minimize:} \quad c_1 x_1 + c_2 x_2 + \tfrac{1}{2}(x_1^2 + x_2^2)
$$

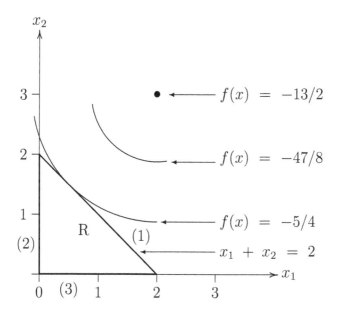

Figure 4.5. Geometry of Example 4.7.

$$\text{subject to:} \quad x_1 + \quad x_2 \leq b, \quad (1)$$
$$- \quad x_1 \quad\quad \leq 0, \quad (2)$$
$$- \quad x_2 \leq 0. \quad (3)$$

Let $c = (c_1, c_2)'$, $\hat{c} = (-2, -3)'$, and $\hat{b} = 2$. The above problem is obtained from the previous by replacing $\hat{c}$ with c and $\hat{b}$ with b. We investigate the behavior of the optimal solution as it depends on b and c as b and c are changed a small amount from $\hat{b}$ and $\hat{c}$, respectively. Such small changes have the effect of a parallel shift to the line $x_2 + x_2 = b$ and a shift to the center of the circular level sets of the objective function. It seems reasonable that only the first constraint will be active at the optimal solution for the new problem. To verify this we use the optimality conditions. Letting $u_2 = u_3 = 0$, the remaining conditions are

$$x_1 \quad\quad + u_1 = -c_1,$$
$$x_2 + u_1 = -c_2,$$
$$x_1 + x_2 \quad\quad = b.$$

These are readily solved to obtain

$$x_1(b, c) = -c_1 + \tfrac{1}{2}(c_1 + c_2 + b),$$
$$x_2(b, c) = -c_2 + \tfrac{1}{2}(c_1 + c_2 + b),$$
$$u_1(b, c) = \quad\quad -\tfrac{1}{2}(c_1 + c_2 + b),$$

where the notation $x_1(b\,,\ c)$, $x_2(b\,,\ c)$, and $u_1(b\,,\ c)$ is used to denote explicit dependence on b and c. For all $(b\,,\ c)$ sufficiently close to $(\hat{b}\,,\ \hat{c})$ it is easy to see that $x(b\,,\ c)$ is feasible and $u_1(b\,,\ c)\ >\ 0$. For all such $(b\,,\ c)$, $x(b\,,\ c)$ satisfies the optimality conditions and is therefore an optimal solution. Furthermore, the above shows explicitly that $x(b\,,\ c)$ and $u_1(b\,,\ c)$ are linear functions of b and c. This property holds for more general problems and will be demonstrated shortly.

Substitution of $x(b\,,\ c)$ into the objective function gives the optimal objective function value as an explicit function of b and c:

$$f(x(b\,,\ c))\ =\ \tfrac{1}{4}(c_1\ +\ c_2\ +\ b)^2\ -\ \tfrac{1}{2}(c_1^2\ +\ c_2^2)\,.$$

Differentiation gives

$$\frac{\partial f}{\partial b}(x(b\,,\ c))\ =\ \tfrac{1}{2}(c_1\ +\ c_2\ +\ b)\,,$$

$$\frac{\partial f}{\partial c_1}(x(b\,,\ c))\ =\ -c_1\ +\ \tfrac{1}{2}(c_1\ +\ c_2\ +\ b)\,,$$

$$\frac{\partial f}{\partial c_2}(x(b\,,\ c))\ =\ -c_2\ +\ \tfrac{1}{2}(c_1\ +\ c_2\ +\ b)\,.$$

Comparing these with the above solution for $x(b\,,\ c)$ and $u_1(b\,,\ c)$ shows that

$$\frac{\partial f}{\partial b}(x(b\,,\ c))\ =\ -u_1(b\,,\ c)\,,$$

$$\frac{\partial f}{\partial c_1}(x(b\,,\ c))\ =\ x_1(b\,,\ c)\,,$$

$$\frac{\partial f}{\partial c_2}(x(b\,,\ c))\ =\ x_2(b\,,\ c)\,.$$

These relationships between the partial derivatives of $f(x(b\,,\ c))$ with respect to b and the components of c and $u(b\,,\ c)$ and $x(b\,,\ c)$ hold for more general problems. $\qquad\qquad\qquad\qquad\qquad\qquad\qquad\Diamond$

Returning to our model QP

$$\min\ \{\ c'x\ +\ \tfrac{1}{2}x'Cx\ |\ Ax\ \leq\ b\ \}\qquad\qquad(4.44)$$

we formulate and prove more general versions of the properties established for Example 4.6. Let $x(b\,,\ c)$ denote an optimal solution for this problem and $u(b\,,\ c)$ an associated vector of multipliers, as explicit functions of b and c. In general, $x(b\,,\ c)$ and $u(b\,,\ c)$ may not be uniquely determined. However, the assumptions made in the hypothesis of the following theorem ensure uniqueness.

Theorem 4.17 *Let $x(b \, , \, c)$ be an optimal solution for*

$$\min \{ c'x + \tfrac{1}{2}x'Cx \mid Ax \leq b \}$$

as an explicit function of b and c. Let $u(b \, , \, c)$ denote a multiplier vector associated with $x(b \, , \, c)$. Suppose that $\hat{b}$ and $\hat{c}$ are specified m- and n-vectors, respectively, such that the gradients of linearly independent, $\hat{u} = u(\hat{b} \, , \, \hat{c})$ satisfies the strict complementary slackness condition, and $s'Cs > 0$ for all $s \neq 0$ satisfying $a_i's = 0$ for all i with $a_i'\hat{x} = \hat{b}_i$. Then there are nonnull neighborhoods about $\hat{b}$ and $\hat{c}$ such that for all b and c in these neighborhoods

(a) $x(b \, , \, c)$ *and* $u(b \, , \, c)$ *are uniquely determined linear functions of b and c,*

(b) $\dfrac{\partial f}{\partial c_i}(x(b \, , \, c)) = x_i(b \, , \, c)$ *for all* $i = 1, \ldots, n$,

(c) $\dfrac{\partial f}{\partial b_i}(x(b \, , \, c)) = -u_i(b \, , \, c)$ *for all* $i = 1, \ldots, m$,

where b_i, c_i, $x_i(b \, , \, c)$, and $u_i(b \, , \, c)$ denote the ith components of b, c, $x(b \, , \, c)$, and $u(b \, , \, c)$, respectively.

Proof: By reordering the constraints if necessary, we may assume without loss of generality that

$$a_i'\hat{x} = \hat{b}_i \, , \quad i = 1, \ldots, k \, , \quad \text{and}$$
$$a_i'\hat{x} < \hat{b}_i \, , \quad i = k+1, \ldots, m \, .$$

Let $A_1' = [a_1, \ldots, a_k]$, $b_1 = (b_1, \ldots, b_k)'$, $\hat{b}_1 = (\hat{b}_1, \ldots, \hat{b}_k)'$, $u_1 = (u_1, \ldots, u_k)'$, and $\hat{u}_1 = (\hat{u}_1, \ldots, \hat{u}_k)'$. Consider the system of linear equations

$$\begin{bmatrix} C & A_1' \\ A_1 & 0 \end{bmatrix} \begin{bmatrix} x \\ u_1 \end{bmatrix} = \begin{bmatrix} -c \\ b_1 \end{bmatrix}, \tag{4.45}$$

where 0 denotes a $(k \, , \, k)$ matrix of zeros. Because $(\hat{x} \, , \, \hat{u})$ satisfy the optimality conditions, this system has solution $x = \hat{x}, u_1 = \hat{u}_1$ when $c = \hat{c}$ and $b_1 = \hat{b}_1$. Let

$$H = \begin{bmatrix} C & A_1' \\ A_1 & 0 \end{bmatrix}$$

denote the coefficient matrix of the system. From the assumptions in the statement of the theorem it can be shown that H is nonsingular (Exercise 3.21). We partition H^{-1} as follows:

$$H^{-1} = \begin{bmatrix} M_1 & M_2 \\ M_3 & M_4 \end{bmatrix},$$

where M_1, M_2, M_3, and M_4 are (n , n), (n , k), (k , n), and (k , k) matrices, respectively. Because H is symmetric, so too is H^{-1}. Therefore M_1 and M_4 are both symmetric and $M_2' = M_3$. Using this partitioned form of H^{-1}, the solution of (4.45) is written

$$x = M_2 b_1 - M_1 c, \qquad (4.46)$$

$$u_1 = M_4 b_1 - M_3 c. \qquad (4.47)$$

Let $u = (u_1 , 0 , 0, \ldots, 0)'$. Then for any b and c, $x = x(b , c)$ and $u = u(b , c)$ defined by (4.46) and (4.47) are linear functions of b and c. Because of (4.45), any such x and u satisfy part, but in general not all, of the optimality conditions for (4.44). In particular, for some values of b and c one or more constraints active at $\hat{x}$ may be violated at $x(b , c)$ or one or more components of $u_1(b , c)$ may be negative. However, since each component of $\hat{u}_1$ is strictly positive and $a_1' \hat{x} < \hat{b}_i$, $i = k + 1, \ldots, m$, there are nonnull neighborhoods of $\hat{b}$ and $\hat{c}$ such that for all b and c in these neighborhoods, all constraints inactive at $\hat{x}$ are also inactive at $x(b , c)$ and all components of $u_1(b , c)$ are positive. It follows that for all such b and c, $x(b , c)$ is optimal with multiplier vector $u(b , c)$. Because of the assumption in the statement of the theorem, it follows from Theorem 4.16 that $x(b , c)$ is the unique optimal solution for (4.44) and furthermore that $u(b , c)$ is uniquely determined by (4.47). This proves part (a) of the theorem.

To demonstrate the remainder of the theorem, let $x = x(b , c)$ and $u_1 = u_1(b , c)$ be defined by (4.46) and (4.47), respectively, where b and c are sufficiently close to $\hat{b}$ and $\hat{c}$, respectively, such that constraints $k + 1, \ldots, m$ are inactive at $x(b , c)$ and each component of $u_1(b , c)$ is strictly positive. We have from (4.45) that $Cx = -c - A_1' u_1$ and $A_1 x = b_1$ so that $x'Cx = -c'x - u_1' A_1 x = -c'x - u_1' b_1$. Therefore

$$c'x + \tfrac{1}{2} x'Cx = \tfrac{1}{2}(c'x - b_1' u_1) .$$

Substitution of x and u_1 from (4.46) and (4.47), respectively, gives the objective function value explicitly in terms of b and c:

$$f(x(b , c)) = b_1' M_3 c - \tfrac{1}{2} c' M_1 c - \tfrac{1}{2} b_1' M_4 b_1 ,$$

where we have used the fact that $M_2' = M_3$. Using the symbols ∇_{b_1} and ∇_c to denote gradient vectors with respect to b_1 and c, respectively, we calculate

$$\nabla_c f(x(b , c)) = M_2 b_1 - M_1 c \quad \text{and}$$
$$\nabla_{b_1} f(x(b , c)) = -M_4 b_1 + M_3 c.$$

The remainder of the theorem follows by comparing these formulae with (4.46) and (4.47), and the observation that

$$\frac{\partial f}{\partial b_i}(x(b , c)) = 0 = u_i(b , c) \text{ for } i = k + 1, \ldots, m ,$$

because $f(x(b , c))$ does not depend on the last $m - k$ components of b. $\square$

Theorem 4.17 requires the assumption that $x(b , c)$ be the unique optimal solution. In order to appreciate this assumption, consider the problem of Example 4.6 further. We previously observed that all points on the line segment joining $(2 , 0)'$ and $(0 , 2)'$ are optimal. In particular, $x_0 = (1 , 1)'$ is optimal. Suppose now that c_1 is allowed to vary a small amount from -6. If $c_1 < -6$, but arbitrarily close to -6, the optimal solution "jumps" to $(2 , 0)'$ where constraint (3) becomes active. This may be verified by observing that at $(2 , 0)'$ the optimality conditions are

$$\begin{bmatrix} 2 & 2 \\ 2 & 2 \end{bmatrix}\begin{bmatrix} x_1 \\ x_2 \end{bmatrix} + u_1 \begin{bmatrix} 2 \\ 2 \end{bmatrix} + u_3 \begin{bmatrix} 0 \\ -1 \end{bmatrix} = \begin{bmatrix} -c_1 \\ 6 \end{bmatrix} ,$$

$$x_1 + x_2 = 2 ,$$
$$x_2 = 0 ,$$
$$u_2 = 0 .$$

These have solution $x = (2 , 0)'$, $u_1 = (-c_2 - 2)/2$, $u_2 = 0$, and $u_3 = (-6 - c_1)$, so that $u_3 \geq 0$ provided that $c_1 \leq -6$.

Similarly, if $c_1 > -6$ but arbitrarily close to -6, the optimal solution "jumps" to $(0 , 2)'$ with $u_1 = 1$, $u_2 = c_1 + 6$, and $u_3 = 0$, so that $u_2 \geq 0$ provided that $c_1 \geq -6$.

4.5 Exercises

4.1 Use Theorem 4.3 to find an optimal solution for

minimize: $\quad x_1 - 2x_2 + 2x_3 + \frac{1}{2}(x_1^2 + 2x_2^2 + 3x_3^2)$
subject to: $\quad - x_1 + x_2 - x_3 \leq 1.$ (1)

Hint: This problem has just one constraint. If it is inactive at the optimal solution, then the optimal solution must be the unconstrained minimum of the objective function. Obtain this point first, show that it violates the constraint and conclude that the constraint must be active. Note that a generalization of this problem is given in Exercise 4.5.

4.2 Consider the problem of minimizing each of the objective functions

(a) $4x_1^2 + 9x_2^2$,

(b) $-36x_1 + 54x_2 + 4x_1^2 + 9x_2^2 - 12x_1x_2$,

(c) $-64x_1 - 36x_2 + 4x_1^2 + 9x_2^2$,

(d) $-128x_1 - x_2 + 16x_1^2$,

$$
\begin{array}{rrrrll}
\text{subject to:} & - & x_1 & & \le & 0, & (1) \\
& - & x_1 & + \quad x_2 & \le & 3, & (2) \\
& - & x_1 & + \quad 3x_2 & \le & 13, & (3) \\
& & 2x_1 & + \quad x_2 & \le & 16, & (4) \\
& & 2x_1 & - \quad 3x_2 & \le & 8, & (5) \\
& & - & x_2 & \le & 0, & (6) \\
& - & 2x_1 & - \quad 3x_2 & \le & -6. & (7)
\end{array}
$$

By sketching the level sets of the objective function in each case, determine which constraints are active at the optimal solution. Use Theorem 4.3 to precisely determine the optimal solution. In each case, determine whether or not the optimal solution is unique and, if not, determine all optimal solutions. State the dual problem and give an optimal solution for it.

4.3 Calculate $u_1, \ldots, u_4$ for each of the quasistationary points of Example 4.1. Verify that x_6 is the only quasistationary point for which all of the u_i are nonnegative.

4.4 Consider the quadratic programming problem

$$
\begin{array}{rrrrll}
\text{minimize:} & - & 8x_1 - 4x_2 + & x_1^2 + 2x_2^2 & & \\
\text{subject to:} & & x_1 + 2x_2 & \le 6, & (1) \\
& & 2x_1 + \quad x_2 & \le 6, & (2) \\
& - & x_1 & \le 0, & (3) \\
& & - \quad x_2 & \le 0. & (4)
\end{array}
$$

(a) Draw a graph of the feasible region and sketch level sets for the objective function. Locate an approximation to the optimal solution and show graphically that the optimality conditions are satisfied at it. Which constraint(s) is(are) active?

(b) Use the optimality conditions to find the optimal solution for it.

(c) State the dual of the problem and determine an optimal solution for it.

(d) Is the optimal solution for the primal uniquely determined? Why?

(e) Let $b = (6, 6, 0, 0)'$ denote the right-hand side of the problem's constraints. Suppose that b is replaced with $b + t(1, -1, 0, 0)'$, where t is a nonnegative scalar parameter. Show graphically the behavior of the optimal solution, $x(t)$, as t is increased a small amount from 0. Determine $x(t)$ analytically as an explicit function of t. For what range of t does $x(t)$ remain optimal? What happens as t is increased further?

4.5 Suppose that $\gamma_1, \ldots, \gamma_n > 0$, $a_1, \ldots, a_n$, $c_1, \ldots, c_n$ and b are given scalars, and that $\sum_{i=1}^{n} (a_i c_i / \gamma_i) < -b$. Consider the problem

$$\text{minimize:} \quad c_1 x_1 + \cdots + c_n x_n + \tfrac{1}{2}(\gamma_1 x_1^2 + \cdots + \gamma_n x_n^2)$$
$$\text{subject to:} \quad a_1 x_1 + \cdots + a_n x_n \leq b. \quad (1)$$

(a) Use Theorem 4.3 to find an optimal solution. (Note that a special case of this problem is considered in Exercise 4.1.)

(b) State the dual problem and find an optimal solution for it.

(c) By direct computation, verify parts (a), (b), and (c) of Theorem 4.17 for this problem.

4.6 Verify the final entry of Table 4.1(a) and both entries of Table 4.1(b).

4.7 Verify the remaining entries in Tables 4.2(a) and (b).

4.8 Let $R = \{x \mid A_1 x \leq b_1, A_2 x = b_2\}$. Suppose that x_0 and x_1 are both in R. Show that $\sigma x_0 + (1 - \sigma)x_1$ is also in R for all σ with $0 \leq \sigma \leq 1$.

4.9 Suppose that x_0 is an optimal solution for

$$\min \{ c'x + \tfrac{1}{2}x'Cx \mid a_i'x \leq b_i, \ i = 1, \ldots, m \}$$

and that there is $u = (u_1, \ldots, u_m)'$ satisfying

$$-g(x_0) = u_1 a_1 + \cdots + u_m a_m,$$
$$u_i(a_i'x_0 - b_i) = 0, \quad i = 1, \ldots, m.$$

Does this imply that $u \geq 0$?

4.10 The following uniqueness result is known for linear programming problems [Theorem 3.12, Best and Ritter, [6]].

Let x_0 be an optimal solution for

$$\text{minimize}: \quad c'x$$
$$\text{subject to}: \quad a_i'x \leq b_i, \quad i = 1, \ldots, m,$$
$$a_i'x = b_i, \quad i = m+1, \ldots, m+r.$$

If x_0 is a nondegenerate extreme point and x_0 satisfies the strict complementary slackness condition, then x_0 is the unique optimal solution.

With the additional assumption that the gradients of those constraints active at x_0 are linearly independent, show that Theorem 4.16 reduces to the above when $C = 0$.

4.11 Let A and C be (m , n) and (n , n) matrices, respectively, with C positive semidefinite. Show that solving min $\{ c'x + \frac{1}{2}x'Cx \mid Ax \geq b , x \geq 0 \}$ is equivalent to finding a pair of $(m + n)$-vectors, w and z, satisfying

$$w = Mz + q , \quad w \geq 0 , \quad z \geq 0 , \quad \text{and} \quad w'z = 0 ,$$

where
$$M = \begin{bmatrix} C & -A' \\ A & 0 \end{bmatrix} \quad \text{and} \quad q = \begin{bmatrix} c \\ -b \end{bmatrix} .$$

This latter problem is called the linear complementarity problem.

Chapter 5

QP Solution Algorithms

In this chapter, we develop a basic QP algorithm. The development is in two parts. We first assume that an initial feasible point is available and develop an algorithm which, in a finite number of steps, will either determine an optimal solution for the QP or determine that the problem is unbounded from below. This is done in Section 5.1.

In Section 5.2, we address the problem of finding an initial feasible point. We formulate this as a linear programming problem for which an initial feasible point is available by inspection. Since an LP is a special case of a QP, the LP can be solved using the QP algorithm of Section 5.1. The complete solution procedure then consists of two passes of the QP algorithm. The first finds an initial feasible point and the second, beginning with the initial feasible point obtained by the first, solves the QP.

In Section 5.3, we develop a variation of the basic QP algorithm that was developed in Section 5.1. The modification substantially reduces the total number of iterations required to solve a QP.

Each of the two QP algorithms developed in this chapter require that all quasistationary determined by the algorithms be nondegenerate in order to guarantee finite termination. In Section 5.4, we will formulate a method to resolve degeneracy. The two QP solution algorithms presented in this chapter can then use this method to achieve finite termination without the assumption that each quasistationary is nondegenerate.

Throughout this chapter, we will assume that the objective function of the QP is convex.

5.1   A Basic QP Algorithm: Algorithm 3

We begin by developing a solution procedure for a quadratic programming problem having only inequality constraints. Consider the model problem

$$\min \left\{ \, c'x \, + \, \tfrac{1}{2}x'Cx \mid a_i'x \, \le \, b_i \, , \;\; i \, = \, 1,\ldots,m \, \right\} \tag{5.1}$$

and let

$$R \, = \, \left\{ \, x \mid a_i'x \, \le \, b_i \, , \;\; i \, = \, 1,\ldots,m \, \right\}$$

denote the feasible region for it. An overview of the algorithm is as follows. At each iteration, certain of the **inequality** constraints will be forced, temporarily, to hold as **equality** constraints. The indices of these constraints will be referred to as the *active set*. The algorithm will consist of two parts. The first will begin with an arbitrary feasible point and, by adding constraints to the active set, will determine a quasistationary point with an improved objective function value. The second part will test for optimality. If the optimality criterion is not satisfied, a single constraint will be deleted from the active set. This will lead to an improved objective function value. Control then passes back to the first part to find another quasistationary point with reduced objective function value and so on. Repeating gives a sequence of quasistationary points with strictly decreasing objective function values. No associated quasistationary set can ever be repeated and in a finite number of steps an optimal solution will be obtained.

The first part of the algorithm is developed as follows. Assume that an $x_0 \, \in \, R$ is known. To begin, we require any subset of those constraints active at x_0 for which the gradients are linearly independent. For simplicity, suppose that this subset is the first ρ constraints. The initial active set is thus $\{ \, 1 \, , \, 2, \ldots, \rho \, \}$. We formulate a variation of Algorithm 2 which will determine a quasistationary point $\hat{x}$ for (5.1) with the properties

$$f(\hat{x}) \, \le \, f(x_0)$$

and

$$a_i'\hat{x} \, = \, b_i \, , \;\; i \, = \, 1 \, , \, 2 \, , , \ldots, \rho \, .$$

Constraints other than the first ρ may be active at $\hat{x}$ but in particular, the first ρ will be active.

Let

$$D_0' \, = \, [\, a_1, \ldots, a_\rho \, , \, d_{\rho+1}, \ldots, d_n \,] \; ,$$

where $d_{\rho+1}, \ldots, d_n$ are any n-vectors such that $a_1, \ldots, a_\rho, d_{\rho+1}, \ldots, d_n$ are linearly independent. Consider the effect of applying Algorithm 2 to

$$\min \left\{ \, c'x \, + \, \tfrac{1}{2}x'Cx \mid a_i'x \, = \, b_i \, , \;\; i \, = \, 1 \, , \, 2, \ldots, \rho \, \right\} \tag{5.2}$$

with initial data x_0, D_0^{-1}, and $J_0 = \{1, 2, \ldots, \rho, 0, \ldots, 0\}$. Because (5.2) does not account for the remaining $m - \rho$ constraints of (5.1), the points constructed by Algorithm 2 may be infeasible for (5.1). In order to ensure feasibility, we introduce the following modification to Algorithm 2. Let s_j denote the search direction at a typical iteration j and let $\hat{\sigma}_j$ denote the maximum feasible step size for s_j. Then $x_j - \sigma s_j \in R$ for all σ with $0 \le \sigma \le \hat{\sigma}_j$ and

$$
\begin{aligned}
\hat{\sigma}_j &= \min \left\{ \frac{a_i' x_j - b_i}{a_i' s_j} \mid \text{all } i = \rho + 1, \ldots, m \text{ with } a_i' s_j < 0 \right\} \\
&= \frac{a_l' x_j - b_l}{a_l' s_j},
\end{aligned}
\tag{5.3}
$$

where l is the index of the restricting constraint. Assume that $s_j' C s_j > 0$ so that the optimal step size is given by

$$
\tilde{\sigma}_j = \frac{g_j' s_j}{s_j' C s_j}.
$$

If $\tilde{\sigma}_j \le \hat{\sigma}_j$, then setting $\sigma_j = \tilde{\sigma}_j$ and $x_{j+1} = x_j - \sigma_j s_j$ results in $x_{j+1} \in R$ with no change to Algorithm 2 or the active set. If $\tilde{\sigma}_j > \hat{\sigma}_j$, then using the step size $\tilde{\sigma}_j$ would result in constraint l being violated. However since $g_j' s_j > 0$, $f(x_j - \sigma s_j)$ is a strictly decreasing function of σ for $0 \le \sigma \le \tilde{\sigma}_j$ and therefore the greatest feasible reduction in f is obtained by setting $x_{j+1} = x_j - \sigma_j s_j$ with $\sigma_j = \hat{\sigma}_j$. In summary, we continue by setting

$$
\sigma_j = \min \{ \tilde{\sigma}_j, \hat{\sigma}_j \}.
$$

The manner in which we continue depends on whether $\sigma_j = \tilde{\sigma}_j$ (Case 1) or $\sigma_j = \hat{\sigma}_j$ (Case 2).

Case 1: $\sigma_j = \tilde{\sigma}_j$
In this case, $x_j - \tilde{\sigma}_j s_j$ just happens to be feasible. We continue with the usual updating for Algorithm 2 and proceed to the next iteration.

Case 2: $\sigma_j = \hat{\sigma}_j$
In this case, constraint l becomes active at x_{j+1}. We proceed by augmenting the active set with l giving the new active set $\{1, 2, \ldots, \rho, l\}$. This means that we add the constraint $a_l' x = b_l$ to those of (5.2) and proceed by applying Algorithm 2 to the revised problem

$$
\min \{ c'x + \tfrac{1}{2} x' C x \mid a_i' x = b_i, \ i = 1, 2, \ldots, \rho \text{ and } i = l \}.
\tag{5.4}
$$

In order to apply Algorithm 2 to (5.4), we need an appropriate D_j^{-1} and an index set J_{j+1}. These can be obtained from D_j^{-1} and J_j as follows. Suppose that

$$D'_j = [\, a_1, \ldots, \, a_\rho \, , \, Cc_{\rho+1}, \ldots, Cc_{\rho+j} \, , \, d_{\rho+j+1}, \ldots, d_n \,]$$

and that

$$D_j^{-1} = [\, c_1, \ldots, \, c_\rho \, , \quad c_{\rho+1}, \ldots, \quad c_{\rho+j} \, , \quad c_{\rho+j+1}, \ldots, c_n \,].$$

Because of the simplifying order assumption on the columns of D'_j,

$$J_j = \{\, 1, \, 2, \ldots, \rho \, , \, -1, \ldots, -1 \, , \, 0, \ldots, 0 \,\}.$$

In keeping with this ordering assumption, s_j must be parallel to $c_{\rho+j+1}$. That is, either $s_j = +c_{\rho+j+1}$ or $s_j = -c_{\rho+j+1}$. By definition of $\hat{\sigma}_j$, $a'_l s_j < 0$ and therefore $a'_l c_{\rho+j+1} \neq 0$. It follows from Procedure Φ that replacing row $\rho + j + 1$ of D_j with a_l will give a nonsingular matrix. Therefore we can proceed by setting

$$D_{j+1}^{-1} = \Phi(D_j^{-1} \, , \, a_l \, , \, \rho + j + 1).$$

Letting $D_{j+1}^{-1} = [\hat{c}_1, \ldots, \hat{c}_n \,]$, Procedure Φ gives the updated conjugate direction columns of D_{j+1}^{-1} as

$$\hat{c}_i = c_i - \frac{a'_l c_i}{a'_l c_{\rho+j+1}} \, c_{\rho+j+1} \, , \quad i = \rho + 1, \ldots, \rho + j \, .$$

There is no reason to expect that a_l will be orthogonal to any of $c_{\rho+1}, \ldots, c_{\rho+j}$ and therefore the conjugate direction columns of D_j^{-1} are destroyed by this update. This must be reflected in the new index set and so we set

$$J_{j+1} = \{\, 1 \, , \, 2, \ldots, \rho \, , \, 0 \, , \, 0, \ldots, \, 0 \, , \, l \, , \, 0 \, , \, 0, \ldots, 0 \,\} \, ,$$

which completes Case 2.

In the discussion prior to Cases 1 and 2, we assumed that $s'_j C s_j > 0$. If $s'_j C s_j = 0$, then

$$f(x_j - \sigma s_j) \to -\infty \text{ as } \sigma \to +\infty.$$

In this case, it is convenient to write $\tilde{\sigma}_j = +\infty$ by which we mean that the objective function can be decreased indefinitely by taking σ sufficiently large. The situation is analogous for the maximum feasible step size. By writing

$$\hat{\sigma}_j = \frac{a'_l x_j - b_l}{a'_l s_j} \, ,$$

we implicitly assume that there is at least one i with $1 \leq i \leq m$ and $a'_i s_j < 0$. This need not be so. If it is not; i.e., $a'_i s_j \geq 0$ for $i = 1, \ldots, m$, then $x_j - \sigma s_j \in R$ for all $\sigma \geq 0$ and we indicate this by setting $\hat{\sigma}_j = +\infty$. If both $\tilde{\sigma}_j = +\infty$ and $\hat{\sigma}_j = +\infty$, then our original QP (5.1) is unbounded from below and the algorithm should terminate with this information. Otherwise, at least one of $\tilde{\sigma}_j$ and $\hat{\sigma}_j$ is finite and setting $\sigma_j = \min\{\tilde{\sigma}_j, \hat{\sigma}_j\}$, the analysis continues as before with either Case 1 or Case 2.

Suppose that the modified Algorithm 2 is continued. Case 1 or Case 2 will apply at each iteration. Case 1 can occur successively for at most k iterations, where k is n less the number of constraints in the current active set. Case 2 can occur at most $n - \rho$ times since at each occurrence, the number of constraints in the active set increases by exactly one. Therefore the modified Algorithm 2 must terminate in a finite number of steps with either the information that (5.1) is unbounded from below, or an optimal solution x_j for

$$\min\{c'x + \tfrac{1}{2}x'Cx \mid a_i'x = b_i, \text{ all } i \in I_j\}$$

which also satisfies $x_j \in R$, where I_j is the set of indices of constraints in the jth active set. Considering the latter possibility further, we have shown that x_j is a quasistationary point with $f(x_j) \leq f(x_0)$. This completes the first part of the algorithm.

The second part of the algorithm deals with the quasistationary point found by the first part. Suppose that x_j is such a quasistationary point and that (for simplicity) the active constraints are $1, 2, \ldots, \rho$. Then x_j is feasible for (5.1) and an optimal solution for

$$\min\{c'x + \tfrac{1}{2}x'Cx \mid a_i'x = b_i, \; i = 1, \ldots, \rho\}. \tag{5.5}$$

Let D_j' be obtained from the first part of the algorithm and assume for simplicity that the columns of D_j' are ordered so that

$$D_j' = [a_1, \ldots, a_\rho, Cc_{\rho+1}, \ldots, Cc_{\rho+j}, d_{\rho+j+1}, \ldots, d_n].$$

Let

$$D_j^{-1} = [c_1, \ldots, c_n].$$

Because x_j is optimal for (5.5), it satisfies the optimality conditions for it:

$$g_j = u_1 a_1 + \cdots + u_\rho a_\rho,$$

where $g_j = g(x_j)$. From Theorem 3.14, the multipliers $u_1, \ldots, u_\rho$ may be obtained from g_j and D_j^{-1} according to

$$u_i = -g_j' c_i, \; i = 1, \ldots, \rho.$$

By defining

$$u_{\rho+1} = \cdots = u_m = 0,$$

we see that x_j satisfies all of the optimality conditions for (5.1) with the possible exception of $u \geq 0$. Considering this further, let k be such that

$$g_j'c_k = \max \{ g_j'c_i \mid i = 1, \ldots, \rho \} . \tag{5.6}$$

If $g_j'c_k \leq 0$ then $u \geq 0$, all of the optimality conditions for (5.1) are satisfied, and Theorem 4.3 asserts that x_j is an optimal solution for (5.1). Thus $g_j'c_k \leq 0$ is an optimality test for the algorithm.

Suppose next that $g_j'c_k > 0$. We claim that setting $s_j = c_k$ and proceeding with the step size and update of D_j^{-1} just described for the first part of the algorithm will reduce the objective function and cause constraint k to become inactive. To see this, first observe that by Taylor's series

$$f(x_j - \sigma s_j) = f(x_j) - \sigma g_j's_j + \tfrac{1}{2}\sigma^2 s_j'C s_j .$$

Since $g_j's_j = g_j'c_k > 0$, it follows that a small increase in σ will decrease the objective function. Furthermore, by definition of the inverse matrix, $a_k'c_k = 1$, so that

$$a_k'(x_j - \sigma s_j) = b_k - \sigma .$$

Therefore

$$a_k'(x_j - \sigma s_j) \leq b_k ;$$

i.e., constraint k becomes inactive at $x_j - \sigma s_j$, for all $\sigma > 0$. Proceeding as in the first part of the algorithm, we compute $\sigma_j = \min \{ \tilde{\sigma}_j , \hat{\sigma}_j \}$, update D_j^{-1} accordingly, and set $x_{j+1} = x_j - \sigma_j s_j$. Constraint k, which was active at iteration j, has become inactive at iteration $j + 1$ and is removed from the active set for iteration $j + 1$. Continuing, we return to the first part of the algorithm to determine a quasistationary point with an objective function value smaller than $f(x_{j+1})$.

It should be noted that the first part of the algorithm **adds** constraints to the active set until a quasistationary point is obtained. Only at this point is a single constraint **dropped** from the active set by the second part of the algorithm.

Let $j_1 , j_2 , \ldots$ be the iterations for which $x_{j_1} , x_{j_2} , \ldots$ are quasistationary points. By construction,

$$f(x_0) \geq f(x_1) \geq f(x_2) \geq \cdots .$$

If each $\sigma_{j_i} > 0$, then $f(x_{j_i+1}) < f(x_{j_i})$ for $i = 1 , 2 , \ldots$. Consequently,

$$f(x_{j_1}) > f(x_{j_2}) > \cdots > f(x_{j_i}) > \cdots ,$$

so that no quasistationary set can ever be repeated. Since (5.1) possesses only finitely many quasistationary sets, the algorithm must terminate in a finite number of steps. Since termination occurs only when the optimality conditions are satisfied, the final point determined by the algorithm must indeed be an optimal solution for (5.1).

We next introduce a simple modification to account for explicit equality constraints. Suppose that our model problem is

$$\left. \begin{array}{rll} \text{minimize}: & c'x + \frac{1}{2}x'\Sigma x & \\ \text{subject to}: & a_i'x \leq b_i, & i = 1, \ldots, m, \\ & a_i'x = b_i, & i = m+1, \ldots, m+r, \end{array} \right\} \tag{5.7}$$

which has r equality constraints. Assume that $a_{m+1}, \ldots, a_{m+r}$ are linearly independent and that a feasible point x_0 is available. Since both x_0 and the optimal solution satisfy the equality constraints, the only modification required is to start with $m + 1, \ldots, m + r$ in the active set and make sure that they remain there throughout the algorithm. Thus an equality constraint is never a candidate to be dropped from the active set in the second part of the algorithm. Furthermore, the multipliers for the equality constraints need not be computed nor considered in the optimality test since from the second entry of Table 4.1, the multipliers for equality constraints are not constrained in sign.

At various points in the development of the algorithm, we have introduced simplifying order assumptions on the columns of D_j'. As with Algorithm 2, these assumptions may be removed by introducing an ordered index set J_j. At iteration j, let

$$\begin{aligned} D_j' &= [\, d_{1j}, \ldots, d_{nj} \,] \,, \\ D_j^{-1} &= [\, c_{1j}, \ldots, c_{nj} \,] \,, \quad \text{and} \\ J_j &= \{\, \alpha_{1j}, \ldots, \alpha_{nj} \,\} \,, \end{aligned}$$

where for $i = 1, \ldots, n$,

$$\alpha_{ij} = \begin{cases} -1 \,, & \text{if } d_{ij} = Cc_{ij} \,, \\ k > 0 \,, & \text{if } d_{ij} = a_k \,, \\ 0 \,, & \text{otherwise.} \end{cases}$$

Thus $\{\, c_{ij} \mid \text{all } i = 1, \ldots, n \text{ with } \alpha_{ij} = -1 \,\}$ are conjugate direction columns of D_j^{-1}, $\{\, c_{ij} \mid \text{all } i = 1, \ldots, n \text{ with } 1 \leq \alpha_{ij} \leq m \,\}$ are those columns of D_j^{-1} associated with active inequality constraints, $\{\, c_{ij} \mid \text{all } i = 1, \ldots, n \text{ with } m + 1 \leq \alpha_{ij} \leq m + r \,\}$ are those columns of D_j^{-1} associated with equality constraints and $\{\, c_{ij} \mid \text{all } i = 1, \ldots, n \text{ with } \alpha_{ij} = 0 \,\}$ are the remaining, or "free," columns of D_j^{-1}.

In terms of the index set, the first part of the algorithm differs from Algorithm 2 only in the step size calculation, where in addition to the calculation of the optimal step size, the maximum feasible step size is also calculated. For the second part of the algorithm, (5.6) becomes: let k be the smallest index such that

$$g_j'c_{kj} = \max \{ g_j'c_{ij} \mid \text{all } i = 1, \ldots, n \text{ with } 1 \leq \alpha_{ij} \leq m \} .$$

If $g_j'c_{kj} \leq 0$, x_j is optimal. Otherwise we continue by setting $s_j = c_{kj}$. The phrase "the smallest index" means that in the event of ties for the maximum, choose k to be the smallest index among those tied for the maximum. We also resolve ties in the computation of the maximum feasible step size using a similar rule so that (5.3) becomes: let l be the smallest index such that

$$\hat{\sigma}_j = \frac{a_l'x_j - b_l}{a_l's_j} .$$

These two tie-resolving rules are arbitrary at this point. However, we will subsequently demonstrate their utility in resolving a technical difficulty caused by degeneracy. Assembling the above ideas gives Algorithm 3.

ALGORITHM   3

Model Problem:

$$
\begin{aligned}
\text{minimize:} \quad & c'x + \tfrac{1}{2}x'Cx \\
\text{subject to:} \quad & a_i'x \leq b_i, \quad i = 1, \ldots, m, \\
& a_i'x = b_i, \quad i = m + 1, \ldots, m + r.
\end{aligned}
$$

Initialization:
Start with any feasible point x_0, $J_0 = \{ \alpha_{10}, \ldots, \alpha_{n0} \}$, and D_0^{-1}, where $D_0' = [d_1, \ldots, d_n]$ is nonsingular, $d_i = a_{\alpha_{i0}}$ for all i with $1 \leq \alpha_{i0} \leq m + r$, and, each of $m + 1, \ldots, m + r$ is in J_0. Compute $f(x_0) = c'x_0 + \tfrac{1}{2}x_0'Cx_0$, $g_0 = c + Cx_0$, and set $j = 0$.

Step 1:   Computation of Search Direction s_j.
Let $D_j^{-1} = [c_{1j}, \ldots, c_{nj}]$ and $J_j = \{ \alpha_{1j}, \ldots, \alpha_{nj} \}$. If there is at least one i with $\alpha_{ij} = 0$, go to Step 1.1. Otherwise, go to Step 1.2.

Step 1.1:
Determine the smallest index k such that

$$| g_j'c_{kj} | = \max \{ | g_j'c_{ij} | \mid \text{all } i \text{ with } \alpha_{ij} = 0 \} .$$

If $g_j'c_{kj} = 0$, go to Step 1.2. Otherwise, set $s_j = c_{kj}$ if $g_j'c_{kj} > 0$ and $s_j = -c_{kj}$ if $g_j'c_{kj} < 0$. Go to Step 2.

Step 1.2:

If there is no i with $1 \leq \alpha_{ij} \leq m$, then stop with optimal solution x_j. Otherwise, determine the smallest index k such that

$$g_j' c_{kj} = \max \{ g_j' c_{ij} \mid \text{all } i \text{ with } 1 \leq \alpha_{ij} \leq m \} .$$

If $g_j' c_{kj} \leq 0$, then stop with optimal solution x_j. Otherwise, set $s_j = c_{kj}$ and go to Step 2.

Step 2:   **Computation of Step Size σ_j.**
Compute $s_j' C s_j$. If $s_j' C s_j = 0$, set $\tilde{\sigma}_j = +\infty$. Otherwise, set

$$\tilde{\sigma}_j = \frac{g_j' s_j}{s_j' C s_j}.$$

If $a_i' s_j \geq 0$ for $i = 1, \ldots, m$, set $\hat{\sigma}_j = +\infty$. Otherwise, compute the smallest index l and $\hat{\sigma}_j$ such that

$$\hat{\sigma}_j = \frac{a_l' x_j - b_l}{a_l' s_j} = \min \left\{ \frac{a_i' x_j - b_i}{a_i' s_j} \mid \text{all } i \notin J_j \text{ with } a_i' s_j < 0 \right\} .$$

If $\tilde{\sigma}_j = \hat{\sigma}_j = +\infty$, print x_j, s_j, and the message "the objective function is unbounded from below" and stop. Otherwise, set $\sigma_j = \min \{ \tilde{\sigma}_j , \hat{\sigma}_j \}$ and go to Step 3.

Step 3:   **Update.**
Set $x_{j+1} = x_j - \sigma_j s_j$, $g_{j+1} = c + C x_{j+1}$, and $f(x_{j+1}) = c' x_{j+1} + \frac{1}{2} x_{j+1}' C x_{j+1}$. If $\tilde{\sigma}_j \leq \hat{\sigma}_j$, go to Step 3.1. Otherwise, go to Step 3.2.

Step 3.1:
Set $d_j = (s_j' C s_j)^{-1/2} C s_j$, $D_{j+1}^{-1} = \Phi_1(D_j^{-1} , d_j , k , J_j)$, and $J_{j+1} = \{ \alpha_{1,j+1}, \ldots, \alpha_{n,j+1} \}$, where

$$\alpha_{i,j+1} = \alpha_{ij} , \quad i = 1, \ldots, n , \quad i \neq k ,$$
$$\alpha_{k,j+1} = -1 .$$

Replace j with $j + 1$ and go to Step 1.

Step 3.2:
Set $D_{j+1}^{-1} = \Phi(D_j^{-1}, a_l, k)$ and $J_{j+1} = \{ \alpha_{1,j+1}, \ldots, \alpha_{n,j+1} \}$, where

$$\alpha_{i,j+1} = \alpha_{ij} , \quad \text{for all } i \text{ with } \alpha_{ij} \geq 0 , \ i \neq k ,$$
$$\alpha_{i,j+1} = 0 , \quad \text{for all } i \text{ with } \alpha_{ij} = -1,$$
$$\alpha_{k,j+1} = l .$$

Replace j with $j + 1$ and go to Step 1.

We illustrate Algorithm 3 by applying it to the following two examples. The problem of the first example (Example 5.1) was used in Example 4.1 to illustrate quasistationary points.

Example 5.1

$$
\begin{aligned}
\text{minimize:} \quad & - 10x_1 - 4x_2 + x_1^2 + x_2^2 \\
\text{subject to:} \quad & x_1 \qquad\qquad \leq \quad 4, \quad (1) \\
& \qquad - x_2 \leq -1, \quad (2) \\
& - x_1 \qquad\quad \leq -1, \quad (3) \\
& \qquad\quad x_2 \leq \quad 3. \quad (4)
\end{aligned}
$$

Here

$$
c = \begin{bmatrix} -10 \\ -4 \end{bmatrix} \quad \text{and} \quad C = \begin{bmatrix} 2 & 0 \\ 0 & 2 \end{bmatrix} .
$$

Initialization:

$$
x_0 = \begin{bmatrix} 1 \\ 3 \end{bmatrix} , \quad J_0 = \{ 3 , 4 \} , \quad D_0^{-1} = \begin{bmatrix} -1 & 0 \\ 0 & 1 \end{bmatrix} , \quad f(x_0) = -12 ,
$$

$$
g_0 = \begin{bmatrix} -8 \\ 2 \end{bmatrix} , \quad j = 0 .
$$

Iteration $\underline{0}$

Step 1:   Transfer to Step 1.2.

Step 1.2: $g_0' c_{10} = \max \{ 8 , 2 \} = 8 , \quad k = 1 , \quad s_0 = \begin{bmatrix} -1 \\ 0 \end{bmatrix} .$

Step 2: $s_0' C s_0 = 2 ,$

$$
\tilde{\sigma}_0 = \frac{8}{2} = 4 ,
$$

$$
\hat{\sigma}_0 = \min \left\{ \frac{-3}{-1} , - , - , - \right\} = 3 , \quad l = 1 ,
$$

$$
\sigma_0 = \min \{ 4 , 3 \} = 3 .
$$

Step 3: $x_1 = \begin{bmatrix} 1 \\ 3 \end{bmatrix} - 3 \begin{bmatrix} -1 \\ 0 \end{bmatrix} = \begin{bmatrix} 4 \\ 3 \end{bmatrix} , \quad g_1 = \begin{bmatrix} -2 \\ 2 \end{bmatrix} ,$

$$f(x_1) = -27 . \text{ Transfer to Step 3.2.}$$

Step 3.2: $D_1^{-1} = \begin{bmatrix} 1 & 0 \\ 0 & 1 \end{bmatrix}$, $J_1 = \{ 1 , 4 \}$, $j = 1$.

Iteration 1

Step 1: Transfer to Step 1.2.

Step 1.2: $g_1' c_{21} = \max \{ -2 , 2 \} = 2$, $k = 2$, $s_1 = \begin{bmatrix} 0 \\ 1 \end{bmatrix}$.

Step 2: $s_1' C s_1 = 2$,

$$\tilde{\sigma}_1 = \frac{2}{2} = 1 ,$$

$$\hat{\sigma}_1 = \min \left\{ - , \frac{-2}{-1} , - , - \right\} = 2 , \ l = 2 ,$$

$$\sigma_1 = \min \{ 1 , 2 \} = 1 .$$

Step 3: $x_2 = \begin{bmatrix} 4 \\ 3 \end{bmatrix} - \begin{bmatrix} 0 \\ 1 \end{bmatrix} = \begin{bmatrix} 4 \\ 2 \end{bmatrix}$, $g_2 = \begin{bmatrix} -2 \\ 0 \end{bmatrix}$,

$$f(x_2) = -28 . \text{ Transfer to Step 3.1.}$$

Step 3.1: $d_1 = \begin{bmatrix} 0 \\ 2/\sqrt{2} \end{bmatrix}$,

$$D_2^{-1} = \begin{bmatrix} 1 & 0 \\ 0 & 1/\sqrt{2} \end{bmatrix} , \ J_2 = \{ 1 , -1 \} , \ j = 2 .$$

Iteration 2

Step 1: Transfer to Step 1.2.

Step 1.2: $g_2' c_{12} = \max \{ -2 , - \} = -2$, $k = 1$.

$g_2' c_{12} \leq 0$; stop with optimal solution $x_2 = (4 , 2)'$.

$\Diamond$

The iterates of Example 5.1 are shown in Figure 5.1. The algorithm is initiated at x_0. Algorithm 3 recognizes that x_0 is a quasistationary point (in this case an extreme point) from the fact that all elements of J_0 are strictly positive. Control then passes to Step 1.2 for the optimality test. This is not passed and the algorithm proceeds by deleting constraint (3) ($\alpha_{10} = 3$) from the active set. The optimal step size would take the next iterate outside of R. The restricting constraint is constraint (1). Then x_1 is also an extreme point. Again, control transfers to Step 1.2 for the optimality test. This fails and constraint (4) is dropped from the active set. In this case, the optimal step size is not limited by any constraint. x_2 is, in fact, the optimal solution although Algorithm 3 does not know it yet. Because the optimal step size is used, the updating uses Procedure Φ_1 to perform a conjugate direction update. Control then passes to Step 1. Because J_2 includes no zeros, Algorithm 3 realizes that x_2 is a quasistationary point and proceeds to the optimality test in Step 1.2. This is passed, and Algorithm 3 terminates with optimal solution x_2.

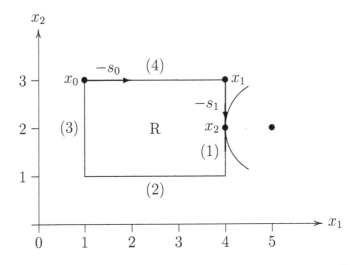

Figure 5.1.   Progress of Algorithm 3 in solving Example 5.1.

A Matlab program, Alg3.m, which implements Algorithm 3 is given in Figure 5.4, Section 5.5. Note that the computations for Step 1 are performed in a separate Matlab function called Step1.m (Figure 5.5). Similarly, the computations for Step 2 are given in the separate Matlab function called Step2.m (Figure 5.6). Both of these functions are called by Alg3.m. The reason for doing this is that Step 1 and Step 2 are common to Algorithms 3, 4 and 10 and coding them as separate routines avoids repetition.

When Algorithm 3 terminates with an optimal solution to the primal problem, it is useful to determine an optimal solution to the dual problem. This is done using the function GetDual.m (Figure 5.7). This is invoked at several points in Alg3.m.

After Algorithm 3 terminates with optimal solutions for the primal and dual problems, it is useful to calculate the primal and dual feasibility errors. This is done in checkKKT.m (Figure 5.8).

A Matlab program which sets up the data for Example 5.1 and the results of applying Alg3.m to it are shown in Figures 5.9 and 5.10, respectively, in Section 5.5.

A second example of Algorithm 3 is given in

Example 5.2

$$
\begin{array}{lrcrcrcl}
\text{minimize:} & x_1 & - & 2x_2 & - & x_3 & + & \tfrac{1}{2}(x_1^2 + 2x_2^2 + 4x_3^2 - 4x_2x_3) \\
\text{subject to:} & x_1 & + & x_2 & + & x_3 & \leq & 10, \quad (1) \\
& 2x_1 & & & + & 3x_3 & \leq & 11, \quad (2) \\
& -2x_1 & + & 2x_2 & - & 5x_3 & \leq & -13, \quad (3) \\
& -x_1 & & & & & \leq & 0, \quad (4) \\
& & & -x_2 & & & \leq & 0, \quad (5) \\
& & & & & -x_3 & \leq & 0. \quad (6)
\end{array}
$$

Here

$$
c = \begin{bmatrix} 1 \\ -2 \\ -1 \end{bmatrix} \quad \text{and} \quad C = \begin{bmatrix} 1 & 0 & 0 \\ 0 & 2 & -2 \\ 0 & -2 & 4 \end{bmatrix} .
$$

Initialization:

$$
x_0 = \begin{bmatrix} 0 \\ 0 \\ 3 \end{bmatrix} , \quad J_0 = \{ 4, 5, 0 \} , \quad D_0^{-1} = \begin{bmatrix} -1 & 0 & 0 \\ 0 & -1 & 0 \\ 0 & 0 & 1 \end{bmatrix} ,
$$

$$
f(x_0) = 15 , \quad g_0 = \begin{bmatrix} 1 \\ -8 \\ 11 \end{bmatrix} , \quad j = 0 .
$$

Iteration 0

Step 1:   Transfer to Step 1.1.

Step 1.1: $| g_0' c_{30} | = \max \{ - , - , | 11 | \} = 11 , \quad k = 3 ,$

$$
s_0 = \begin{bmatrix} 0 \\ 0 \\ 1 \end{bmatrix} .
$$

Step 2: $s_0' C s_0 = 4 ,$

$$\tilde{\sigma}_0 = \frac{11}{4} ,$$

$$\hat{\sigma}_0 = \min \left\{ - , - , \frac{-2}{-5} , - , - , \frac{-3}{-1} \right\} = \frac{2}{5} , \quad l = 3 ,$$

$$\sigma_0 = \min \left\{ \frac{11}{4} , \frac{2}{5} \right\} = \frac{2}{5} .$$

Step 3: $\quad x_1 = \begin{bmatrix} 0 \\ 0 \\ 3 \end{bmatrix} - \frac{2}{5} \begin{bmatrix} 0 \\ 0 \\ 1 \end{bmatrix} = \begin{bmatrix} 0 \\ 0 \\ 13/5 \end{bmatrix} , \quad g_1 = \begin{bmatrix} 1 \\ -36/5 \\ 47/5 \end{bmatrix} ,$

$$f(x_1) = \frac{273}{25} . \text{ Transfer to Step 3.2.}$$

Step 3.2: $\quad D_1^{-1} = \begin{bmatrix} -1 & 0 & 0 \\ 0 & -1 & 0 \\ 2/5 & -2/5 & -1/5 \end{bmatrix} , \quad J_1 = \{ 4 , 5 , 3 \} ,$

$$j = 1 .$$

Iteration 1

Step 1:   Transfer to Step 1.2.

Step 1.2: $\quad g_1' c_{21} = \max \left\{ \frac{69}{25} , \frac{86}{25} , \frac{-47}{25} \right\} = \frac{86}{25} , \quad k = 2 ,$

$$s_1 = \begin{bmatrix} 0 \\ -1 \\ -2/5 \end{bmatrix} .$$

Step 2: $\quad s_1' C s_1 = \frac{26}{25} ,$

$$\tilde{\sigma}_1 = \frac{86/25}{26/25} = \frac{43}{13} ,$$

$$\hat{\sigma}_1 = \min \left\{ \frac{-37/5}{-7/5} , \frac{-16/5}{-6/5} , - , - , - , - \right\} = \frac{8}{3} ,$$

$$l = 2 ,$$

$$\sigma_1 = \min \left\{ \frac{43}{13} , \frac{8}{3} \right\} = \frac{8}{3} .$$

Step 3: $\quad x_2 = \begin{bmatrix} 0 \\ 0 \\ 13/5 \end{bmatrix} - \frac{8}{3} \begin{bmatrix} 0 \\ -1 \\ -2/5 \end{bmatrix} = \begin{bmatrix} 0 \\ 8/3 \\ 11/3 \end{bmatrix} ,$

$$g_2 = \begin{bmatrix} 1 \\ -4 \\ 25/3 \end{bmatrix} ,$$

$$f(x_2) = \frac{49}{9} . \text{ Transfer to Step 3.2.}$$

Step 3.2: $D_2^{-1} = \begin{bmatrix} -1 & 0 & 0 \\ 2/3 & 5/6 & 1/2 \\ 2/3 & 1/3 & 0 \end{bmatrix}$, $J_2 = \{4, 2, 3\}$,

$$j = 2 .$$

Iteration 2

Step 1: Transfer to Step 1.2.

Step 1.2: $g_2' c_{12} = \max \left\{ \frac{17}{9} , \frac{-5}{9} , -2 \right\} = \frac{17}{9} , \quad k = 1 ,$

$$s_2 = \begin{bmatrix} -1 \\ 2/3 \\ 2/3 \end{bmatrix} .$$

Step 2: $s_2' C s_2 = \frac{17}{9} ,$

$$\tilde{\sigma}_2 = \frac{17/9}{17/9} = 1 ,$$

$$\hat{\sigma}_2 = \min \left\{ - , - , - , - , \frac{-8/3}{-2/3} , \frac{-11/3}{-2/3} \right\} = 4 ,$$

$$l = 5 ,$$

$$\sigma_2 = \min \{ 1 , 4 \} = 1 .$$

Step 3: $x_3 = \begin{bmatrix} 0 \\ 8/3 \\ 11/3 \end{bmatrix} - \begin{bmatrix} -1 \\ 2/3 \\ 2/3 \end{bmatrix} = \begin{bmatrix} 1 \\ 2 \\ 3 \end{bmatrix}$, $g_3 = \begin{bmatrix} 2 \\ -4 \\ 7 \end{bmatrix}$,

$$f(x_3) = \frac{9}{2} . \text{ Transfer to Step 3.1.}$$

Step 3.1: $d_2 = \begin{bmatrix} -3/\sqrt{17} \\ 0 \\ 4/\sqrt{17} \end{bmatrix} ,$

$$D_3^{-1} = \begin{bmatrix} -3/\sqrt{17} & 4/17 & 0 \\ 2/\sqrt{17} & 23/24 & 1/2 \\ 2/\sqrt{17} & 3/17 & 0 \end{bmatrix} , \quad J_3 = \{ -1, 2, 3 \} ,$$

$j = 3$.

Iteration 3

Step 1:   Transfer to Step 1.2.

Step 1.2: $g_3' c_{23} = \max \{ -, -1, -2 \} = -1 , \quad k = 2$.

$g_3' c_{23} \leq 0$;   stop with optimal solution $x_3 = (1, 2, 3)'$.

$\diamond$

The results of applying Alg3.m to Example 5.2 are shown in Figures 5.11 and 5.12.

In the following theorem we will establish the main properties of Algorithm 3. In order to prove finite termination, we will assume that every quasistationary point determined by Algorithm 3 is nondegenerate. This is analogous to the assumption of nondegenerate extreme points for linear programming.

Theorem 5.1 *Let Algorithm 3 be applied to the model problem*

$$\min \{ c'x + \tfrac{1}{2} x' C x \mid A_1 x \leq b_1 , \quad A_2 x = b_2 \}$$

beginning with an arbitrary feasible point x_0 and let x_1, $x_2, \ldots, x_j$, $\ldots$ be the sequence of iterates so obtained. If each quasistationary point obtained by Algorithm 3 is nondegenerate, then Algorithm 3 terminates after a finite number of steps with either an optimal solution x_j or the information that the problem is unbounded from below. In the former case, the multipliers for the active constraints are given by $u_{\alpha_{ij}} = -g_j' c_{ij}$ for all i with $1 \leq \alpha_{ij} \leq m + r$, where $D_j^{-1} = [c_{1j}, \ldots, c_{nj}]$ and $J_j = \{ \alpha_{1j}, \ldots, \alpha_{nj} \}$, and the multipliers for the remaining constraints have value zero.

Proof: By construction, at each iteration j

$$a_i' x_j = b_i , \quad \text{for all } i \in J_j ,$$

and

$$a_i' x_j \leq b_i , \quad \text{for all } i \notin J_j ,$$

so that each x_j is feasible. Furthermore, letting $D_j' = [d_{1j}, \ldots, d_{nj}]$ it follows from the initial data assumptions, Step 3, and Lemma 3.2 that

$$d_{ij} = \begin{cases} a_{\alpha_{ij}} , & \text{for all } i \text{ with } 1 \leq \alpha_{ij} \leq m + r, \\ C c_{ij} , & \text{for all } i \text{ with } \alpha_{ij} = -1 , \end{cases} \tag{5.8}$$

and

$$g_j'c_{ij} = 0, \quad \text{for all } i \text{ with } \alpha_{ij} = -1. \tag{5.9}$$

Consider the possibility that termination occurs at Step 2 with the message that "the objective function is unbounded from below." From Step 1 $g_j's_j > 0$ and since $\tilde{\sigma}_j = +\infty$, $s_j'Cs_j = 0$. With Taylor's series, this implies that

$$f(x_j - \sigma s_j) = f(x_j) - \sigma g_j's_j$$

is a strictly decreasing function of σ and therefore $f(x_j - \sigma s_j) \to -\infty$ as $\sigma \to +\infty$. Since $\hat{\sigma}_j = +\infty$, $x_j - \sigma s_j \in R$ for all $\sigma \geq 0$. Therefore the objective function is indeed unbounded from below.

Suppose next that termination occurs with "optimal solution x_j." From Lemma 3.1,

$$-g_j = \sum_{i=1}^{n} (-g_j'c_{ij})d_{ij}. \tag{5.10}$$

From Step 1.1, termination with "optimal solution x_j" implies that either there is no i with $\alpha_{ij} = 0$ or that

$$g_j'c_{ij} = 0, \quad \text{for all } i \text{ with } \alpha_{ij} = 0. \tag{5.11}$$

Using (5.8), (5.9), and (5.11) in (5.10) gives

$$-g_j = \sum_{\alpha_{ij} = -1} (-g_j'c_{ij})d_{ij} + \sum_{\alpha_{ij} = 0} (-g_j'c_{ij})d_{ij} + \sum_{1 \leq \alpha_{ij} \leq m+r} (-g_j'c_{ij})d_{ij}$$

$$= \sum_{1 \leq \alpha_{ij} \leq m} (-g_j'c_{ij})a_{\alpha_{ij}} + \sum_{m+1 \leq \alpha_{ij} \leq m+r} (-g_j'c_{ij})a_{\alpha_{ij}}. \tag{5.12}$$

Termination in Step 1.2 implies that either there is no i with $1 \leq \alpha_{ij} \leq m$ or that

$$\max \{ g_j'c_{ij} \mid \text{all } i \text{ with } 1 \leq \alpha_{ij} \leq m \} \leq 0. \tag{5.13}$$

The latter possibility implies that

$$g_j'c_{ij} \leq 0, \quad \text{for all } i \text{ with } 1 \leq \alpha_{ij} \leq m. \tag{5.14}$$

Define the $(m + r)$-vector u according to $u_i = 0$ for all $i \notin J_j$ and

$$u_{\alpha_{ij}} = -g_j'c_{ij}, \quad \text{for all } i \text{ with } 1 \leq \alpha_{ij} \leq m + r.$$

Then (5.14) together with the definition of u imply that

$$u_i \geq 0 , \quad \text{for } i = 1, \ldots, m ,$$
$$u_i(a_i'x_j - b_i) = 0 , \quad \text{for } i = 1, \ldots, m ,$$

and from (5.12),

$$-g_j = \sum_{i=1}^{m+r} u_i a_i .$$

The optimality conditions for the model problem are thus satisfied and it follows from Theorem 4.5 that x_j is indeed an optimal solution.

To prove finite termination, we first observe that no constraint is dropped from the active set until a quasistationary point is reached. Each iteration prior to the determination of a quasistationary point either constructs an additional conjugate direction or adds a new constraint to the active set. Thus, beginning with an arbitrary feasible point a quasistationary point will be located in finitely many iterations. Let the iterations corresponding to quasistationary points be $j_1 < j_2 < \cdots < j_i$. Since $f(x_0) \geq f(x_1) \geq \cdots$, it follows that if each $\sigma_{j_i} > 0$ then

$$f(x_{j_i+1}) < f(x_{j_i}) .$$

That is, the subsequence of objective function values for the quasistationary points will be strictly decreasing. This means that no quasistationary set will ever be repeated. Since the problem possesses only finitely many quasistationary sets, termination in a finite number of iterations is then assured.

It remains to consider the case of a quasistationary point x_j for which $\sigma_j = 0$. We will show that after a finite number of steps a strict decrease in the objective function is obtained so that the previous argument can be used to guarantee finite termination of the algorithm. Renumbering the constraints, if necessary, we suppose for convenience that constraints $1 , 2, \ldots, \rho$ are in the active set at iteration j and that their gradients occur in natural order as the first ρ columns of D_j'. Again, for simplicity, assume that $k = \rho$ and that $l = \rho + 1$, where k and l are the indices of Steps 1.2 and 2, respectively, of Algorithm 3. Then

$$\begin{aligned}
s_j &= c_{\rho+j} , \quad a_{\rho+1}'x_j = b_{\rho+1}, \\
a_{\rho+1}'s_j &< 0, \\
D_j' &= [\, a_1, \ldots, a_\rho , d_{\rho+1,j}, \ldots, d_{nj} \,], \quad \text{and,} \\
D_{j+1}' &= [\, a_1, \ldots, a_\rho , a_{\rho+1} , d_{\rho+1,j}, \ldots, d_{nj} \,].
\end{aligned} \tag{5.15}$$

Because x_j is a quasistationary point, $g_j'c_{ij} = 0$ for $i = \rho + 1, \ldots, n$ so that from Lemma 3.1

$$g_j = \sum_{i=1}^{\rho} (g_j'c_{ij})a_i . \tag{5.16}$$

Furthermore

$$g'_j s_j = g'_j c_{\rho+j} > 0 . \tag{5.17}$$

Let

$$a_{\rho+1} = \lambda_1 a_1 + \cdots + \lambda_\rho a_\rho + \lambda_{\rho+1} d_{\rho+1,j} + \cdots + \lambda_n d_{nj} . \tag{5.18}$$

Multiplying this last by $c_{\rho+j}$ gives $\lambda_\rho = a_{\rho+1} c_{\rho+j}$ which with (5.15) shows that $\lambda_\rho < 0$. Solving (5.18) for a_ρ gives

$$a_\rho = \frac{1}{\lambda_\rho} (a_{\rho+1} - \lambda_1 a_1 - \cdots - \lambda_{\rho-1} a_{\rho-1} - \lambda_{\rho+1} d_{\rho+1,j} - \cdots - \lambda_n d_{nj}) .$$

Substitution of this into (5.16) gives

$$g_j = v_1 a_1 + \cdots + v_{\rho-1} a_{\rho-1} + v_\rho a_{\rho+1} + v_{\rho+1} d_{\rho+1,j} + \cdots + v_n d_{nj} , \tag{5.19}$$

where $v_\rho = g'_j c_{\rho j}/\lambda_\rho$, $v_i = g'_j c_{ij} - \lambda_i g'_j c_{\rho j}/\lambda_\rho$ for $i = 1, 2, \ldots, \rho - 1$ and $v_i = -\lambda_i g'_j c_{ij}/\lambda_\rho$ for $i = \rho + 1, \ldots, n$. Because x_j is a quasistationary point, the assumption in the statement of the theorem asserts that $a_1, \ldots, a_\rho$, $a_{\rho+1}$ must be linearly independent. This together with (5.18) implies that at least one of $\lambda_{\rho+1}, \ldots, \lambda_n$ is nonzero. With (5.17) this in turn implies that at least one of $v_{\rho+1}, \ldots, v_n$ is nonzero. This gives the critical result that s_{j+1} will be chosen according to Step 1.1. If $\sigma_{j+1} = 0$, then because s_{j+1} is determined by Step 1.1 the number of active constraints in J_{j+2} will be one greater than in J_j. Suppose that a zero step size occurs for the next several iterations. Each time the search direction comes from Step 1.1, the number of constraints in the active set increases by one for the next iteration. Each time the search direction comes from Step 1.2, the number of constraints in the active set increases by one, two iterations later (as we have just argued). By hypothesis, the gradients of those constraints active at x_j are linearly independent and therefore there are at most n of them. Therefore, after a finite number of steps a positive step size and a strict decrease in the objective function must be obtained and this completes the proof of the theorem. $\square$

Theorem 5.1 together with the Strong Duality Theorem (Theorem 4.9) show that an optimal solution for the dual problem may be obtained from the final data of Algorithm 3. The critical quantities are $g'_j c_{ij}$ for all i with $1 \leq \alpha_{ij} \leq m + r$. For the active inequality constraints, these quantities have already been computed in obtaining k during Step 1.2 just prior to termination and they need not be recomputed. For the equality constraints, these inner products must be computed directly. For Example 5.1, an optimal solution

for the dual is $x_2 = (4 , 2)'$ together with $u_0 = (2 , 0 , 0 , 0)'$. For
Example 5.2, an optimal solution for the dual is $x_3 = (1 , 2 , 3)'$ together
with $u_0 = (0 , 1 , 2 , 0 , 0 , 0)'$.

The assumption of nondegenerate quasistationary points is required in
Theorem 5.1 to guarantee finite termination. This assumption will be removed
in Section 5.4.

5.2   Determination of an Initial Feasible Point

Algorithm 3 must be initiated with a feasible point x_0 and corresponding data
J_0 and D_0^{-1}. Determination of an initial feasible point for a QP is identical to
that for an LP. A procedure for doing this is given in Sections 2.5 and 2.6 of
Best and Ritter [1985]. We summarize the procedure in this section.

First consider the linear programming problem

$$\left. \begin{array}{ll} \text{minimize :} & c'x \\ \text{subject to :} & a_i'x \leq b_i, \quad i = 1, \ldots, m, \\ & a_i'x = b_i, \quad i = m+1, \ldots, m+r. \end{array} \right\} \tag{5.20}$$

Because $C = 0$ is positive semidefinite, (5.20) is a special case of the model
problem for Algorithm 3. Therefore Algorithm 3 can be used to solve an LP.
However, because the model problem is now an LP rather than a QP, some
simplification can be made. In Step 2, $s_j'Cs_j$ will always be zero and thus
$\tilde{\sigma}_j = +\infty$. In Step 3, $g_{j+1} = c$ and need not be evaluated. Similarly, g_j may
be replaced with c in Steps 1.1 and 1.2. Because $\tilde{\sigma}_j = +\infty$, Step 3.1 will
never be taken and Step 3.2 can be incorporated into Step 3. Incorporating
these changes gives

LP ALGORITHM 3

Model Problem:

$$\begin{array}{ll} \text{minimize:} & c'x \\ \text{subject to:} & a_i'x \leq b_i, \quad i = 1, \ldots, m, \\ & a_i'x = b_i, \quad i = m + 1, \ldots, m + r. \end{array}$$

Initialization:

Start with any feasible point x_0, $J_0 = \{ \alpha_{10}, \ldots, \alpha_{n0} \}$, and D_0^{-1}, where
$D_0' = [d_1, \ldots, d_n]$ is nonsingular, $d_i = \alpha_{i0}$ for all i with $1 \leq \alpha_{i0} \leq m + r$,
and, each of $m + 1, \ldots, m + r$ is in J_0. Compute $c'x_0$ and set $j = 0$.

Step 1: **Computation of Search Direction** s_j.
Let $D_j^{-1} = [\,c_{1j}, \ldots, c_{nj}\,]$ and $J_j = \{\,\alpha_{1j}, \ldots, \alpha_{nj}\,\}$. If there is at least one i with $\alpha_{ij} = 0$, go to Step 1.1. Otherwise, go to Step 1.2.

Step 1.1:
Determine the smallest index k such that

$$|\,c'c_{kj}\,| = \max \{\,|\,c'c_{ij}\,| \mid \text{all } i \text{ with } \alpha_{ij} = 0\,\}.$$

If $c'c_{kj} = 0$, go to Step 1.2. Otherwise, set $s_j = c_{kj}$ if $c'c_{kj} > 0$ and $s_j = -c_{kj}$ if $c'c_{kj} < 0$. Go to Step 2.

Step 1.2:
If there is no i with $1 \leq \alpha_{ij} \leq m$, then stop with optimal solution x_j. Otherwise, determine the smallest index k such that

$$c'c_{kj} = \max \{\,c'c_{ij} \mid \text{all } i \text{ with } 1 \leq \alpha_{ij} \leq m\,\}.$$

If $c'c_{kj} \leq 0$, then stop with optimal solution x_j. Otherwise, set $s_j = c_{kj}$ and go to Step 2.

Step 2: **Computation of Step Size** σ_j.
If $a_i's_j \geq 0$ for $i = 1, \ldots, m$, print the message "the objective function is unbounded from below" and stop. Otherwise, compute the smallest index l and σ_j such that

$$\sigma_j = \frac{a_l'x_j - b_l}{a_l's_j} = \min \left\{ \frac{a_i'x_j - b_i}{a_i's_j} \mid \text{all } i \notin J_j \text{ with } a_i's_j < 0 \right\},$$

and go to Step 3.

Step 3: **Update.**
Set $x_{j+1} = x_j - \sigma_j s_j$ and compute $c'x_{j+1}$. Set $D_{j+1}^{-1} = \Phi(D_j^{-1}, a_l, k)$ and $J_{j+1} = \{\,\alpha_{1,j+1}, \ldots, \alpha_{n,j+1}\,\}$, where

$$\alpha_{i,j+1} = \alpha_{ij}, \quad \text{for all } i \text{ with } i \neq k,$$
$$\alpha_{k,j+1} = l.$$

Replace j with $j + 1$ and go to Step 1.

Not surprisingly, the above LP Algorithm 3 is equivalent to Algorithm 3 of Best and Ritter [1985] for the solution of a linear programming problem.[1] We thus have the capability of solving a linear programming problem using

---

[1]Note that Algorithm 3 of Best and Ritter [1985] uses the additional symbol v_i for $c'c_{ij}$. In Algorithm 3 of Section 5.1 here, all computations are formulated directly in terms of $g_j'c_{ij}$.

either Algorithm 3 with $C = 0$ explicitly or LP Algorithm 3 with $C = 0$ implicitly.

A Matlab computer program (LPAlg3) is given in Figure 5.13 for the solution of a Linear Programming problem.

Having digressed to discuss solution procedures for a linear programming problem, we now return to the problem of determining initial data x_0, D_0^{-1}, and J_0 for Algorithm 3 so that it may be applied to our model quadratic programming problem

$$\left. \begin{array}{ll} \text{minimize}: & c'x + \frac{1}{2}x'Cx \\ \text{subject to}: & a_i'x \leq b_i, \quad i = 1, \ldots, m, \\ & a_i x = b_i, \quad i = m+1, \ldots, m+r. \end{array} \right\} \tag{5.21}$$

It is convenient to assume that $b_{m+1}, \ldots, b_{m+r}$ are all nonnegative. This assumption may be satisfied by multiplying both sides of an equality constraint by -1, when necessary. Having done this, set

$$d = -\sum_{i=m+1}^{m+r} a_i .$$

The <u>initial point problem</u> or <u>phase 1 problem</u> for (5.21) is

$$\left. \begin{array}{ll} \text{minimize}: & d'x + \alpha \\ \text{subject to}: & a_i'x - \alpha \leq b_i, \quad i = 1, \ldots, m, \\ & a_i'x \quad\quad \leq b_i, \quad i = m+1, \ldots, m+r, \\ & \quad\quad -\alpha \leq 0. \end{array} \right\} \tag{5.22}$$

The relevant properties of the initial point problem are summarized as follows.

1. The initial point problem (5.22) is an LP in $n + 1$ variables (α is a scalar) and $m + r + 1$ inequality constraints ($-\alpha \leq 0$ is constraint $m + r + 1$).

2. A feasible solution for (5.22) is $x_0 = 0$ together with any α_0 satisfying

$$\alpha_0 \geq \max \{ 0, -b_i \mid i = 1, \ldots, m \} .$$

   (We have assumed that $b_i \geq 0$, $i = m + 1, \ldots, m + r$.)

3. The initial point problem may be solved with initial data as in property 2 using either Algorithm 3 (with $C = 0$ explicitly) or LP Algorithm 3 (with $C = 0$ implicitly). An optimal solution exists because for any point $(x', \alpha)'$ feasible for (5.22),

$$d'x + \alpha = -\sum_{i=m+1}^{m+r} a_i'x + \alpha \geq -\sum_{i=m+1}^{m+r} b_i .$$

Let x^* , α^* be an optimal solution for the initial point problem. There are two cases to be considered.

Case 1: $\quad d'x^* + \alpha^* > - \displaystyle\sum_{i=m+1}^{m+r} b_i$

This implies that (5.21) has no feasible solution.

Case 2: $\quad d'x^* + \alpha^* = - \displaystyle\sum_{i=m+1}^{m+r} b_i$

In this case, $\alpha^* = 0$ and x^* is feasible for (5.21).

If Case 1 applies, nothing remains to be done. Suppose then, that Case 2 applies. In addition to the feasible point $x_0 = x^*$, Algorithm 3 requires a matrix D_0^{-1} and an index set J_0 with the property that each of the numbers $m + 1, \ldots, m + r$ is in J_0. D_0^{-1} and J_0 may be obtained from the final data in solving the initial point problem as follows. Let $\hat{x}_0$, $\hat{J}_0$, and $\hat{D}_0^{-1}$ denote this final data. We use the "hat" notation to emphasize that $\hat{x}_0$, $\hat{J}_0$, and $\hat{D}_0^{-1}$ are $(n + 1)$-dimensional quantities. Suppose that the kth element of $\hat{J}_0$ is $m + r + 1$ [i.e., the kth column of $\hat{D}_0'$ is the gradient of the $(m + r + 1)$st constraint $-\alpha \leq 0$]. Then we obtain x_0 from $\hat{x}_0$ by deleting the last element, J_0 from $\hat{J}_0$ by striking out the kth element, and D_0^{-1} from $\hat{D}_0^{-1}$ by deleting the last row and the kth column. Using the initial data x_0, J_0, and D_0^{-1}, we can now apply Algorithm 3 to (5.21). A complete procedure for solving the QP

$$\left. \begin{array}{ll} \text{minimize}: & c'x + \frac{1}{2}x'Cx \\ \text{subject to}: & a_i'x \leq b_i, \qquad i = 1, \ldots, m, \\ & a_ix = b_i, \qquad i = m+1, \ldots, m+r, \end{array} \right\} \qquad (5.23)$$

can be described as follows.

Phase 1:

Formulate the initial point problem (5.22) for (5.23). Solve it using either Algorithm 3 with $C = 0$ or LP Algorithm 3. Let $\hat{x}_0$, $\hat{J}_0$, and $\hat{D}_0^{-1}$ be the final data thus obtained.

Phase 2:

If (5.23) has a feasible solution, obtain x_0, J_0, and D_0^{-1} from $\hat{x}_0$, $\hat{J}_0$, and $\hat{D}_0^{-1}$, respectively, as described above and then apply Algorithm 3 with these initial data to (5.23).

It remains to show that $\hat{J}_0$ contains each of the numbers $m + 1, \ldots, m + r$, $m + r + 1$. This is indeed the case if the gradients a_i of those constraints that are active at x_0 are linearly independent. Let LP Algorithm 3, applied to

(5.22), terminate with

$$\hat{x}_0 = \begin{bmatrix} x_0 \\ 0 \end{bmatrix}, \quad \hat{J}_0 = \{ \alpha_{10}, \ldots, \alpha_{n0}, \ \alpha_{n+1,0} \}, \quad \text{and} \quad \hat{D}_0^{-1}.$$

We use the following notation:

$$\hat{c} = \begin{bmatrix} d \\ 1 \end{bmatrix} \quad \text{and} \quad \hat{D}_0' = \begin{bmatrix} d_1, \ldots, d_n, & d_{n+1} \\ \delta_i, \ldots, \delta_n, & \delta_{n+1} \end{bmatrix},$$

where $d_i = a_{\alpha_{i0}}$ for all i with $1 \le \alpha_{i0} \le m + r$,

$$\delta_i = 0, \quad \text{for all } i \text{ with } m + 1 \le \alpha_{i0} \le m + r, \tag{5.24}$$

and $d_i = 0, \delta_i = -1$ if $\alpha_{i0} = m + r + 1$. With

$$\hat{D}_0^{-1} = [\, \hat{c}_{10}, \ldots, \hat{c}_{n0}, \ \hat{c}_{n+1,0} \,] \tag{5.25}$$

and

$$v_i = \hat{c}' \hat{c}_{i0}, \quad i = 1, \ldots, n + 1$$

we have

$$\hat{c} = \sum_{i=1}^{n+1} v_i \begin{bmatrix} d_i \\ \delta_i \end{bmatrix}$$

or

$$d = \sum_{i=1}^{n+1} v_i d_i \tag{5.26}$$

and

$$1 = \sum_{i=1}^{n+1} v_i \delta_i. \tag{5.27}$$

Observing that $d_i = 0$ if $\alpha_{i0} = m + r + 1$,

$$v_i = 0, \quad \text{for all } i \text{ with } \alpha_{i0} = 0, \tag{5.28}$$

and using the definition of d, we obtain from (5.26)

$$- \sum_{k=m+1}^{m+r} a_k = \sum_{1 \le \alpha_{i0} \le m+r} v_i a_{\alpha_{i0}}. \tag{5.29}$$

Because x_0 is a feasible solution for (5.21), every a_i occurring on either side of (5.29) is the gradient of a constraint which is active at x_0. By assumption, these gradients are linearly independent. This implies that each of the vectors $a_{m+1}, \ldots, a_{m+r}$ occurs on the right-hand side of (5.29); that is,

$$m + i \in \hat{J}_0 , \quad \text{for } i = 1, \ldots, r .$$

Furthermore, $v_i = -1$ for all i with $m + 1 \le \alpha_{i0} \le m + r$ and

$$v_i = 0 , \quad \text{for all } i \text{ with } 1 \le \alpha_{i0} \le m . \tag{5.30}$$

Using (5.24), (5.28), and (5.30), we note that $v_i \delta_i = 0$ for all i with $0 \le \alpha_{i0} \le m + r$. Thus it follows from (5.27) that $\alpha_{i0} = m + r + 1$ for some $1 \le i \le n + 1$ (i.e., $m + r + 1$ is in $\hat{J}_0$).

To conclude this section we discuss briefly the case that at least one of the numbers $m + 1, \ldots, m + r$, $m + r + 1$ is not in $\hat{J}_0$.

First suppose that $m + r + 1 \notin \hat{J}_0$; that is, the gradient of the constraint $-\alpha \le 0$ is not a column of $\hat{D}'_0$. Using the same notation for $\hat{D}'_0$ as before, we can write $(0' , -1)'$ as

$$\begin{bmatrix} 0 \\ -1 \end{bmatrix} = \sum_{i=1}^{n+1} \lambda_i \begin{bmatrix} d_i \\ \delta_i \end{bmatrix} . \tag{5.31}$$

Because of (5.24) there is at least one i with $0 \le \alpha_{i0} \le m$ and $\lambda_i \ne 0$. Now determine k such that

$$\mid (\hat{c}_{k0})_{n+1} \mid = \max \{ \mid (\hat{c}_{i0})_{n+1} \mid \mid \text{ all } i \text{ with } 0 \le \alpha_{i0} \le m \} .$$

Thus $(\hat{c}_{k0})_{n+1} \ne 0$ because it follows from (5.25) and (5.31) that

$$\hat{c}'_{i0} \begin{bmatrix} 0 \\ 1 \end{bmatrix} = \lambda_i , \quad i = 1, \ldots, n + 1 .$$

Thus we can use Procedure Φ to determine the matrix

$$\Phi \left(\hat{D}_0^{-1} , \begin{bmatrix} 0 \\ -1 \end{bmatrix} , k \right) .$$

Replacing $\hat{D}_0^{-1}$ with this new matrix if necessary, we may assume that the vector $(0' , -1)'$ is one of the columns of $\hat{D}'_0$.

Next suppose that there is a ρ such that $m + 1 \le \rho \le m + r$ and $\rho \notin \hat{J}_0$. Let

$$a_\rho = \sum_{i=1}^{n+1} \lambda_i \begin{bmatrix} a_\rho \\ 0 \end{bmatrix} .$$

If $\lambda_i \ne 0$ for at least one i with $0 \le \alpha_{i0} \le m$, determine k such that

$$\mid \lambda_k \mid = \max \{ \mid \lambda_i \mid \mid \text{ all } i \text{ with } 0 \le \alpha_{i0} \le m \} .$$

Then $\lambda_k \neq 0$ and

$$\Phi\left(\hat{D}_0^{-1}, \begin{bmatrix} a_\rho \\ 0 \end{bmatrix}, k\right)$$

is well-defined. Now let $\lambda_i = 0$ for all i with $0 \leq \alpha_{i0} \leq m$. Then

$$a_\rho \in \text{span } \{ a_{\alpha_{i0}} \mid \text{all } i \text{ with } m + 1 \leq \alpha_{i0} \leq m + r \}.$$

It follows that the equation $a_\rho' x = b_\rho$ is superfluous and can be deleted.

Continuing this process and deleting equations if necessary, we obtain a matrix $\tilde{D}_0^{-1}$ and a corresponding index set $\tilde{J}_0$ which contains the indices of all remaining equality constraints. Thus we can replace $\hat{J}_0$ and $\hat{D}_0^{-1}$ by $\tilde{J}_0$ and $\tilde{D}_0^{-1}$, respectively.

We illustrate the Phase 1–Phase 2 procedure in the following two examples.

Example 5.3

$$
\begin{array}{rlcrl}
\text{minimize:} & -10x_1 & - & 4x_2 & | & x_1^2 + x_2^2 \\
\text{subject to:} & x_1 & & & \leq & 4, \quad (1) \\
& & - & x_2 & \leq & -1, \quad (2) \\
& - & x_1 & & \leq & -1, \quad (3) \\
& & & x_2 & \leq & 3. \quad (4)
\end{array}
$$

Here

$$n = 2, \quad m = 4, \quad r = 0, \quad c = \begin{bmatrix} -10 \\ -4 \end{bmatrix}, \quad \text{and} \quad C = \begin{bmatrix} 2 & 0 \\ 0 & 2 \end{bmatrix}.$$

The Phase 1 problem for this example is

$$
\begin{array}{rlcrl}
\text{minimize:} & & & \alpha & & \\
\text{subject to:} & x_1 & - & \alpha & \leq & 4, \quad (1) \\
& - x_2 & - & \alpha & \leq & -1, \quad (2) \\
& - x_1 & - & \alpha & \leq & -1, \quad (3) \\
& x_2 & - & \alpha & \leq & 3, \quad (4) \\
& & - & \alpha & \leq & 0. \quad (5)
\end{array}
$$

Note that $d = 0$ because $r = 0$. We solve the Phase 1 problem using LP Algorithm 3 as follows.

Initialization:

$$x_0 = \begin{bmatrix} 0 \\ 0 \end{bmatrix}, \quad \alpha_0 = \max\{0, -4, 1, 1, -3\} = 1, \quad J_0 = \{0, 0, 0\},$$

$$D_0^{-1} = \begin{bmatrix} 1 & 0 & 0 \\ 0 & 1 & 0 \\ 0 & 0 & 1 \end{bmatrix}, \quad c = \begin{bmatrix} d \\ 1 \end{bmatrix} = \begin{bmatrix} 0 \\ 0 \\ 1 \end{bmatrix}, \quad c'\begin{bmatrix} x_0 \\ \alpha_0 \end{bmatrix} = 1, \quad j = 0.$$

Iteration 0

Step 1: Transfer to Step 1.1.

Step 1.1: $|c'c_{30}| = \max\{|0|, |0|, |1|\} = 1, \ k = 3,$

$$s_0 = \begin{bmatrix} 0 \\ 0 \\ 1 \end{bmatrix}.$$

Step 2: $\sigma_0 = \min\left\{\dfrac{-5}{-1}, \dfrac{0}{-1}, \dfrac{0}{-1}, \dfrac{-4}{-1}, \dfrac{-1}{-1}\right\} = 0, \ l = 2,$

Step 3: $\begin{bmatrix} x_1 \\ \alpha_1 \end{bmatrix} = \begin{bmatrix} 0 \\ 0 \\ 1 \end{bmatrix}, \ c'\begin{bmatrix} x_1 \\ \alpha_1 \end{bmatrix} = 1,$

$$D_1^{-1} = \begin{bmatrix} 1 & 0 & 0 \\ 0 & 1 & 0 \\ 0 & -1 & -1 \end{bmatrix}, \ J_1 = \{0, 0, 2\}, \ j = 1.$$

Iteration 1

Step 1: Transfer to Step 1.1.

Step 1.1: $|c'c_{21}| = \max\{|0|, |-1|, -\} = 1, \ k = 2,$

$$s_1 = \begin{bmatrix} 0 \\ -1 \\ 1 \end{bmatrix}.$$

Step 2: $\sigma_1 = \min\left\{\dfrac{-5}{-1}, -, \dfrac{0}{-1}, \dfrac{-4}{-2}, \dfrac{-1}{-1}\right\} = 0, \ l = 3,$

Step 3: $\begin{bmatrix} x_2 \\ \alpha_2 \end{bmatrix} = \begin{bmatrix} 0 \\ 0 \\ 1 \end{bmatrix}, \ c'\begin{bmatrix} x_2 \\ \alpha_2 \end{bmatrix} = 1,$

$$D_2^{-1} = \begin{bmatrix} 1 & 0 & 0 \\ 1 & 1 & -1 \\ -1 & -1 & 0 \end{bmatrix}, \ J_2 = \{0, 3, 2\},$$

$j = 2.$

Iteration 2

Step 1:   Transfer to Step 1.1.

Step 1.1: $|\, c' c_{12} \,| \;=\; \max \{\, |\, -1 \,|\, ,\, -\, ,\, -\, \} \;=\; 1\, ,\quad k = 1\, ,$

$$s_2 \;=\; \begin{bmatrix} -1 \\ -1 \\ 1 \end{bmatrix}.$$

Step 2: $\sigma_2 \;=\; \min \left\{\, \dfrac{-5}{-2}\, ,\, -\, ,\, -\, ,\, \dfrac{-4}{-2}\, ,\, \dfrac{-1}{-1}\, \right\} \;=\; 1\, ,\quad l = 5\, ,$

Step 3: $\begin{bmatrix} x_3 \\ \alpha_3 \end{bmatrix} \;=\; \begin{bmatrix} 0 \\ 0 \\ 1 \end{bmatrix} - \begin{bmatrix} -1 \\ -1 \\ 1 \end{bmatrix} \;=\; \begin{bmatrix} 1 \\ 1 \\ 0 \end{bmatrix}\, ,\quad c' \begin{bmatrix} x_3 \\ \alpha_3 \end{bmatrix} \;=\; 0\, ,$

$$D_3^{-1} \;=\; \begin{bmatrix} 1 & -1 & 0 \\ 1 & 0 & -1 \\ -1 & 0 & 0 \end{bmatrix}\, ,\quad J_3 = \{\, 5\, ,\, 3\, ,\, 2\, \}\, ,\quad j = 3\, .$$

Iteration 3

Step 1:   Transfer to Step 1.2.

Step 1.2: $c' c_{23} \;=\; \max \{\, -1\, ,\, 0\, ,\, 0\, \} \;=\; 0\, ,\quad k = 2\, .$

$c' c_{23} \leq 0$; stop with optimal solution x_3 and α_3 .

We obtain the initial data D_0^{-1}, J_0, and x_0 for the Phase 2 problem by striking out row 3 and column 1 of D_3^{-1}, the first element of J_3, and the last component of $(x_3'\, ,\, \alpha_3)'$ above, respectively. The Phase 2 problem is now solved by applying Algorithm 3.

Initialization:

$$x_0 \;=\; \begin{bmatrix} 1 \\ 1 \end{bmatrix}\, ,\quad J_0 = \{\, 3\, ,\, 2\, \}\, ,\quad D_0^{-1} \;=\; \begin{bmatrix} -1 & 0 \\ 0 & -1 \end{bmatrix}\, ,\quad f(x_0) = -12\, ,$$

$$g_0 \;=\; \begin{bmatrix} -8 \\ -2 \end{bmatrix}\, ,\quad j = 0\, .$$

Iteration 0

Step 1:   Transfer to Step 1.2.

Step 1.2: $g_0' c_{10} = \max \{ 8 , 2 \} = 8 , \ k = 1 , \ s_0 = \begin{bmatrix} -1 \\ 0 \end{bmatrix} .$

Step 2: $s_0' C s_0 = 2 ,$

$$\tilde{\sigma}_0 = \frac{8}{2} = 4 ,$$

$$\hat{\sigma}_0 = \min \left\{ \frac{-3}{-1} , - , - , - \right\} = 3 , \ l = 1 ,$$

$$\sigma_0 = \min \{ 4 , 3 \} = 3 .$$

Step 3: $x_1 = \begin{bmatrix} 1 \\ 1 \end{bmatrix} - 3 \begin{bmatrix} -1 \\ 0 \end{bmatrix} = \begin{bmatrix} 4 \\ 1 \end{bmatrix} , \ g_1 = \begin{bmatrix} -2 \\ -2 \end{bmatrix} ,$

$f(x_1) = -27 .$ Transfer to Step 3.2.

Step 3.2: $D_1^{-1} = \begin{bmatrix} 1 & 0 \\ 0 & -1 \end{bmatrix} , \ J_1 = \{ 1 , 2 \} , \ j = 1 .$

Iteration 1

Step 1:   Transfer to Step 1.2.

Step 1.2: $g_1' c_{21} = \max \{ -2 , 2 \} = 2 , \ k = 2 , \ s_1 = \begin{bmatrix} 0 \\ -1 \end{bmatrix} .$

Step 2: $s_1' C s_1 = 2 ,$

$$\tilde{\sigma}_1 = \frac{2}{2} = 1 ,$$

$$\hat{\sigma}_1 = \min \left\{ - , - , - , \frac{-2}{-1} \right\} = 2 , \ l = 4 ,$$

$$\sigma_1 = \min \{ 1 , 2 \} = 1 .$$

Step 3: $x_2 = \begin{bmatrix} 4 \\ 1 \end{bmatrix} - \begin{bmatrix} 0 \\ -1 \end{bmatrix} = \begin{bmatrix} 4 \\ 2 \end{bmatrix} , \ g_2 = \begin{bmatrix} -2 \\ 0 \end{bmatrix} ,$

$f(x_2) = -28 .$ Transfer to Step 3.1.

Step 3.1: $d_1 = \begin{bmatrix} 0 \\ -2/\sqrt{2} \end{bmatrix} ,$

$$D_2^{-1} = \begin{bmatrix} 1 & 0 \\ 0 & -1/\sqrt{2} \end{bmatrix} , \quad J_2 = \{ 1 , -1 \} , \quad j = 2 .$$

Iteration 2

Step 1:   Transfer to Step 1.2.

Step 1.2: $g_2' c_{12} = \max \{ -2 , - \} = -2 , \quad k = 1 .$

$g_2' c_{12} \leq 0$; stop with optimal solution $x_2 = (4 , 2)'$.

This completes Example 5.3.   ◇

A computer program which implements the above Phase 1–Phase 2 procedure, (P1P2.m), is given in Figure 5.14, Section 5.5. Data for solving Example 5.3 by P1P2.m is given in eg5p3.m (Figure 5.15) and the output is given in Figure 5.16. Both figures are in Section 5.5.

We next give a second illustration of the Phase 1–Phase 2 procedure by using it to solve a 3-dimensional problem whose constraints include an equality constraint.

Example 5.4

$$\begin{aligned} \text{minimize:} \quad & -2x_2 + 6x_3 + \tfrac{1}{2}(2x_1^2 + 3x_2^2 + 4x_3^2 + 4x_1x_2) \\ \text{subject to:} \quad & 2x_1 + x_2 + x_3 \leq 10, \quad (1) \\ & x_1 \qquad\quad + x_3 \leq 5, \quad (2) \\ & x_2 + x_3 \leq 5, \quad (3) \\ & x_1 + 2x_2 + x_3 \leq 9, \quad (4) \\ & x_1 + x_2 + x_3 = 6. \quad (5) \end{aligned}$$

Here

$$n = 3 , \quad m = 4 , \quad r = 1 , \quad c = \begin{bmatrix} 0 \\ -2 \\ 6 \end{bmatrix} , \quad \text{and} \quad C = \begin{bmatrix} 2 & 2 & 0 \\ 2 & 3 & 0 \\ 0 & 0 & 4 \end{bmatrix} .$$

The Phase 1 problem for this example is

$$\begin{aligned} \text{minimize:} \quad & -x_1 - x_2 - x_3 + \alpha \\ \text{subject to:} \quad & 2x_1 + x_2 + x_3 - \alpha \leq 10, \quad (1) \\ & x_1 \qquad\quad + x_3 - \alpha \leq 5, \quad (2) \\ & x_2 + x_3 - \alpha \leq 5, \quad (3) \end{aligned}$$

$$x_1 + 2x_2 + x_3 - \alpha \leq 9, \quad (4)$$
$$x_1 + x_2 + x_3 \leq 6, \quad (5)$$
$$- \alpha \leq 6. \quad (6)$$

We solve the Phase 1 problem using LP Algorithm 3 as follows.

Initialization:

$$x_0 = \begin{bmatrix} 0 \\ 0 \\ 0 \end{bmatrix}, \quad \alpha_0 = \max \{ 0, -10, -5, -5, -9 \} = 0,$$

$$J_0 = \{ 0, 0, 0, 0 \}, \quad D_0^{-1} = \begin{bmatrix} 1 & 0 & 0 & 0 \\ 0 & 1 & 0 & 0 \\ 0 & 0 & 1 & 0 \\ 0 & 0 & 0 & 1 \end{bmatrix},$$

$$c = \begin{bmatrix} d \\ 1 \end{bmatrix} = \begin{bmatrix} -1 \\ -1 \\ -1 \\ 1 \end{bmatrix}, \quad c' \begin{bmatrix} x_0 \\ \alpha_0 \end{bmatrix} = 0, \quad j = 0.$$

Iteration 0

Step 1: Transfer to Step 1.1.

Step 1.1: $| c'c_{10} | = \max \{ | -1 |, | -1 |, | -1 |, | 1 | \} = 1,$

$$k = 1,$$

$$s_0 = \begin{bmatrix} -1 \\ 0 \\ 0 \\ 0 \end{bmatrix}.$$

Step 2: $\sigma_0 = \min \left\{ \dfrac{-10}{-2}, \dfrac{-5}{-1}, -, \dfrac{-9}{-1}, \dfrac{-6}{-1}, - \right\} = 5,$

$$l = 2,$$

Step 3: $\begin{bmatrix} x_1 \\ \alpha_1 \end{bmatrix} = \begin{bmatrix} 0 \\ 0 \\ 0 \\ 0 \end{bmatrix} - 5 \begin{bmatrix} -1 \\ 0 \\ 0 \\ 0 \end{bmatrix} = \begin{bmatrix} 5 \\ 0 \\ 0 \\ 0 \end{bmatrix}, \quad c' \begin{bmatrix} x_1 \\ \alpha_1 \end{bmatrix} = -5,$

$$D_1^{-1} = \begin{bmatrix} 1/2 & -1/2 & -1/2 & 1/2 \\ 0 & 1 & 0 & 0 \\ 0 & 0 & 1 & 0 \\ 0 & 0 & 0 & 1 \end{bmatrix}, \quad J_1 = \{\, 1\,,\, 0\,,\, 0\,,\, 0\,\},$$

$$j = 1 \,.$$

Iteration 1

Step 1:   Transfer to Step 1.1.

Step 1.1: $|\, c'c_{21}\,| = \max \left\{\, -\,,\, |\,\dfrac{-1}{2}\,|\,,\, |\,\dfrac{-1}{2}\,|\,,\, |\,\dfrac{1}{2}\,|\, \right\} = \dfrac{1}{2}\,,$

$k = 2\,,$

$$s_1 = \begin{bmatrix} 1/2 \\ -1 \\ 0 \\ 0 \end{bmatrix} \,.$$

Step 2: $\sigma_1 = \min \left\{\, -\,,\, -\,,\, \dfrac{-5}{-1}\,,\, \dfrac{-4}{-3/2}\,,\, \dfrac{-1}{-1/2}\,,\, -\, \right\} = 2\,,$

$l = 5\,,$

Step 3: $\begin{bmatrix} x_2 \\ \alpha_2 \end{bmatrix} = \begin{bmatrix} 5 \\ 0 \\ 0 \\ 0 \end{bmatrix} - 2 \begin{bmatrix} 1/2 \\ -1 \\ 0 \\ 0 \end{bmatrix} = \begin{bmatrix} 4 \\ 2 \\ 0 \\ 0 \end{bmatrix}, \quad c' \begin{bmatrix} x_2 \\ \alpha_2 \end{bmatrix} = -6,$

$$D_2^{-1} = \begin{bmatrix} 1 & -1 & 0 & 1 \\ -1 & 2 & -1 & -1 \\ 0 & 1 & 1 & 0 \\ 0 & 0 & 0 & 1 \end{bmatrix}, \quad J_2 = \{\, 1\,,\, 5\,,\, 0\,,\, 0\,\},$$

$$j = 2 \,.$$

Iteration 2

Step 1:   Transfer to Step 1.1.

Step 1.1: $|\, c'c_{42}\,| = \max \{\, -\,,\, -\,,\, |\,0\,|\,,\, |\,1\,|\, \} = 1\,,\quad k = 4\,,$

$$s_2 = \begin{bmatrix} 1 \\ -1 \\ 0 \\ 1 \end{bmatrix}.$$

Step 2: $\sigma_2 = \min \left\{ -, -, \dfrac{-3}{-2}, \dfrac{-1}{-2}, -, \dfrac{0}{-1} \right\} = 0, \; l = 6,$

Step 3: $\begin{bmatrix} x_3 \\ \alpha_3 \end{bmatrix} = \begin{bmatrix} 4 \\ 2 \\ 0 \\ 0 \end{bmatrix}, \; c' \begin{bmatrix} x_3 \\ \alpha_3 \end{bmatrix} = -6,$

$$D_3^{-1} = \begin{bmatrix} 1 & -1 & 0 & -1 \\ -1 & 2 & -1 & 1 \\ 0 & 0 & 1 & 0 \\ 0 & 0 & 0 & -1 \end{bmatrix}, \; J_3 = \{\, 1\,,\, 5\,,\, 0\,,\, 6 \,\},$$

$$j = 3.$$

Iteration 3

Step 1:   Transfer to Step 1.1.

Step 1.1: $|\, c' c_{33} \,| = \max \{\, -\,,\, -\,,\, |\, 0\, |\,,\, - \,\} = 0, \; k = 3.$

             Transfer to Step 1.2.

Step 1.2: $c' c_{13} = \max \{\, 0\,,\, -1\,,\, -\,,\, -1 \,\} = 0, \; k = 1.$

             $c' c_{13} \leq 0$; stop with optimal solution x_3 and α_3 .

We obtain the initial data D_0^{-1}, J_0, and x_0 for the Phase 2 problem by striking out row 4 and column 4 of D_3^{-1}, the last element of J_3, and the last component of $(x_3', \alpha_3)'$ above, respectively. The Phase 2 problem is now solved by applying Algorithm 3.

Initialization:

$$x_0 = \begin{bmatrix} 4 \\ 2 \\ 0 \end{bmatrix}, \; J_0 = \{\, 1\,,\, 5\,,\, 0 \,\}, \; D_0^{-1} = \begin{bmatrix} -1 & -1 & 0 \\ 1 & 2 & -1 \\ 0 & 0 & 1 \end{bmatrix},$$

$$f(x_0) = 34, \; g_0 = \begin{bmatrix} 12 \\ 12 \\ 6 \end{bmatrix}, \; j = 0.$$

Iteration 0

Step 1:   Transfer to Step 1.1.

Step 1.1: $| g_0' c_{30} | \ = \ \max \{ \, - \, , \, - \, , \, | -6 \, | \, \} \ = \ 6 \, , \quad k \ = \ 3 \, ,$

$$s_0 \ = \ \begin{bmatrix} 0 \\ 1 \\ -1 \end{bmatrix} .$$

Step 2: $s_0' C s_0 \ = \ 7 \, ,$

$$\tilde{\sigma}_0 \ = \ \frac{6}{7} \, ,$$

$$\hat{\sigma}_0 \ = \ \min \left\{ \, - \, , \, \frac{-1}{1} \, , \, - \, , \, - \, \right\} \ = \ 1 \, , \quad l \ = \ 2 \, ,$$

$$\sigma_0 \ = \ \min \left\{ \, \frac{6}{7} \, , \, 1 \, \right\} \ = \ \frac{6}{7} \, .$$

Step 3: $x_1 \ = \ \begin{bmatrix} 4 \\ 2 \\ 0 \end{bmatrix} - \frac{6}{7} \begin{bmatrix} 0 \\ 1 \\ -1 \end{bmatrix} \ = \ \begin{bmatrix} 4 \\ 8/7 \\ 6/7 \end{bmatrix} \, , \quad g_1 \ = \ \begin{bmatrix} 72/7 \\ 66/7 \\ 66/7 \end{bmatrix} \, ,$

$$f(x_1) \ = \ \frac{220}{7} \, . \quad \text{Transfer to Step 3.1.}$$

Step 3.1: $d_0 \ = \ \begin{bmatrix} 2/\sqrt{7} \\ 3/\sqrt{7} \\ -4/\sqrt{7} \end{bmatrix} \, ,$

$$D_1^{-1} \ = \ \begin{bmatrix} 1 & -1 & 0 \\ -6/7 & 10/7 & 1/\sqrt{7} \\ -1/7 & 4/7 & -1/\sqrt{7} \end{bmatrix} \, , \quad J_1 \ = \ \{ \, 1 \, , \, 5 \, , \, -1 \, \} \, ,$$

$$j \ = \ 1 \, .$$

Iteration 1

Step 1:   Transfer to Step 1.2.

Step 1.2: $g_1' c_{11} \ = \ \max \left\{ \, \frac{6}{7} \, , \, - \, , \, - \, \right\} \ = \ \frac{6}{7} \, , \quad k \ = \ 1 \, ,$

$$s_1 \ = \ \begin{bmatrix} 1 \\ -6/7 \\ -1/7 \end{bmatrix} .$$

Step 2: $s_1' C s_1 = \dfrac{6}{7}$,

$$\tilde{\sigma}_1 = \frac{6/7}{6/7} = 1 \ ,$$

$$\hat{\sigma}_1 = \min \left\{ - \ , \ - \ , \ \frac{-3}{-1} \ , \ \frac{-13/7}{-6/7} \right\} = \frac{13}{6} \ , \ l = 4 \ ,$$

$$\sigma_1 = \min \left\{ 1 \ , \ \frac{13}{6} \right\} = 1 \ .$$

Step 3: $x_2 = \begin{bmatrix} 4 \\ 8/7 \\ 6/7 \end{bmatrix} - \begin{bmatrix} 1 \\ -6/7 \\ -1/7 \end{bmatrix} = \begin{bmatrix} 3 \\ 2 \\ 1 \end{bmatrix}$, $g_2 = \begin{bmatrix} 10 \\ 10 \\ 10 \end{bmatrix}$,

$f(x_2) = 31$. Transfer to Step 3.1.

Step 3.1: $d_1 = \begin{bmatrix} 2\sqrt{7}/7\sqrt{6} \\ -4\sqrt{7}/7\sqrt{6} \\ -4\sqrt{7}/7\sqrt{6} \end{bmatrix}$,

$$D_2^{-1} = \begin{bmatrix} \sqrt{7}/\sqrt{6} & 2/3 & 0 \\ -\sqrt{6}/\sqrt{7} & 0 & 1/\sqrt{7} \\ -1/\sqrt{42} & 1/3 & -1/\sqrt{7} \end{bmatrix} , \ J_2 = \{ -1 \ , \ 5 \ , \ -1 \ \},$$

$j = 2$.

Iteration 2

Step 1: Transfer to Step 1.2.

Step 1.2: There is no i with $1 \leq \alpha_{i2} \leq m$; stop with optimal solution
$x_2 = (3 \ , \ 2 \ , \ 1)'$.

This completes Example 5.4. $\Diamond$

Example 5.4 has been solved using the Matlab program P1P2.m (Figure 5.14). The data setup and summary of results are shown in Figures 5.17 and 5.18. All figures are in Section 5.5

5.3 An Efficient QP Algorithm: Algorithm 4

Each iteration of Algorithm 3 which results in a new active constraint causes the current set of conjugate directions to be destroyed. Subsequent iterations must build a new set of conjugate directions, one per iteration, which are orthogonal to the gradient of the new active constraint. In this section we show how to transform the current set of conjugate directions into a second set of conjugate directions, all but one of which are orthogonal to the gradient of the new active constraint. This capability of "saving" conjugate directions will reduce the number of iterations required to solve a QP.

The situation we are interested in is illustrated in

Example 5.5

$$
\text{minimize:} \quad \tfrac{1}{2}(9x_1^2 + 8x_2^2 + 7x_3^2 + 5x_4^2 + 4x_5^2 + 3x_6^2 + x_7^2 + x_8^2 + x_9^2 + x_{10}^2)
$$

$$
\begin{aligned}
\text{subject to:} \quad & x_1 + x_2 + x_3 + x_4 + x_5 + x_6 \\
& \qquad + x_7 + x_8 + x_9 + x_{10} \le 34, \quad (1) \\
& \qquad\qquad\qquad\qquad\quad - x_{10} \le -3.6, \quad (2) \\
& \qquad\qquad - x_8 \qquad\qquad\qquad \le -6.13, \quad (3) \\
& -x_1 - x_2 - x_3 - x_4 - x_5 - x_6 \\
& \qquad - x_7 - x_8 - x_9 - x_{10} \le -9.8, \quad (4) \\
& \qquad - x_3 - x_4 \qquad\qquad\qquad \le -0.06. \quad (5)
\end{aligned}
$$

This problem is somewhat large for hand calculation. It has been solved using the Matlab computer program Alg3. The input data for it is shown Figure 5.19 with the final optimization summarized in Figure 5.20 (both figures are in Section 5.5). The initial data used was $D_0^{-1} = I$, $J_0 = \{0, 0, \ldots, 0\}$, and $x_0 = (-99, -98, -97, -96, -95, -94, -93, -92, -91, 35)'$.

The key thing to observe from the output file (Figure 5.20) is that Algorithm 3 requires 25 iterations to solve it. This problem has $n = 10$ variables and if it were not for the constraints, the problem would be solved in 10 iterations.

The situation in Example 5.5 may arise in general. Suppose that a 100-dimensional problem begins with an x_0 for which no constraints are active. Suppose that the first 99 iterations produce 99 conjugate directions and at iteration 100, a new constraint becomes active. The 99 conjugate directions are then destroyed by the updating. Suppose that iterations 101 to 198 again construct conjugate directions which are orthogonal to the gradient of the single active constraint and then at iteration 199, a new constraint becomes active. The updating then destroys the 98 conjugate directions. And so on. Suppose in this case that at iteration 100, it were possible to transform the 99 conjugate directions into another set of 99 conjugate directions such that 98 of the new conjugate directions were orthogonal to the gradient of the single active constraint. Then the normal updating with Procedure Φ would leave 98 of the new conjugate directions unchanged and the subsequent 98 conjugate direction constructing iterations would be replaced by a single iteration. We will now discuss in detail how such a transformation may be performed.

Suppose at some iteration of Algorithm 3 that D and D^{-1} have the following form (we suppress subscripts as much as possible);

$$D' = [\, a_1, \ldots, \, a_p \,, \, d_{p+1}, \ldots, d_{p+\nu} \,, \, Cc_{p+\nu+1}, \ldots, Cc_n \,]$$

and

$$D^{-1} = [\, c_1, \ldots, \, c_p \,, \, c_{p+1}, \ldots, c_{p+\nu} \,, \, c_{p+\nu+1}, \ldots, c_n \,] \, .$$

Then constraints $1, \ldots, p$ are active, D^{-1} has ν free columns and $n - p - \nu$ conjugate direction columns. Suppose that x_0 is the current iterate, $s_0 = \pm c_{p+1}$, $\sigma_0 = \hat{\sigma}_0$, and l is the index of the new active constraint. In addition, assume that

$$\left. \begin{array}{ll} a_l'c_{p+\nu+1} & \neq \quad 0 \\ a_l'c_i & = \quad 0, \quad i \; = \; p + \nu + 2, \ldots, n. \end{array} \right\} \tag{5.32}$$

If we set

$$\tilde{D}' = [\, a_1, \ldots, a_p \,, \, d_{p+1}, \ldots, d_{p+\nu} \,, \, a_l \,, \, Cc_{p+\nu+2}, \ldots, Cc_n \,]$$

then because $a_l'c_{p+\nu+1} \neq 0$,

$$\begin{aligned} \tilde{D}^{-1} &= \Phi(D^{-1} \,, \, a_l \,, \, p + \nu + 1) \\ &= [\, \tilde{c}_1, \ldots, \tilde{c}_n \,] \, , \end{aligned}$$

where

$$\tilde{c}_i = c_i - \frac{a_l'c_i}{a_l'c_{p+\nu+1}} c_{p+\nu+1} \,, \quad i = 1, \ldots, n \,, \; i \neq p + \nu + 1 \,,$$

$$\tilde{c}_{p+\nu+1} = \frac{1}{a_l'c_{p+\nu+1}} c_{p+\nu+1} \, .$$

Because we have assumed that (5.32) holds, $\tilde{c}_i = c_i$ for $i = p + \nu + 2, \ldots, n$ so that $\tilde{D}$ and $\tilde{D}^{-1}$ are related as

$$\tilde{D}' = [\, a_1, \ldots, a_p \,, \, d_{p+1}, \ldots, d_{p+\nu} \,, \, a_l \,, \, Cc_{p+\nu+2}, \ldots, Cc_n \,]$$

and

$$\tilde{D}^{-1} = [\, \tilde{c}_1, \ldots, \, \tilde{c}_p \,, \, \tilde{c}_{p+1}, \ldots, \tilde{c}_{p+\nu} \,, \, \tilde{c}_{p+\nu+1} \,, \, c_{p+\nu+2}, \ldots, c_n \,] \, .$$

Furthermore, since $s_0 = \pm c_{p+1}$ it follows by definition of the inverse matrix that

$$s_0'Cc_i = 0 \,, \quad i = p + \nu + 1, \ldots, n \,,$$

so assuming that $g_0' c_{p+\nu+2} = \cdots = g_0' c_n = 0$, it follows from Taylor's series that

$$g_1' c_i = g_0' c_i - \sigma_0 s_0' C c_i = 0 , \quad i = p + \nu + 2, \ldots, n .$$

This means that the last $n - p - \nu - 1$ columns of D^{-1} are unchanged and remain conjugate direction columns in $\tilde{D}^{-1}$.

In order to save all but one conjugate direction we have assumed that (5.32) is satisfied. Of course there is no reason to expect that this will be true in general. However, suppose instead that vectors $\hat{c}_{p+\nu+1}, \ldots, \hat{c}_n$ are available (we'll show how to construct them shortly) having the properties

$$\hat{c}_i' C \hat{c}_j = 0 , \text{ for } p + \nu + 1 \leq i, j \leq n \text{ and } i \neq j, \tag{5.33}$$

$$\hat{c}_i' C \hat{c}_i = 1 , \text{ for } i = p + \nu + 1, \ldots, n, \tag{5.34}$$

$$\text{span } \{ \hat{c}_{p+\nu+1}, \ldots, \hat{c}_n \} = \text{span } \{ c_{p+\nu+1}, \ldots, c_n \}, \tag{5.35}$$

$$a_l' \hat{c}_i = 0 , \text{ for } i = p + \nu + 2, \ldots, n, \tag{5.36}$$

$$a_l' \hat{c}_{p+\nu+1} \neq 0. \tag{5.37}$$

That is, $\hat{c}_{p+\nu+1}, \ldots, \hat{c}_n$ is a set of normalized conjugate directions [(5.33) and (5.34)] which span the same space as $c_{p+\nu+1}, \ldots, c_n$ [(5.35)], the last $n - p - \nu - 1$ of them are orthogonal to a_l [(5.36)], and $\hat{c}_{p+\nu+1}$ is not orthogonal to a_l [(5.37)]. If we then define

$$\hat{D} = [a_1, \ldots, a_p , d_{p+1}, \ldots, d_{p+\nu} , C\hat{c}_{p+\nu+1}, \ldots, C\hat{c}_n] ,$$

it then follows from (5.33), (5.34), and (5.35) that

$$\hat{D}^{-1} = [c_1, \ldots, c_p , c_{p+1}, \ldots, c_{p+\nu} , \hat{c}_{p+\nu+1}, \ldots, \hat{c}_n] .$$

Furthermore, $g_0' c_i = 0$ for $i = p + \nu + 1, \ldots, n$ which, together with (5.35) imply

$$g_0' \hat{c}_i = 0 , \quad i = p + \nu + 1, \ldots, n .$$

Therefore, if $\hat{c}_{p+\nu+1}, \ldots, \hat{c}_n$ were available, we could replace $c_{p+\nu+1}, \ldots, c_n$ with them giving $\hat{D}^{-1}$ and then set

$$D_1^{-1} = \Phi(\hat{D}^{-1} , a_l , p + \nu + 1) .$$

In fact, we can even do better. Since the conjugate direction columns are unchanged by this update, it is computationally more efficient to set

$$D_1^{-1} = \Phi_1(\hat{D}^{-1} , a_l , p + \nu + 1 , J_0) .$$

This brings us to the central problem. Given a_l and a set of normalized conjugate directions $c_{p+\nu+1}, \ldots, c_n$, find $\hat{c}_{p+\nu+1}, \ldots, \hat{c}_n$ satisfying (5.33) to (5.36).

Since this problem is independent of the properties of D and D^{-1}, we reformulate it in the following way. First observe that from (5.35), in order that (5.36) and (5.37) be satisfied it is necessary that

$$a'_l c_i \neq 0 , \quad \text{for at least one } i \text{ with } p + \nu + 1 \leq i \leq n .$$

Now let $p_1, \ldots, p_k$ be given normalized conjugate directions. Then

$$p'_i C p_j = 0 , \quad \text{for } 1 \leq i , j \leq k \text{ and } i \neq j , \quad \text{and}$$
$$p'_i C p_i = 1 , \quad \text{for } i = 1, \ldots, k .$$

In this, $p_1, \ldots, p_k$ play the role of $c_{p+\nu+1}, \ldots, c_n$ above. The normalized conjugate direction conditions can be represented more compactly by the equivalent matrix condition $P'CP = I$, where $P = [\, p_1, \ldots, p_k \,]$ and I denotes the (k , k) identity matrix. Given some n-vector d with $P'd \neq 0$ (previously we used a_l for d) we seek k n-vectors $\hat{p}_1, \ldots, \hat{p}_k$ satisfying $\hat{P}'C\hat{P} = I$ where $\hat{P} = [\, \hat{p}_1, \ldots, \hat{p}_k \,]$, $\hat{P}'d = \theta e_\nu$ where e_ν denotes the νth unit vector and θ is a nonzero scalar, and span $\{\, p_1, \ldots, p_k \,\} = $ span $\{\, \hat{p}_1, \ldots, \hat{p}_k \,\}$. This last condition is equivalent to $\hat{P} = PQ$, where Q is a nonsingular (k , k) matrix.

The critical problem can now be summarized as follows. Given an (n , k) matrix P and an n-vector

$$P'CP = I , \tag{5.38}$$
$$P'd \neq 0 , \tag{5.39}$$

find a (k , k) matrix Q such that

$$Q \text{ is nonsingular} , \tag{5.40}$$

and with

$$\hat{P} = PQ , \tag{5.41}$$

$\hat{P}$ is to satisfy

$$\hat{P}'C\hat{P} = I \tag{5.42}$$

and

$$\hat{P}'d = \theta e_\nu, \tag{5.43}$$

for some scalar $\theta \neq 0$ and some ν with $1 \leq \nu \leq k$. Equation (5.42) is a reformulation of (5.33) and (5.34). Equations (5.40) and (5.41) are equivalent to (5.35). Equation (5.43) is equivalent to (5.36) and (5.37).

Now the critical thing is to find an appropriate matrix Q. Substitution of (5.41) into (5.42) and use of (5.38) gives

$$
\begin{aligned}
\hat{P}'C\hat{P} &= (PQ)'C(PQ) \\
&= Q'P'CPQ \\
&= Q'Q \\
&= I .
\end{aligned}
$$

Given $P'CP = I$, we have shown that

$$\hat{P}'C\hat{P} = I \text{ if and only if } Q'Q = I. \tag{5.44}$$

A matrix Q satisfying $Q'Q = I$ is said to be <u>orthogonal</u>. It is easily shown that an orthogonal matrix is nonsingular and that $Q^{-1} = Q'$ (Exercise 5.4). A particular orthogonal matrix is the <u>Householder matrix</u>

$$Q = I - 2ww' ,$$

where w is any k-vector with $\| w \| = 1.$[2] Equations (5.40), (5.41), and (5.42) are automatically satisfied by any orthogonal matrix. By choosing w in the definition of the Householder matrix in an appropriate way, we will also be able to satisfy (5.43). First we establish the basic properties of a Householder matrix and give a method of constructing one which has certain properties.

Lemma 5.1 *Let u and v be any k-vectors such that $\| u \| = \| v \|$ and $u \neq v$. Define $w = (v - u)/\| v - u \|$ and $Q = I - 2ww'$. Then*

    (a) $Q' = Q$,

    (b) $Q'Q = I$,

    (c) $Qu = v$.

Proof: (a) By definition,

$$Q' = I - 2(ww')' = I - 2ww' = Q .$$

(b) From part (a), Q is symmetric so

$$
\begin{aligned}
Q'Q &= (I - 2ww')(I - 2ww') \\
&= I - 4ww' + 4ww'ww' \\
&= I - 4ww' + 4w(w'w)w' .
\end{aligned}
$$

---

[2]$\| w \|$ means the Euclidean norm of w and is defined by $\| w \| = (w_1^2 + \cdots + w_k^2)^{1/2}$ where $w = (w_1, \ldots, w_k)'$. Note that $\| w \|^2 = w'w$.

By construction, $\| w \| = 1$; that is, $w'w = 1$, from which it follows that $Q'Q = I$.

(c)
$$
\begin{aligned}
Qu &= u - 2ww'u \\
&= u - \frac{2(v - u)(v - u)'u}{\| v - u \|^2} \\
&= u - \frac{2(v - u)'u}{\| v - u \|^2}(v - u).
\end{aligned}
$$

Because $v'v = u'u$, it follows that

$$
\begin{aligned}
\| v - u \|^2 &= v'v - 2u'v + u'u \\
&= -2u'v + 2u'u \\
&= -2(v - u)'u,
\end{aligned}
$$

and therefore,

$$
Qu = u + v - u = v,
$$

as required. $\square$

In the previous lemma, each pair of vectors u and v defines a new Householder matrix. We next show that a particular choice of u and v produces the additional properties that we desire. Let P, $\hat{P}$, d, and ν be as in (5.38) to (5.43) and Q as in Lemma 5.1. Since Q is symmetric, $\hat{P}'d = QP'd$. The requirement (5.43) that $\hat{P}'d = \theta e_\nu$ will then be satisfied provided that

$$
Q(P'd) = \theta e_\nu .
$$

According to Lemma 5.1, this in turn will be satisfied provided that we choose

$$
u = P'd \text{ and } v = \theta e_\nu . \tag{5.45}
$$

Because the lemma requires $\| u \| = \| v \|$, we choose

$$
\theta = \| P'd \| . \tag{5.46}
$$

Choosing u and v according to (5.45) and (5.46) will satisfy the requirements of Lemma 5.1 with the possible exception of one case. Suppose that

$$
\left.\begin{aligned}
p_i'd &= 0 , \text{ for } i = 1,\ldots,k \text{ and } i \neq \nu , \text{ and} \\
p_\nu'd &> 0 .
\end{aligned}\right\} \tag{5.47}
$$

In this case, $v = u$ so that Lemma 5.1 no longer applies. In terms of modifying Algorithm 3, this situation can arise if either there is only one conjugate direction ($k = \nu = 1$) or there are many and all but one of them are orthogonal to the gradient of the new active constraint. In this special case, we could proceed by updating with Procedure Φ_1 and all conjugate directions (if there are any other than p_ν) would be saved. However, we prefer to avoid special cases and to keep the statement of the modified algorithm as simple as possible. Therefore we treat the case of (5.47) as follows. Lemma 5.1 requires only that $u \neq v$ and that $\| u \| = \| v \|$, and this is satisfied by setting $u = P'd$ and $v = -\| u \| e_\nu$. That is, we need only change the sign of v. In general then, we set

$$\mu = \begin{cases} 1 \,, & \text{if } \| u \| e_\nu \neq u \,, \\ -1 \,, & \text{otherwise,} \end{cases}$$

and

$$w = \frac{\mu \| u \| e_\nu - u}{\| \mu \| u \| e_\nu - u \|}$$

and then proceed with $Q = I - 2ww'$ as in Lemma 5.1.

In the context of Lemma 5.1, it was appropriate to use the symbols "u" and "v" to emphasize the critical relationship $Qu = v$. Now however, we are close to interfacing the results of Lemma 5.1 with Algorithm 3 and in the latter context we have used the symbol "u" to denote the dual variables. In order to avoid confusion in what follows we shall replace the symbol "u" of Lemma 5.1 with "y." With this change of notation, the results of the analysis so far are summarized in

Lemma 5.2 *Let* $p_1, \ldots, p_k$ *and* d *be given* n-*vectors and let* $P = [\, p_1, \ldots, p_k \,]$. *Suppose that* $P'CP = I$ *and* $P'd \neq 0$. *Let* ν *be a given index with* $1 \leq \nu \leq k$ *and set*

$$y = (d'p_1, \ldots, d'p_k)' \,,$$
$$\mu = 1 \text{ if } \| y \| e_\nu \neq y \,, \text{ and } -1 \text{ otherwise,}$$
$$w = \frac{\mu \| y \| e_\nu - y}{\| \mu \| y \| e_\nu - y \|} \,,$$
$$Q = I - 2ww' \,, \text{ and}$$
$$[\, \hat{p}_1, \ldots, \hat{p}_k \,] = [\, p_1, \ldots, p_k \,] Q \,.$$

Then

    (a) $\hat{p}_i = p_i - 2w_i p$, *for* $i = 1, \ldots, k$ *with* $p = w_1 p_1 + \cdots + w_k p_k$,

    (b) $\hat{p}_i' C \hat{p}_j = p_i' C p_j$, *for* $1 \leq i, j \leq k$,

    (c) $d' \hat{p}_i = 0$, *for* $i = 1, \ldots, k$, $i \neq \nu$, $d' \hat{p}_\nu \neq 0$,

    (d) *span* $\{ \hat{p}_1, \ldots, \hat{p}_k \} =$ *span* $\{ p_1, \ldots, p_k \}$.

Proof:

(a) $[\hat{p}_1, \ldots, \hat{p}_k] = [p_1, \ldots, p_k] - 2[p_1, \ldots, p_k]ww'$

$\qquad\qquad = [p_1, \ldots, p_k] - 2(w_1 p_1 + \cdots + w_k p_k)w'$

$\qquad\qquad = [p_1, \ldots, p_k] - 2pw'$,

from which part (a) follows.

(b) This follows from (5.44).

(c) This follows from Lemma 5.1(c).

(d) This follows from the fact that Q is nonsingular. $\qquad\square$

Analogous to Procedures Φ and Φ_1, it is helpful to formulate a procedure which performs the updating according to Lemma 5.2. This is done as

Procedure Φ_2.

Let $D' = [d_1, \ldots, d_n]$, $D^{-1} = [c_1, \ldots, c_n]$, and $J = \{\alpha_1, \ldots, \alpha_n\}$ be given with $d_i = Cc_i$ for all i with $\alpha_i = -1$. Let d and ν be a given n-vector and an index, respectively, with $d'c_\nu \neq 0$. By definition of the inverse matrix, $\{c_i \mid \text{all } i \text{ with } \alpha_i = -1\}$ form a set of normalized conjugate directions. Procedure Φ_2 uses a Householder transformation to compute a second set of normalized conjugate directions $\{\hat{c}_i \mid \text{all } i \text{ with } \alpha_i = -1\}$ which span the same space as the first set. The second set has the additional property that $d'c_i = 0$ for all i with $\alpha_i = -1$ and $i \neq \nu$. Let $\hat{D}' = [\hat{d}_1, \ldots, \hat{d}_n]$ and $\hat{D}^{-1} = [\hat{c}_1, \ldots, \hat{c}_n]$. Then $\hat{d}_i = d_i$ for all i with $\alpha_i \neq -1$, $\hat{d}_i = C\hat{c}_i$ for all i with $\alpha_i = -1$, and $\hat{c}_i = c_i$ for all i with $\alpha_i \neq -1$. We use the notation $\Phi_2(D_j^{-1}, a_l, \nu, J_j)$ to denote $\hat{D}^{-1}$. An explicit way to compute the columns of $\hat{D}^{-1}$ is as follows. For $i = 1, \ldots, n$ compute

$$y_i = \begin{cases} d'c_i , & \text{if } \alpha_i = -1 , \\ 0, & \text{otherwise.} \end{cases}$$

Compute

$$\mu = 1 \text{ if } \| y \| e_\nu \neq y , \text{ and } -1 , \text{ otherwise,}$$

$$w = \frac{\mu \| y \| e_\nu - y}{\| \mu \| y \| e_\nu - y \|} ,$$

$$p = \sum_{\alpha_i = -1} w_i c_i ,$$

$$\hat{c}_i = c_i - 2w_i p , \text{ for all } i \text{ with } \alpha_i = -1 ,$$

$$\hat{c}_i = c_i , \text{ for all } i \text{ with } \alpha_i \neq -1 .$$

We illustrate the computations of Procedure Φ_2 in

Example 5.6

Let $n = 4$, $C = I$, $D^{-1} = I$, $J = \{-1, -1, -1, -1\}$, $d = (1, 1, 1, 1)'$, and $\nu = 1$. We wish to replace all four columns of D^{-1} with four new ones which also form a set of normalized conjugate directions for C and for which the last three new columns are also orthogonal to d. Procedure Φ_2 proceeds as follows.

$$y = (1, 1, 1, 1)', \quad \| y \| = 2,$$

$$\mu = 1,$$

$$w = (1/2, -1/2, -1/2, -1/2)',$$

$$p = \frac{1}{2}\begin{bmatrix} 1 \\ 0 \\ 0 \\ 0 \end{bmatrix} - \frac{1}{2}\begin{bmatrix} 0 \\ 1 \\ 0 \\ 0 \end{bmatrix} - \frac{1}{2}\begin{bmatrix} 0 \\ 0 \\ 1 \\ 0 \end{bmatrix} - \frac{1}{2}\begin{bmatrix} 0 \\ 0 \\ 0 \\ 1 \end{bmatrix} = \begin{bmatrix} 1/2 \\ -1/2 \\ -1/2 \\ -1/2 \end{bmatrix},$$

$$\hat{c}_1 = \begin{bmatrix} 1 \\ 0 \\ 0 \\ 0 \end{bmatrix} - 2(\frac{1}{2})\begin{bmatrix} 1/2 \\ -1/2 \\ -1/2 \\ -1/2 \end{bmatrix} = \begin{bmatrix} 1/2 \\ 1/2 \\ 1/2 \\ 1/2 \end{bmatrix},$$

$$\hat{c}_2 = \begin{bmatrix} 0 \\ 1 \\ 0 \\ 0 \end{bmatrix} - 2(\frac{-1}{2})\begin{bmatrix} 1/2 \\ -1/2 \\ -1/2 \\ -1/2 \end{bmatrix} = \begin{bmatrix} 1/2 \\ 1/2 \\ -1/2 \\ -1/2 \end{bmatrix},$$

$$\hat{c}_3 = \begin{bmatrix} 0 \\ 0 \\ 1 \\ 0 \end{bmatrix} - 2(\frac{-1}{2})\begin{bmatrix} 1/2 \\ -1/2 \\ -1/2 \\ -1/2 \end{bmatrix} = \begin{bmatrix} 1/2 \\ -1/2 \\ 1/2 \\ -1/2 \end{bmatrix},$$

$$\hat{c}_4 = \begin{bmatrix} 0 \\ 0 \\ 0 \\ 1 \end{bmatrix} - 2(\frac{-1}{2})\begin{bmatrix} 1/2 \\ -1/2 \\ -1/2 \\ -1/2 \end{bmatrix} = \begin{bmatrix} 1/2 \\ -1/2 \\ -1/2 \\ 1/2 \end{bmatrix},$$

$$\hat{D}^{-1} = \begin{bmatrix} 1/2 & 1/2 & 1/2 & 1/2 \\ 1/2 & 1/2 & -1/2 & -1/2 \\ 1/2 & -1/2 & 1/2 & -1/2 \\ 1/2 & -1/2 & -1/2 & 1/2 \end{bmatrix}.$$

It is straightforward to verify that $\hat{c}_1$, $\hat{c}_2$, $\hat{c}_3$, and $\hat{c}_4$ form a normalized set of conjugate directions for C, that d is orthogonal to each of $\hat{c}_1$, $\hat{c}_2$, $\hat{c}_3$, and $\hat{c}_4$, and that $d'\hat{c}_1 (= 2)$ is nonzero. $\diamondsuit$

A Matlab program which implements Procedure Φ_2 is given in Figure 5.21.

A Matlab program which sets up the data for Example 5.6 is given in Figure 5.22. The output is given in Figure 5.23 and is in agreement with the results of Example 5.6.

We now incorporate Procedure Φ_2 into Algorithm 3. The initialization as well as Steps 1 and 2 remain unchanged. If $\tilde{\sigma}_j \leq \hat{\sigma}_j$ in Step 3, we proceed to update with $d_j = (s_j' C s_j)^{-1/2} C s_j$ as in Algorithm 3 so that Step 3.1 is unchanged. In Algorithm 3, Step 3.2 is used if $\tilde{\sigma}_j > \hat{\sigma}_j$. Now we split Step 3.2 into two parts. If either there are no conjugate directions or there are some and they are all orthogonal to a_l then (in the new Step 3.2) we obtain D_{j+1}' from D_j' by replacing column k with a_l. We could obtain D_{j+1}^{-1} from D_j^{-1} using Procedure Φ. However, since we already know that $a_l' c_{ij} = 0$ for all i with $\alpha_{ij} = -1$, it is computationally more efficient to use Procedure Φ_1. Furthermore, since no conjugate directions are destroyed we obtain J_{j+1} from J_j only by changing α_{kj} to l.

The remaining case is that of at least one conjugate direction which is orthogonal to a_l and this is handled by the new Step 3.3. The first requirement of this step is to determine ν. The index of any conjugate direction column of D_j^{-1} which makes a nonzero inner product with a_l will suffice. However, it seems appropriate to choose the index for which the inner product is largest (in absolute value); that is, we choose the smallest index ν such that

$$| a_l' c_{\nu j} | = \max \{ \, | \, a_l' c_{ij} \, | \, | \, \text{all } i \text{ with } \alpha_{ij} = -1 \, \} \, .$$

Next we apply Procedure Φ_2 to modify the conjugate direction columns giving

$$\hat{D}_{j+1}^{-1} = \Phi_2(D_j^{-1} , a_l , \nu , J_j) \, .$$

The remaining columns are then updated using Φ_1:

$$D_{j+1}^{-1} = \Phi_1(\hat{D}_{j+1}^{-1} , a_l , \nu , J_j) \, .$$

Finally, the new index set is obtained from the old by changing the νth element to l and the kth to zero. Making these changes in Algorithm 3 gives

ALGORITHM   4

Model Problem:

$$
\begin{aligned}
\text{minimize:} \quad & c'x + \tfrac{1}{2} x' C x \\
\text{subject to:} \quad & a_i' x \leq b_i, \quad i = 1, \ldots, m, \\
& a_i' x = b_i, \quad i = m + 1, \ldots, m + r.
\end{aligned}
$$

Initialization:

Start with any feasible point x_0, $J_0 = \{ \alpha_{10}, \ldots, \alpha_{n0} \}$, and D_0^{-1}, where $D_0' = [d_1, \ldots, d_n]$ is nonsingular, $d_i = a_{\alpha_{i0}}$ for all i with $1 \leq \alpha_{i0} \leq m + r$, and, each of $m + 1, \ldots, m + r$ is in J_0. Compute $f(x_0) = c'x_0 + \frac{1}{2}x_0'Cx_0$, $g_0 = c + Cx_0$, and set $j = 0$.

Step 1:   **Computation of Search Direction s_j.**
Same as Step 1 of Algorithm 3.

Step 2:   **Computation of Step Size σ_j .**
Same as Step 2 of Algorithm 3.

Step 3:   **Update.**
Set $x_{j+1} = x_j - \sigma_j s_j$, $g_{j+1} = c + Cx_{j+1}$, and $f(x_{j+1}) = c'x_{j+1} + \frac{1}{2}x_{j+1}'Cx_{j+1}$. If $\tilde{\sigma}_j \leq \hat{\sigma}_j$, go to Step 3.1. Otherwise, go to Step 3.2.

Step 3.1:
Set $d_j = (s_j'Cs_j)^{-1/2}Cs_j$, $D_{j+1}^{-1} = \Phi_1(D_j^{-1} , d_j , k , J_j)$, and $J_{j+1} = \{ \alpha_{1,j+1}, \ldots, \alpha_{n,j+1} \}$, where

$$\alpha_{i,j+1} = \alpha_{ij} , \quad i = 1, \ldots, n , i \neq k ,$$
$$\alpha_{k,j+1} = -1 .$$

Replace j with $j + 1$ and go to Step 1.

Step 3.2:
If there is at least one i with $\alpha_{ij} = -1$ and $a_l'c_{ij} \neq 0$, go to Step 3.3. Otherwise, set $D_{j+1}^{-1} = \Phi_1(D_j^{-1} , a_l , k , J_j)$ and $J_{j+1} = \{ \alpha_{1,j+1}, \ldots, \alpha_{n,j+1} \}$, where
$$\alpha_{i,j+1} = \alpha_{ij} , \quad i = 1, \ldots, n , i \neq k ,$$
$$\alpha_{k,j+1} = l .$$

Replace j with $j + 1$ and go to Step 1.

Step 3.3:
Compute the smallest index ν such that

$$| a_l'c_{\nu j} | = \max \{ | a_l'c_{ij} | | \text{ all } i \text{ with } \alpha_{ij} = -1 \} .$$

Set
$$\hat{D}_{j+1}^{-1} = \Phi_2(D_j^{-1} , a_l , \nu , J_j) ,$$
$$D_{j+1}^{-1} = \Phi_1(\hat{D}_{j+1}^{-1} , a_l , \nu , J_j) ,$$

and $J_{j+1} = \{ \alpha_{1,j+1}, \ldots, \alpha_{n,j+1} \}$, where

$$\alpha_{i,j+1} = \alpha_{ij} , \quad i = 1, \ldots, n , i \neq \nu , i \neq k ,$$
$$\alpha_{\nu,j+1} = l ,$$
$$\alpha_{k,j+1} = 0 .$$

Replace j with $j + 1$ and go to Step 1.

A Matlab program which implements Algorithm 4 is given in Figure 5.24, Section 5.5.

We illustrate Algorithm 4 by applying it to the following example.

Example 5.7

$$\text{minimize:} \quad \tfrac{1}{2}(x_1^2 + x_2^2 + x_3^2)$$
$$\text{subject to:} \quad 3x_1 + 4x_2 + x_3 \leq -1. \quad (1)$$

Here

$$c = \begin{bmatrix} 0 \\ 0 \\ 0 \end{bmatrix} \quad \text{and} \quad C = \begin{bmatrix} 1 & 0 & 0 \\ 0 & 1 & 0 \\ 0 & 0 & 1 \end{bmatrix}.$$

Initialization:

$$x_0 = \begin{bmatrix} -2 \\ -2 \\ -2 \end{bmatrix}, \quad J_0 = \{\, 0\,,\, 0\,,\, 0\, \}, \quad D_0^{-1} = \begin{bmatrix} 1 & 0 & 0 \\ 0 & 1 & 0 \\ 0 & 0 & 1 \end{bmatrix}, \quad f(x_0) = 6,$$

$$g_0 = \begin{bmatrix} -2 \\ -2 \\ -2 \end{bmatrix}, \quad j = 0.$$

Iteration 0

Step 1:   Transfer to Step 1.1.

Step 1.1: $|\, g_0' c_{10}\,| = \max\{\,|-2|\,,\, |-2|\,,\, |-2|\,\} = 2, \quad k = 1,$

$$s_0 = \begin{bmatrix} -1 \\ 0 \\ 0 \end{bmatrix}.$$

Step 2: $s_0' C s_0 = 1,$

$$\tilde{\sigma}_0 = \frac{2}{1} = 2$$

$$\hat{\sigma}_0 = \min\left\{\, \frac{-15}{-3}\, \right\} = 5, \quad l = = 1,$$

$$\sigma_0 = \min\{\, 2\,,\, 5\, \} = 2.$$

Step 3: $x_1 = \begin{bmatrix} 0 \\ -2 \\ -2 \end{bmatrix},$

$f(x_1) = 4$. Transfer to Step 3.1.

Step 3.1: $d_0 = \begin{bmatrix} -1 \\ 0 \\ 0 \end{bmatrix}$,

$$D_1^{-1} = \begin{bmatrix} -1 & 0 & 0 \\ 0 & 1 & 0 \\ 0 & 0 & 1 \end{bmatrix} , \quad J_1 = \{ -1 , 0 , 0 \} , \quad j = 1 .$$

Iteration 1

Step 1:   Transfer to Step 1.1.

Step 1.1: $| g_1' c_{21} | = \max \{ - , | -2 | , | -2 | \} = 2 , \quad k = 2 ,$

$$s_1 = \begin{bmatrix} 0 \\ -1 \\ 0 \end{bmatrix} .$$

Step 2: $s_1' C s_1 = 1$,

$$\tilde{\sigma}_1 = \frac{2}{1} = 2 ,$$

$$\hat{\sigma}_1 = \min \left\{ \frac{-9}{-4} \right\} = \frac{9}{4} , \quad l = 1 ,$$

$$\sigma_1 = \min \left\{ 2 , \frac{9}{4} \right\} = 2 .$$

Step 3: $x_2 = \begin{bmatrix} 0 \\ -2 \\ -2 \end{bmatrix} - 2 \begin{bmatrix} 0 \\ -1 \\ 0 \end{bmatrix} = \begin{bmatrix} 0 \\ 0 \\ -2 \end{bmatrix} , \quad g_2 = \begin{bmatrix} 0 \\ 0 \\ -2 \end{bmatrix} ,$

$f(x_2) = 2$. Transfer to Step 3.1.

Step 3.1: $d_1 = \begin{bmatrix} 0 \\ -1 \\ 0 \end{bmatrix}$,

$$D_2^{-1} = \begin{bmatrix} -1 & 0 & 0 \\ 0 & -1 & 0 \\ 0 & 0 & 1 \end{bmatrix} , \quad J_2 = \{ -1 , -1 , 0 \} , \quad j = 2 .$$

Iteration 2

Step 1:   Transfer to Step 1.1.

Step 1.1: $|g_2' c_{32}| = \max\{-, -, |-2|\} = 2$, $k = 3$,

$$s_2 = \begin{bmatrix} 0 \\ 0 \\ -1 \end{bmatrix}.$$

Step 2: $s_2' C s_2 = 1$,

$$\tilde{\sigma}_2 = \frac{2}{1} = 2,$$

$$\hat{\sigma}_2 = \min\left\{\frac{-1}{-1}\right\} = 1, \quad l = 1,$$

$$\sigma_2 = \min\{2, 1\} = 1.$$

Step 3: $x_3 = \begin{bmatrix} 0 \\ 0 \\ -2 \end{bmatrix} - \begin{bmatrix} 0 \\ 0 \\ -1 \end{bmatrix} = \begin{bmatrix} 0 \\ 0 \\ -1 \end{bmatrix}$, $g_3 = \begin{bmatrix} 0 \\ 0 \\ -1 \end{bmatrix}$,

$$f(x_3) = \frac{1}{2}. \text{ Transfer to Step 3.2.}$$

Step 3.2: Transfer to Step 3.3.

Step 3.3: $|a_1' c_{22}| = \max\{|-3|, |-4|, -\} = 4$, $\nu = 2$,

$$y = \begin{bmatrix} -3 \\ -4 \\ 0 \end{bmatrix}, \quad \|y\| = 5,$$

$$\mu = 1,$$

$$w = \begin{bmatrix} 1/\sqrt{10} \\ 3/\sqrt{10} \\ 0 \end{bmatrix}, \quad p = \begin{bmatrix} -1/\sqrt{10} \\ -3/\sqrt{10} \\ 0 \end{bmatrix},$$

$$\hat{D}_3^{-1} = \begin{bmatrix} -4/5 & 3/5 & 0 \\ 3/5 & 4/5 & 0 \\ 0 & 0 & 1 \end{bmatrix},$$

$$D_3^{-1} = \begin{bmatrix} -4/5 & 3/25 & -3/25 \\ 3/5 & 4/25 & -4/25 \\ 0 & 0 & 1 \end{bmatrix},$$

$$J_3 = \{-1, 1, 0\}, \quad j = 3.$$

Iteration 3

Step 1:   Transfer to Step 1.1.

Step 1.1: $| g_3' c_{33} | = \max \{ - , - , | -1 | \} = 1 , \quad k = 3 ,$

$$s_3 = \begin{bmatrix} 3/25 \\ 4/25 \\ -1 \end{bmatrix} .$$

Step 2: $s_3' C s_3 = \dfrac{26}{25} ,$

$$\tilde{\sigma}_3 = \frac{1}{26/25} = \frac{25}{26} ,$$

$$\hat{\sigma}_3 = +\infty ,$$

$$\sigma_3 = \min \left\{ \frac{25}{26} , +\infty \right\} = \frac{25}{26} .$$

Step 3: $x_4 = \begin{bmatrix} 0 \\ 0 \\ -1 \end{bmatrix} - \dfrac{25}{26} \begin{bmatrix} 3/25 \\ 4/25 \\ -1 \end{bmatrix} = \begin{bmatrix} -3/26 \\ -2/13 \\ -1/26 \end{bmatrix} ,$

$$g_4 = \begin{bmatrix} -3/26 \\ -2/13 \\ -1/26 \end{bmatrix} ,$$

$f(x_4) = \dfrac{1}{52} .$ Transfer to Step 3.1.

Step 3.1: $d_3 = \begin{bmatrix} 3/5\sqrt{26} \\ 4/5\sqrt{26} \\ -5/\sqrt{26} \end{bmatrix} ,$

$$D_4^{-1} = \begin{bmatrix} -4/5 & 3/26 & 3\sqrt{26} \\ 3/5 & 2/13 & 2\sqrt{2}/5\sqrt{13} \\ 0 & 1/26 & 5/\sqrt{26} \end{bmatrix} ,$$

$J_4 = \{ -1 , 1 , -1 \} , \quad j = 4 .$

Iteration 4

Step 1:   Transfer to Step 1.2.

Step 1.2: $g_4' c_{24} = \max \left\{ - , \dfrac{-1}{26} , - \right\} = \dfrac{-1}{26} ,$

$k = 2 .$

$g_4' c_{24} \leq 0$; stop with optimal solution $x_4 =$ $(-3/26 , -2/13 , -1/26)'$.

A Matlab program, Alg4.m, which implements Algorithm 4 is given in Figure 5.24, Section 5.5. A Matlab program which sets up the data for Example 5.7 and the results of applying Alg4.m to it are shown in Figures 5.25 and 5.26, respectively, Section 5.5.

In Example 5.7, Procedure Φ_2 is used just once; at iteration 2 and in order to save a single conjugate direction. The example provides a good numerical illustration of the computations of the algorithm. It does not, however, show the power of Procedure Φ_2. We further illustrate Algorithm 4 in

Example 5.8

Here, we use Algorithm 4 to solve the problem of Example 5.5. We have previously used Algorithm 3 to solve this same problem with the data formulation shown in Figures 5.19 and 5.20. The analogous files for Algorithm 4 are shown in Figures 5.27 and 5.28. Both algorithms arrive at the same primal and dual solutions, as expected. However, there is one important difference between the two algorithms. Algorithm 3 takes 25 iterations to solve the problem whereas Algorithm 4 takes just 16. The reason for this is as follows. Both algorithms start with no active constraints and the first 9 iterations of the two algorithms are identical. At the next iteration, a previously inactive constraint becomes active. For Algorithm 3, this destroys all of the existing 9 conjugate directions and subsequent iterations are needed to rebuild these. By way of contrast, Algorithm 4 loses just one conjugate direction (which is replaced by the gradient of the new active constraint). Thus, Algorithm 4 is to be preferred over Algorithm 3 in terms of speed of computation and this may be important for large problems.

In order to establish the termination properties of Algorithm 4, we observe that Algorithm 4 differs from Algorithm 3 only in the updating of D_j^{-1} when $\hat{\sigma}_j < \tilde{\sigma}_j$. Suppose that $\hat{\sigma}_j < \tilde{\sigma}_j$. Let $D_j^{-1} = [\, c_{1j}, \ldots, c_{nj} \,]$, $J_j = \{\, \alpha_{1j}, \ldots, \alpha_{nj} \,\}$, and $D_j' = [\, d_{1j}, \ldots, d_{nj} \,]$. Given that

$$d_{ij} = C c_{ij} , \quad \text{for all } i \text{ with } \alpha_{ij} = -1 ,$$

it follows from Lemma 5.2 and Procedure Φ_2 that

$$d_{i,j+1} = C c_{i,j+1} , \quad \text{for all } i \text{ with } \alpha_{i,j+1} = -1 , \tag{5.48}$$

where $D_{j+1}^{-1} = [\, c_{1,j+1}, \dots, c_{n,j+1} \,]$, $J_{j+1} = \{\, \alpha_{1,j+1}, \dots, \alpha_{n,j+1} \,\}$, and $D'_{j+1} = [\, d_{1,j+1}, \dots, d_{n,j+1} \,]$. In addition, suppose that

$$g'_j c_{ij} = 0, \quad \text{for all } i \text{ with } \alpha_{ij} = -1. \tag{5.49}$$

We now show that (5.49) also holds at the next iteration. No matter whether s_j is chosen from Step 1.1 or 1.2, it is true that s_j is parallel to some c_{kj} with $\alpha_{kj} \neq -1$. By definition of the inverse matrix,

$$s'_j C c_{ij} = 0, \quad \text{for all } i \text{ with } \alpha_{ij} = -1. \tag{5.50}$$

From Taylor's series, (5.49), and (5.50) we have

$$\begin{aligned} g'_{j+1} c_{ij} &= g'_j c_{ij} - \sigma_j s'_j C c_{ij} \\ &= 0, \quad \text{for all } i \text{ with } \alpha_{ij} = -1. \end{aligned} \tag{5.51}$$

Letting $\hat{D}_{j+1}^{-1} - [\, \hat{c}_{1,j+1}, \dots, \hat{c}_{n,j\,|\,1} \,]$, it follows from Step 3.3 of Algorithm 4 and Lemma 5.2(d) that

$$\text{span } \{\, \hat{c}_{i,j+1} \mid \text{all } i \text{ with } \alpha_{ij} = -1 \,\} = \text{span } \{\, c_{i,j+1} \mid \text{all } i \text{ with } \alpha_{ij} = -1 \,\}$$

and thus from (5.51) that

$$g'_{j+1} \hat{c}_{ij} = 0, \quad \text{for all } i \text{ with } \alpha_{ij} = -1.$$

But $c_{i,j+1} = \hat{c}_{i,j+1}$ for all i with $\alpha_{ij} = -1$ and $i \neq \nu$. Since $\alpha_{\nu,j+1} \neq -1$, we have shown that

$$g'_{j+1} c_{i,j+1} = 0, \quad \text{for all } i \text{ with } \alpha_{i,j+1} = -1. \tag{5.52}$$

Algorithms 3 and 4 differ only in the updating in the event of a new active constraint. Algorithm 4 saves all but one conjugate direction and after the updating, the critical properties (5.48) and (5.52) are satisfied. It therefore follows that the termination properties of Algorithm 3 formulated in Theorem 5.1 are inherited by Algorithm 4. We have established

Theorem 5.2 *Let Algorithm 4 be applied to the model problem*

$$\min \{\, c'x + \tfrac{1}{2} x'Cx \mid A_1 x \leq b_1, \quad A_2 x = b_2 \,\}$$

beginning with an arbitrary feasible point x_0 and let x_1, $x_2, \dots, x_j$, $\dots$ be the sequence of iterates so obtained. If each quasistationary point obtained by Algorithm 4 is nondegenerate, then Algorithm 4 terminates after a finite number of steps with either an optimal solution x_j or the information that the problem is unbounded from below. In the former case, the multipliers for the active constraints are given by $u_{\alpha_{ij}} = -g'_j c_{ij}$ for all i with $1 \leq \alpha_{ij} \leq m+r$, where $D_j^{-1} = [\, c_{1j}, \dots, c_{nj} \,]$ and $J_j = \{\, \alpha_{1j}, \dots, \alpha_{nj} \,\}$, and the multipliers for the remaining constraints have value zero.

In Section 5.4, it will be shown how to deal with degenerate quasistationary points and thus the assumption required by Theorem 5.2 of "If each quasistationary point obtained by Algorithm 4 is nondegenerate" may be removed and the remainder of Theorem 5.2 holds.

Algorithm 4 is computationally superior to Algorithm 3 and therefore is more suitable to solving practical problems. In particular, in the Phase 1–Phase 2 procedure for solving a QP, Algorithm 4 should be used for the Phase 2 problem.

5.4   Degeneracy and Its Resolution

The assumption of nondegenerate quasistationary points is required in Theorem 5.1 to guarantee finite termination. However, if this assumption is not satisfied, Algorithm 3 may still find an optimal solution in finitely many steps.

Suppose that x_j is a degenerate quasistationary point determined by Algorithm 3. Suppose that $x_j = x_{j+1} = \cdots = x_{j+k}$ and that for each of these k iterations the gradient of the new active constraint is linearly dependent on the gradients of those constraints in the active set. Suppose for simplicity that

$$D'_j = [\, a_1, \ldots, a_\rho \,, \, d_{\rho+1,j}, \ldots, d_{nj} \,] \, ,$$

that constraint ρ is dropped, and that constraint $\rho + 1$ is added during iteration j. Because $a_{\rho+1}$ is linearly dependent on $a_1, \ldots, a_\rho$, the updating of D_j^{-1} via Step 3.2 leaves the last $n - \rho$ columns of D_{j+1}^{-1} identical to those of D_j^{-1}. The last $n - \rho$ elements of J_j may include some -1's and these are changed to 0's in J_{j+1}. However, $g_{j+1} = g_j$ is still orthogonal to the last $n - \rho$ columns of D_{j+1}^{-1} as well as those of D_j^{-1}. Consequently, s_{j+1} is chosen via Step 1.2. Because of the assumed linear dependence, only the first ρ components of subsequent index sets are changed. The last $n - \rho$ elements of each J_{j+i} will all have value zero and the last $n - \rho$ columns of D_{j+i}^{-1} will all be identical to those of D_j^{-1}. It is theoretically possible to have $J_{j+k} = J_{j+1}$. This means that the data $x_{j+k} = x_{j+1}$, $J_{j+k} = J_{j+1}$, and $D_{j+k} = D_{j+1}$ is identical for iterations $j + 1$ and $j + k$. This *cycle* of k iterations would then repeat indefinitely.

In practice, however, cycling has hardly ever been observed. Due to rounding errors, the data used on a computer after k iterations is numerically different even though theoretically identical with the previous data.

The notion of cycling for quadratic programming is very similar to cycling for linear programming. Linear programming algorithms typically work with extreme points and at least n constraints are active at these points. Quadratic programming algorithms work with quasistationary points and these are associated with any number of active constraints. In either case, the possibility of cycling is due to the gradients of active constraints being linearly dependent.

As with linear programming, it is common to state quadratic programming algorithms without provision to avoid cycling. For theoretical arguments, however, especially for proving finite termination, it is important to deal with this problem. We do this in detail in this section.

Without the use of antidegeneracy rules, Algorithm 3 may still terminate after finitely many steps in the presence of degenerate quasistationary points as illustrated in

Example 5.10

$$\text{minimize:} \quad -8x_1 - 6x_2 + x_1^2 + x_2^2$$

$$
\begin{array}{rrcll}
\text{subject to:} & 4x_1 - 3x_2 & \leq & 5, & (1) \\
& 2x_1 - x_2 & \leq & 3, & (2) \\
& 4x_1 - x_2 & \leq & 7, & (3) \\
& -x_1 & \leq & -2, & (4) \\
& x_1 & \leq & 2, & (5) \\
& -x_2 & \leq & -1, & (6) \\
& x_2 & \leq & 5, & (7) \\
& -x_1 & \leq & -1. & (8)
\end{array}
$$

Here

$$c = \begin{bmatrix} -8 \\ -6 \end{bmatrix} \quad \text{and} \quad C = \begin{bmatrix} 2 & 0 \\ 0 & 2 \end{bmatrix}.$$

Initialization:

$$x_0 = \begin{bmatrix} 2 \\ 1 \end{bmatrix}, \quad J_0 = \{0, 0\}, \quad D_0^{-1} = \begin{bmatrix} 1 & 0 \\ 0 & 1 \end{bmatrix}, \quad f(x_0) = -17,$$

$$g_0 = \begin{bmatrix} -4 \\ -4 \end{bmatrix}, \quad j = 0.$$

Iteration 0

Step 1:   Transfer to Step 1.1.

Step 1.1: $|g_0' c_{10}| = \max\{|-4|, |-4|\} = 4, \quad k = 1,$

$$s_0 = \begin{bmatrix} -1 \\ 0 \end{bmatrix}.$$

Step 2: $s_0' C s_0 = 2,$

$$\tilde{\sigma}_0 = \frac{4}{2} = 2,$$

$$\hat{\sigma}_0 = \min\left\{\frac{0}{-4}, \frac{0}{-2}, \frac{0}{-4}, -, \frac{0}{-1}, -, -, -\right\} = 0,$$

$$l = 1,$$

$$\sigma_0 = \min\{2, 0\} = 0.$$

Step 3: $\quad x_1 = \begin{bmatrix} 2 \\ 1 \end{bmatrix}, \quad g_1 = \begin{bmatrix} -4 \\ -4 \end{bmatrix}, \quad f(x_1) = -17.$

Transfer to Step 3.2.

Step 3.2: $\quad D_1^{-1} = \begin{bmatrix} 1/4 & 3/4 \\ 0 & 1 \end{bmatrix}, \quad J_1 = \{1, 0\}, \quad j = 1.$

Iteration 1

Step 1: Transfer to Step 1.1.

Step 1.1: $\quad |g_1'c_{21}| = \max\{-, |-7|\} = 7, \quad k = 2,$

$$s_1 = \begin{bmatrix} -3/4 \\ -1 \end{bmatrix}.$$

Step 2: $\quad s_1'Cs_1 = \dfrac{25}{8},$

$$\tilde{\sigma}_1 = \frac{7}{25/8} = \frac{56}{25},$$

$$\hat{\sigma}_1 = \min\left\{-, \frac{0}{-1/2}, \frac{0}{-2}, -, \frac{0}{-3/4}, -, \frac{-4}{-1}, -\right\} = 0,$$

$$l = 2,$$

$$\sigma_1 = \min\left\{\frac{56}{25}, 0\right\} = 0.$$

Step 3: $\quad x_2 = \begin{bmatrix} 2 \\ 1 \end{bmatrix}, \quad g_2 = \begin{bmatrix} -4 \\ -4 \end{bmatrix}, \quad f(x_2) = -17.$

Transfer to Step 3.2.

Step 3.2: $\quad D_2^{-1} = \begin{bmatrix} -1/2 & 3/2 \\ -1 & 2 \end{bmatrix}, \quad J_2 = \{1, 2\}, \quad j = 2.$

Iteration 2

Step 1: Transfer to Step 1.2.

Step 1.2: $g_2' c_{12} = \max\{6, -14\} = 6$, $k = 1$, $s_2 = \begin{bmatrix} -1/2 \\ -1 \end{bmatrix}$.

Step 2: $s_2' C s_2 = \dfrac{5}{2}$,

$$\tilde{\sigma}_2 = \frac{6}{5/2} = \frac{12}{5},$$

$$\hat{\sigma}_2 = \min\left\{-, -, \frac{0}{-1}, -, \frac{0}{-1/2}, -, \frac{-4}{-1}, -\right\} = 0,$$

$$l = 3,$$

$$\sigma_2 = \min\left\{\frac{12}{5}, 0\right\} = 0.$$

Step 3: $x_3 = \begin{bmatrix} 2 \\ 1 \end{bmatrix}$, $g_3 = \begin{bmatrix} -4 \\ -4 \end{bmatrix}$, $f(x_3) = -17$.

Transfer to Step 3.2.

Step 3.2: $D_3^{-1} = \begin{bmatrix} 1/2 & -1/2 \\ 1 & -2 \end{bmatrix}$, $J_3 = \{3, 2\}$, $j = 3$.

Iteration 3

Step 1: Transfer to Step 1.2.

Step 1.2: $g_3' c_{23} = \max\{-6, 10\} = 10$, $k = 2$, $s_3 = \begin{bmatrix} -1/2 \\ -2 \end{bmatrix}$.

Step 2: $s_3' C s_3 = \dfrac{17}{2}$,

$$\tilde{\sigma}_3 = \frac{10}{17/2} = \frac{20}{17},$$

$$\hat{\sigma}_3 = \min\left\{-, -, -, -, \frac{0}{-1/2}, -, \frac{-4}{-2}, -\right\} = 0,$$

$$l = 5,$$

$$\sigma_3 = \min\left\{\frac{20}{17}, 0\right\} = 0.$$

Step 3: $x_4 = \begin{bmatrix} 2 \\ 1 \end{bmatrix}$, $g_4 = \begin{bmatrix} -4 \\ -4 \end{bmatrix}$, $f(x_4) = -17$.

Transfer to Step 3.2.

Step 3.2: $D_4^{-1} = \begin{bmatrix} 0 & 1 \\ -1 & 4 \end{bmatrix}$, $J_4 = \{\, 3 \,,\, 5 \,\}$, $j = 4$.

Iteration 4

Step 1: Transfer to Step 1.2.

Step 1.2: $g_4' c_{14} = \max \{\, 4 \,,\, -20 \,\} = 4$, $k = 1$, $s_4 = \begin{bmatrix} 0 \\ -1 \end{bmatrix}$.

Step 2: $s_4' C s_4 = 2$,

$$\tilde{\sigma}_4 = \frac{4}{2} = 2 ,$$

$$\hat{\sigma}_4 = \min \left\{\, - \,,\, - \,,\, - \,,\, - \,,\, - \,,\, - \,,\, \frac{-4}{-1} \,,\, - \,\right\} = 4 ,$$

$$l = 7 ,$$

$$\sigma_4 = \min \{\, 2 \,,\, 4 \,\} = 2 .$$

Step 3: $x_5 = \begin{bmatrix} 2 \\ 1 \end{bmatrix} - 2 \begin{bmatrix} 0 \\ -1 \end{bmatrix} = \begin{bmatrix} 2 \\ 3 \end{bmatrix}$, $g_5 = \begin{bmatrix} -4 \\ 0 \end{bmatrix}$,

$f(x_5) = -21$. Transfer to Step 3.1.

Step 3.1: $d_4 = \begin{bmatrix} 0 \\ -2/\sqrt{2} \end{bmatrix}$,

$$D_5^{-1} = \begin{bmatrix} 0 & 1 \\ -1/\sqrt{2} & 0 \end{bmatrix} , \quad J_5 = \{\, -1 \,,\, 5 \,\} , \quad j = 5 .$$

Iteration 5

Step 1: Transfer to Step 1.2.

Step 1.2: $g_5' c_{25} = \max \{\, - \,,\, -4 \,\} = -4$, $k = 2$.

$g_5' c_{25} \leq 0$; stop with optimal solution $x_5 = (2 \,,\, 3)'$.

This completes Example 5.10.

In Example 5.10 constraints (4) and (5) imply that $x_1 = 2$. Thus the feasible region is just the line segment joining $(2 \,,\, 1)'$ and $(2 \,,\, 5)'$ (see Figure 5.2). Algorithm 3 begins at $x_0 = (2 \,,\, 1)'$ with no constraints in the active set.

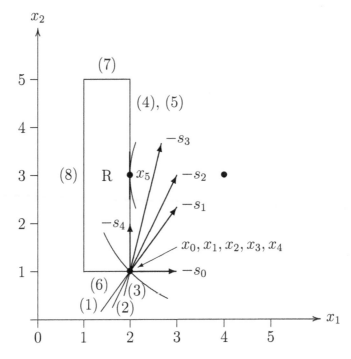

Figure 5.2.   Example 5.10.

Constraints (1), (2), (3), (4), (5), and (6) are active at x_0 but none of these are in the active set. x_0 is an extreme point. Since 6 constraints are active at it, x_0 is a degenerate extreme point and therefore a degenerate quasistationary point. Algorithm 3 must first discover that x_0 is a quasistationary point. It takes two iterations to do this. Since Algorithm 3 does not yet know that x_0 is a quasistationary point, s_0 is computed using Step 1.1. In the calculation of the maximum feasible step size for s_0, constraints (1), (2), (3), and (5) are all tied for the minimum of 0 and the smallest index rule gives $l = 1$. Constraint (1) is added to the active set.

   Algorithm 3 still does not yet know that x_0 is a quasistationary point so s_1 is computed from Step 1.1. In the computation for $\hat{\sigma}_1$, constraints (2), (3), and (5) are tied for the minimum of 0 and the smallest index rule gives $l = 2$. At the beginning of iteration 2, constraints (1) and (2) are in the active set. Algorithm 3 now recognizes that x_0 $(= x_2)$ is a quasistationary point and proceeds with the optimality test in Step 1.2. The test fails and the algorithm proceeds by constructing s_2 to drop constraint (1). Constraints (3) and (5) are tied for the minimum of 0 in the computation of $\hat{\sigma}_2$ and the smallest index rule gives $l = 3$. The net effect of iteration 2 is to replace constraint (1) with constraint (3) in the active set. Similarly, the net effect of iteration 3 is to replace constraint (2) with constraint (5). Finally, at iteration 4 a positive step is taken. The objective function is strictly reduced and indeed, the optimal solution is obtained. This is recognized when the optimality test is performed in Step 1.2 at iteration 5 and Algorithm 3 terminates with the optimal solution.

Example 5.10 illustrates two points. First, Algorithm 3 may spend several iterations at the same point, increasing the number of constraints in the active set until it has been ascertained that the point is in fact a quasistationary point. This is the case for the first two iterations of Example 5.10. The second point is that finite termination may occur even if a degenerate quasistationary point is encountered. Note also that the optimal solution is a degenerate quasistationary point.

All algorithms for the solution of a linear programming problem presented thus far require a nondegeneracy assumption in order to guarantee finite termination. The purpose of the following is to present a modification of the algorithms for which finite termination is guaranteed with no assumptions.

We next give a detailed description of the implementation of Bland's rules [7] which ensures finite termination in the presence of active constraints having linearly dependent gradients. We first formulate Bland's rules for the model problem

$$
\left.
\begin{aligned}
\text{minimize:} \quad & c'x \\
\text{subject to:} \quad & a_i'x \leq b_i, \quad i = 1, \ldots, m, \\
& a_i'x = b_i, \quad i = m+1, \ldots, m+r,
\end{aligned}
\right\} \tag{5.53}
$$

and then generalize these results for quadratic programming algorithms.

The possibility of cycling occurs only if, at some iteration j, $\sigma_j = 0$, and s_j is determined by Step 1.2 of LP Algorithm 3. In this case s_j is rejected. A new search direction is computed by Bland's rules (i.e., $s_j = c_{kj}$), where k is such that

$$
\alpha_{kj} = \min \left\{ \alpha_{ij} \mid \text{all } i \text{ with } c'c_{ij} > 0 \text{ and } 1 \leq \alpha_{ij} \leq m \right\}.
$$

The second part of Bland's rule requires that in the computation of the maximum feasible step size, when there are ties in the computation of it, the smallest index of the restricting constraint be used. In each of Algorithm 3, LP Algorithm 3 and Algorithm 4, this rule is already used.

We refer to the modified algorithm as LP Algorithm 3(NC) because it is based on LP Algorithm 3 and has the no cycling (NC) adaptation of Bland's rules. LP Algorithm 3(NC) can be stated as follows

LP ALGORITHM 3(NC)

Model Problem:
Equation (5.53).

Initialization:
Start with any feasible point x_0, $J_0 = \{ \alpha_{10}, \ldots, \alpha_{n0} \}$, and D_0^{-1}, where $D_0' = [\, d_1, \ldots, d_n \,]$ is nonsingular, $d_i = a_{\alpha_{i0}}$ for all i with $1 \leq \alpha_{i0} \leq m + r$, and, each of $m + 1, \ldots, m + r$ is in J_0. Compute $c'x_0$ and set $j = 0$.

Step 1:   **Computation of Search Direction s_j.**
Same as Step 1 of LP Algorithm 3.

Step 1.1:
Set $\gamma_j = 1$ and continue as in Step 1.1 of Algorithm 3.

Step 1.2:
Set $\gamma_j = 2$ and continue as in Step 1.2 of LP Algorithm 3.

Step 1.3:
Set $\gamma_j = 3$ and compute

$$v_i = c'c_{ij}, \quad \text{for all } i \text{ with } 1 \leq \alpha_{ij} \leq m.$$

If $v_i \leq 0$ for all i with $1 \leq \alpha_{ij} \leq m$, stop with optimal solution x_j. Otherwise, determine k such that

$$\alpha_{kj} = \min \{\, \alpha_{ij} \mid \text{all } i \text{ with } v_i > 0 \text{ and } 1 \leq \alpha_{ij} \leq m \,\}.$$

Set $s_j = c_{kj}$ and go to Step 2.

Step 2:   **Computation of Step Size σ_j.**
Compute σ_j as in Step 2 of LP Algorithm 3.
If $\sigma_j = 0$ and $\gamma_j = 2$, then go to Step 1.3. Otherwise, go to Step 3.

Step 3:   **Update.**
Same as Step 3 of LP Algorithm 3.

The following theorem establishes the main properties of LP Algorithm 3(NC).

Theorem 5.3 *Let LP Algorithm 3(NC) be applied to the model problem (5.53). Then LP Algorithm 3(NC) terminates after a finite number of steps with either an optimal solution or the information that the problem is unbounded from below.*

A proof for Theorem 5.4 is given in Best and Ritter [6].
We next turn to the problem of resolving degeneracy for a QP algorithm. Specifically, we consider the problem of modifying Algorithm 3 so that cycling is guaranteed not to occur. We use the model problem

$$\left.\begin{array}{rl} \text{minimize}: & c'x + \frac{1}{2}x'Cx \\ \text{subject to}: & a_i'x \leq b_i, \quad i = 1, \ldots, m, \\ & a_i'x = b_i, \quad i = m + 1, \ldots, m + r. \end{array}\right\} \qquad (5.54)$$

We first give a detailed statement of the modified QP Algorithm 3 which precludes the possibility of cycling and then establish the anti cycling property. Analogous to the naming of LP Algorithm 3(NC), we name the anti cycling version of Algorithm 3 to be Algorithm 3(NC).

ALGORITHM 3(NC)

Model Problem:
Equation (5.54).

Initialization:
Start with any feasible point x_0, $J_0 = \{\alpha_{10}, \ldots, \alpha_{n0}\}$, and D_0^{-1}, where $D_0' = [d_1, \ldots, d_n]$ is nonsingular, $d_i = a_{\alpha_{i0}}$ for all i with $1 \leq \alpha_{i0} \leq m + r$, and, each of $m + 1, \ldots, m + r$ is in J_0. Compute $f(x_0) = c'x_0 + \frac{1}{2}x_0'Cx_0$ and set $j = 0$.

Step 1: **Computation of Search Direction s_j.**
Same as Step 1 of Algorithm 3.

Step 1.1:
Set $\gamma_j = 1$ and continue as in Step 1.1 of Algorithm 3.

Step 1.2:
Set $\gamma_j = 2$ and continue as in Step 1.2 of Algorithm 3.

Step 1.3:
Set $\gamma_j = 3$ and compute

$$v_i = c'c_{ij}, \quad \text{for all } i \text{ with } 1 \leq \alpha_{ij} \leq m.$$

If $v_i \leq 0$ for all i with $1 \leq \alpha_{ij} \leq m$, stop with optimal solution x_j. Otherwise, determine k such that

$$\alpha_{kj} = \min\{\alpha_{ij} \mid \text{all } i \text{ with } v_i > 0 \text{ and } 1 \leq \alpha_{ij} \leq m\}.$$

Set $s_j = c_{kj}$ and go to Step 2.

Step 2: **Computation of Step Size σ_j.**
Compute σ_j as in Step 2 of Algorithm 3.
If $\sigma_j = 0$ and $\gamma_j = 2$, then go to Step 1.3. Otherwise, go to Step 3.

Step 3: **Update.**
Same as Step 3 of Algorithm 3.

The following theorem establishes the main properties of Algorithm 3(NC).

Theorem 5.4 *Let Algorithm 3(NC) be applied to the model problem (5.54).
Then Algorithm 3(NC) terminates after a finite number of steps with either an
optimal solution or the information that the problem is unbounded from below.*

Proof: Let j be an iteration for which Bland's rules are first invoked and
consider the LP

$$\left. \begin{array}{rl} \text{minimize:} & g_j'x \\ \text{subject to:} & a_i'x \leq b_i, \quad i = 1,\ldots,m, \\ & a_i'x = b_i, \quad i = m+1,\ldots,m+r, \end{array} \right\}, \qquad (5.55)$$

where $g_j = \nabla f(x_j)$. Let LP Algorithm 3(NC) be initiated with the same active J_j as Algorithm 3(NC). Then by construction, both LP Algorithm 3(NC)
and Algorithm 3(NC) will produce identical iterations (determined by Bland's
rules). Therefore, by Theorem 5.4 in a finite number of steps, Algorithm 3(NC)
will determine that either (5.54) is unbounded from below, x_j is optimal for
(5.54) or a point will be determined with an objective function value strictly
smaller than $f(x_j)$. This completes the proof. $\qquad \Box$

It is straightforward to modify Algorithm 4 to preclude the possibility of
cycling. Replace every instance of "Algorithm 3" in Algorithm 3(NC) with
"Algorithm 4" and call the result Algorithm 4(NC).

5.5   Computer Programs

In this section, we formulate Matlab programs for each of the algorithms developed in Chapter 5. In addition, we show the results of applying each algorithm
to the examples used to illustrate each algorithm. In some cases, we develop
some Matlab programs which facilitate the presentation of these results.

```
1  %checkconstraints.m
2  function checkconstraints(A,b,x0,m,r,n)
3  feas = 1;
4  tol = 1.e-8;
5  for i = 1:m
6      sum = 0.;
7      for j = 1:n
8          sum = sum + A(i,j)*x0(j);
9      end
10     if sum - b(i) > tol
11         feas = 0;
12         i
13         msg = 'above constraint is infeasible'
```

```
14        end
15  end
16
17  if r > 0
18       for i = m+1:m+r
19            sum = 0.;
20            for j = 1:n
21                 sum = sum + A(i,j)*x0(j);
22            end
23            if abs(sum-b(i)) > tol
24                 feas = 0;
25                 i
26                 msg = 'above constraint is infeasible'
27            end
28       end
29  end
30  if feas == 0
31       errrmsg = 'Error: initial point is not feasible'
32       soln = 'Press Ctrl+C to continue'
33       pause
34       return
35  end
```

Figure 5.3.   checkconstraints: checkconstraints.m.

Discussion of checkconstraints.m. In Algorithms 3 and 4, an initial feasible point, x_0, is required to start the algorithm. It could happen that an incorrect point is entered. If x_0 violates one or more constraints, the iterations of the algorithm may produce complete nonsense leaving the user wondering what went wrong. This situation can be avoided by running a feasibility check on the initial data. This is precisely what is done by checkconstraints.m. See Figure 5.3.

```
1   %  Alg3.m
2   function [x,J,fx,Dinv,msg,dual,j,alg] = Alg3   ...
3        (c,C,n,m,r,tol,x,J,Dinv,A,b)
4   %  Alg3 implements QP Algorithn 3 to solve a model
5   %  Quadratic Programming Problem.
6
7   %  Initialization
8   alg = 'Algorithm 3';
9   msg = '';
10  dual = [];
11  fx = c'*x + 0.5*( x'*C*x);
12  g = c + C*x;
13  next = 'Step1';
14  for j=0:10000;
15       if strcmp(next,'Step1')
16            [next,k,s,msg,dual,big] = Step1(J,next,g,Dinv,n,m,r,tol);
```

```
17          if strcmp(msg,'Optimal solution obtained')
18              return
19          end
20
21      end
22
23      if strcmp(next,'Step2')
24          [next,sigma,sigtil,sighat,ell,stCs,msg] = ...
25                      Step2(J,m,A,b,C,s,tol,big,x);
26          if strcmp(msg,'Unbounded from below');
27              return
28          end
29      end
30
31      if strcmp(next,'Step3')
32                                              %  Step 3: Update
33          x = x - sigma*s;
34          g = c + C*x;
35          fx = c'*x + 0.5*( x'*C*x);
36          if sigtil ≤ sighat
37              next = 'Step3p1';
38          else
39              next = 'Step3p2';
40          end
41          if strcmp(next,'Step3p1')
42                                              %  Step 3.1
43              d = stCs^(-.5) * C * s;
44              [Dinv] = Phi1(Dinv,d,k,J,n);    % k is from Step 1
45              J(k) = -1;
46              next = 'Step1';
47          end
48          if strcmp(next,'Step3p2')
49                                              % Step 3.2
50              [Dinv] =  Phi(Dinv,A(ell,:)',k,n);
51              for i = 1:n
52                  if J(i) == -1;
53                      J(i) = 0;
54                  end
55              end
56              J(k) = ell;
57              next = 'Step1';
58          end
59      end
60
61  end    % iteration loop for j
62
63  end
```

Figure 5.4.   Algorithm 3: Alg3.m.

Discussion of Alg3.m. Figure 5.4 shows the Matlab code for Algorithm 3 (Alg3.m). Since flow control for Algorithm 3 is a bit more complicated than as for previous algorithms, we control the flow as follows. On line 13, we set the character string next to 'Step1'. Control then proceeds sequentially until it reaches line 15. There, the statement "if strcmp(next,'Step1')" has value 'true' and control passes to the function Step1 in line 16. The details of Step1 are shown in Figure 5.5 and a detailed discussion of Step1 are given immediately below that figure. Matlab uses the convention that output values are placed on the left hand-side of the "=" sign and input quantities are on the right hand-side in line 16. If Step1 determines that the current point is optimal, this is indicated in msg and is checked for in line 17. If the solution is indeed optimal, control returns to the calling program. Otherwise, Step1 returns with next = 'Step2', search direction s (s_j) and k being the index of the corresponding column of Dinv (D_j^{-1}).

Control now continues sequentially to line 23 where next = 'Step2' and in lines 25–25 a call is made to Step2. The details of Step2 are shown in Figure 5.6 and a detailed discussion of Step2 are given immediately below that figure. Step2 computes next = 'Step3', sigma (σ_j), sigtil $(\tilde{\sigma}_j)$, $(\hat{\sigma}_j)$, ell (l). stCs $(s_j'Cs_j)$ and msg. If Step2 determines the problem is unbounded from below, this will be picked up on line 26 and control will transfer back to the calling program. Otherwise x, g and fx are updated on lines 33-35. Lines 36–40 then set next = 'Step3p1' if sigtil $\leq$ sighat and next = 'Step3p2' otherwise.

If next = 'Step3p1' in line 41, the updating proceeds according to Step 3.1 as follows. In line 43, letting d = $(s_j'Cs_j)^{-1/2}Cs_j$, D_j^{-1} is updated in line 44 as $\Phi_1(D_j^{-1}, d, k, n)$ and J is unchanged from its previous value other than the k–th element is changed to -1. Now next is set to 'Step1' and control transfers to that step.

If next = 'Step3p2' in line 48, the updating proceeds according to Step 3.2 as follows. D_j^{-1} is updated in line 50 as $\Phi(D_j^{-1}, a_l, k, n)$, the index set J by changing all elements with value -1 to zero and the k–th element to ell (l). Now next is set to 'Step1' and control transfers to that step.

```
1  %         Step1.m
2  %  This is Step 1 for each of Algorithms 3, 4 and 10.
3  %  It is coded as a function to avoid repetition.
4  function[next,k,s,msg,dual,big] = Step1(J,next,g,Dinv,n,m,r,tol)
5  msg = ' ';   dual = (1.);   s = [1.]; k = 0; big = -100.;
6  %  Step 1: Computation of Search Direction s_j
7            %  nzero = # components of J having value 0
8  nzerolist = find(J == 0);
9  nzero = length(nzerolist);
10 if nzero ≥ 1;
11      next = 'Step1p1';
12 else
```

```
13        next = 'Step1p2';
14   end
15   if strcmp(next,'Step1p1')
16        % Step 1.1
17        dotprod = abs(g'* Dinv(:,nzerolist) );
18        big = max(dotprod);
19        k = nzerolist(find(big==dotprod,1));
20        gjckj = g'* Dinv(:,k);
21        % get the correct sign for gjckj
22        if gjckj > 0.;
23             s = Dinv(:,k);
24        elseif gjckj ≤ 0.
25             s = - Dinv(:,k);
26             end
27        next = 'Step2';
28
29   elseif strcmp(next,'Step1p2')
30        % Step 1.2
31        n12mlist = find(J ≥ 1 & J ≤ m);
32        n12m = length(n12mlist);
33        %  n12m = # components of J having value between
34        %              1 and m
35
36
37        if n12m == 0;
38             % We have an optimal solution
39             msg = 'Optimal solution obtained';
40             [dual] = GetDual(Dinv,J,g,n,m,r);
41             return
42        end
43
44        if n12m ≥ 1
45             dotprod = g'* Dinv(:,n12mlist);
46             big = max(dotprod);
47             k = n12mlist(find(big==dotprod,1));
48        end
49
50        if big ≤ tol
51             % Opimal Solution Obtained
52             msg = 'Optimal solution obtained';
53             [dual] = GetDual(Dinv,J,g,n,m,r);
54             return
55        else
56             s = Dinv(:,k);
57        end
58        next = 'Step2';
59   end      % end of Step1p2
```

Figure 5.5.   Step1 of Algorithms 3, 4 and 10: Step1.m.

Discussion of Step1.m. In the function Step1, in line 8, the set $\{\ i \mid \alpha_i = 0\}$ is computed and stored in nzerolist. The number of elements in nzerolist is computed as nzero. If nzero ≥ 1, line 11 sets next = 'Step1p1'. Otherwise next is set to 'Step1p2'.

If next = 'Step1p1' Step 1.1 proceeds as follows. Line 17 computes dotprod = $\{|g'c_i| \mid$ for all i with $i \in nzerolist\}$. Here, c_i denotes the i-th column of Dinv. Line 18 then computes the biggest of these dot products and Line 19 then calculates the position (k) of this biggest element. Finally, lines 22 - 26 set s to either Dinv(:,k) or -Dinv(:,k), which ever gives the correct sign of the inner product. Note that Dinv(:,k) denotes the k-th column of Dinv. Finally, in line 27 next is set 'Step2'.

On line 29, a check is made for next = 'Step1p2' and if this is true, Step 1 proceeds with Step 1.2. In lines 31 and 32, n12mlist is computed which contains those indices satisfying $1 \leq \alpha_i \leq m$ and n12m is the number of entries in n12mlist. Line 37 checks for n12m being zero and if so (no active inequality constraints) then the current point is optimal. In this case, msg is set appropriately, the dual variables are obtained from GetDual and control is returned to the calling program. In line 44, if the number of active inequality constraints is 1 or more, then dotprod is computed on line 45. The components of dotprod are the inner products of the gradient with all columns of Dinv associated with active inequality constraints. In line 46, the biggest of these is computed as big. Line 47 finds the position (k) of this.

If big $\leq$ tol then the current point is optimal and lines 52–54 return this optimal solution as in lines 39–41. If big ¿ tol, then s = is set to the k–th column of Dinv in line 56, next = 'Step2' and control returns to the calling program.

```
1  %           Step2.m
2  %   This is Step 2 for each of Algorithms 3, 4 and 10.
3  %   It is coded as a function to avoid repetition.
4  function[next,sigma,sigtil,sighat,ell,stCs,msg] = ...
5                        Step2(J,m,A,b,C,s,tol,big,x)
6  msg = ' ';  next = 'Step3'; sigma = 1.;
7  % opimal step size (sigtil)
8  stCs = s'*C*s;
9  if stCs <- 1.e-6
10     sigtil = Inf;
11 else
12     sigtil = big/stCs;
13 end
14
15 % maximum feasible stepsize (sighat)
```

```
16  sighat = Inf;
17  ell = 0;
18  inactive = setdiff(1:m,J);
19  for i=1:length(inactive)
20      As = A(inactive(i),:)*s;
21      if As < -tol
22          test = (A(inactive(i),:)*x - b(inactive(i)))/As;
23          if test < sighat
24              sighat = test;
25              ell = inactive(i);
26          end
27      end
28  end
29  if sighat == Inf && sigtil == Inf
30      msg = 'Unbounded from below';
31      return
32  end
33  sigma = min(sigtil,sighat);
34  next = 'Step3';
```

Figure 5.6.   Step2 of Algorithms 3, 4 and 10 : Step2.m.

Discussion of Step2.m. The function Step2.m is used by Algorithms 3, 4 and 10 to determine the optimal stepsize $\tilde{\sigma}_j$, the maximum feasible stepsize $\hat{\sigma}_j$ and the stepsize σ_j. Lines 8–13 determine the optimal stepsize by computing stCs ($= s_j' C s_j$) If this is small, sigtil ($\tilde{\sigma}_j$) is set to ∞ in line 10. Otherwise, sigtil is set to big/stCs (line 12), where big ($= g_j' c_{kj}$) is available as an input argument for Step2.

The computations for sighat are lines 15–32. On lines 16 and 17, sighat and ell are initialized to ∞ and 0, respectively. Line 18 sets inactive to be the indices of all inactive inequality constraints. The for loop (lines 19–28) examines each of these in turn. Line 20 computes the inner product of the gradient of a typical such constraint with s, the current search direction obtained from the input arguments. If As is negative (line 21), test is computed as a term in the computation of the maximum feasible stepsize. If this term is smaller than the current candidate for sighat, sighat is replaced by test and ell is set to the index of the current inactive constraint.

In line 29, a check is made to see if both $\hat{\sigma}j$ and $\tilde{\sigma}j$ are both equal to ∞. If this is so, msg records this fact and control returns to the calling program. Otherwise sigma is set to the minimum of sigtil and sighat, and, next is set to Step3. Control is then returned to the calling routine.

```
1  %   GetDual.m
2  function [dual] = GetDual(Dinv,J,g,n,m,r)
3  %   GetDual retrieves the dual variables from the QP
4  %   solved by Alg3.  Algorithm 3 may terminate in
```

```
5  %    several places with 'optimal solution obtained'
6  %    and the code for obtaining the optimal dual
7  %    is cenralized in this funcion
8  %
9  for i = 1:(m+r)
10     dual(i) = 0.;
11 end
12 for i=1:n
13     if J(i) ≥ 1
14         dual(J(i)) = - g'*Dinv(:,i);
15     end
16 end
```

Figure 5.7.   GetDual: GetDual.m.

Discussion of GetDual.m . The function *GetDual* (Figure 5.7) obtains an optimal solution for the dual problem according to Theorem 5.1.

```
1  %    checkKKT.m
2  function checkKKT(A,b,x,n,m,r,dual,c,C,fx,j,msg,alg)
3  %   Given a proposed optimal solution for a QP,
4  %   checkKKT evaluates the Karush-Kuhn-Tucker conditions
5  %   at x.
6  if strcmp(msg,'Unbounded from below');
7      iter = j;
8      fprintf('Optimization Summary\n');
9      fprintf('%s\n',msg);
10     fprintf('Algorithm: %s\n',alg);
11     fprintf('Iterations   %d\n',iter);
12     return
13 end
14 iter = j;
15 w1 = 1.;
16 eqnorm = 0.0;
17 for i=1:m
18     sum = - b(i);
19     for j=1:n
20         sum  = sum + A(i,j)*x(j);
21     end
22     y(i) = sum;
23 end
24 ineqmax = max(y);
25 if r ≥ 1
26     for i=m+1:m+r
27         sum = - b(i);
28         for j=1:n
29             sum  = sum + A(i,j)*x(j);
30         end
31         z(i) = sum;
```

```
32      end
33      eqnorm = norm(z);
34  end
35
36  grad = c + C*x;
37  for j=1:m+r
38      if abs(dual(j)) ≥  1.e-6
39          grad = grad + dual(j)*A(j,:)';
40      end
41  end
42  dfnorm = norm(grad);
43
44  primalerr = max(ineqmax,eqnorm);
45
46
47
48  fprintf('Optimization Summary\n');
49  fprintf('%s\n',msg);
50  fprintf('Algorithm: %s\n',alg);
51  fprintf('Optimal objective value   %.8f\n',fx);
52  fprintf('Iterations    %d\n',iter);
53  fprintf('Maximum Primal Error      %.8f\n',primalerr);
54  fprintf('Maximum Dual Error        %.8f\n',dfnorm);
55
56  fprintf('\n\n Optimal Primal Solution\n')
57  for i=1:n
58      fprintf('   %d %12.8f\n',i,x(i))
59  end
60  fprintf('\n\n Optimal Dual Solution\n')
61  for i=1:m+r
62      fprintf('   %d %12.8f\n',i,dual(i))
63  end
```

Figure 5.8.   checkKKT: checkKKT.m.

Discussion of checkKKT.m . When a QP algorithm terminates with what it believes are an optimal solution to the primal and an optimal solution to the dual, the only way we have to check optimality is by examining the potential primal and dual errors. This is done in checkKKT (KKT stands for Karush-Kuhn-Tucker conditions). This is shown in Figure 5.8. The variable ineqmax is the maximum primal feasibility for the inequalities, eqnorm is the maximum error in the equalities (if present) and dfnorm is the maximum error in the equality part of the dual feasibility. Note that these quantities are printed out formatted in lines 40 - 56 using the Matlab printing function 'fprintf'.

```
1  %  eg5p1.m
2  %  Illustrates Algorithm 3
3  n = 2; m = 4; tol = 1.e-6; r = 0; c = [-10 -4]';
```

```
4  A = [1 0; 0 -1; -1 0; 0 1 ]; b = [4 -1 -1 3]'; x = [ 1 3]';
5  Dinv = [-1 0; 0 1]; C = [ 2 0; 0 2]; x = [ 1 3]'; J = [3 4];
6  checkdata(C)
7  checkconstraints(A,b,x,m,r,n)
8  [x,J,fx,Dinv,msg,dual,j,alg] =  ...
9                          Alg3(c,C,n,m,r,tol,x,J,Dinv,A,b);
10 checkKKT(A,b,x,n,m,r,dual,c,C,fx,j,msg,alg);
```

Figure 5.9.   Example 5.1: eg5p1.m.

```
1  eg5p1
2  Optimization Summary
3  Optimal solution obtained
4  Algorithm: Algorithm 3
5  Optimal objective value   -28.00000000
6  Iterations    2
7  Maximum Primal Error      0.00000000
8  Maximum Dual Error        0.00000000
9
10
11   Optimal Primal Solution
12      1    4.00000000
13      2    2.00000000
14
15
16   Optimal Dual Solution
17      1    2.00000000
18      2    0.00000000
19      3    0.00000000
20      4    0.00000000
```

Figure 5.10.   Example 5.1: Output.

Discussion of eg5p1.m and its output. These two files concern Example 5.1 of the text and solve that problem using Alg3. The file eg5p1.m constructs the data for this problem and solves it using Alg3.m. The results, summarized in Figure 5.10 are from checkKKT. Note that the optimal solution so obtained is in agreement with that in Example 5.1 of the text.

```
1  %  eg5p2.m
2  %  Illustrates Algorithm 3
3  n = 3; m = 6; tol = 1.e-6; r = 0; c = [1 -2 -1]';
4  C = [ 1 0 0; 0 2 -2; 0 -2 4]; x = [ 0 0 3]';
5  A = [1 1 1; 2 0 3; -2 2 -5; -1 0 0; 0 -1 0; 0 0 -1];
6  b = [10 11 -13 0 0 0]'; J = [4 5 0];
7  Dinv = [-1 0 0; 0 -1 0; 0 0 1];
```

```
 8  checkdata(C)
 9  checkconstraints(A,b,x,m,r,n)
10  [x,J,fx,Dinv,msg,dual,j,alg] =  ...
11                        Alg3(c,C,n,m,r,tol,x,J,Dinv,A,b);
12  checkKKT(A,b,x,n,m,r,dual,c,C,fx,j,msg,alg);
```

Figure 5.11.   Example 5.2: eg5p2.m.

```
 1  eg5p2
 2  Optimization Summary
 3  Optimal solution obtained
 4  Algorithm: Algorithm 3
 5  Optimal objective value    4.50000000
 6  Iterations    3
 7  Maximum Primal Error       0.00000000
 8  Maximum Dual Error         0.00000000
 9
10
11   Optimal Primal Solution
12      1    1.00000000
13      2    2.00000000
14      3    3.00000000
15
16
17   Optimal Dual Solution
18      1    0.00000000
19      2    1.00000000
20      3    2.00000000
21      4    0.00000000
22      5    0.00000000
23      6    0.00000000
```

Figure 5.12.   Example 5.2: Output.

Discussion of eg5p2.m and its output. These two files are analogs of those in Figure 5.9 and 5.10, respectively. They solve the problem of Example 5.2 using Alg3. Note that the optimal solution so obtained is in agreement with that in Example 5.2 of the text.

```
 1  %  LPAlg3.m
 2  function [x,J,fx,Dinv,msg,dual] = LPAlg3(c,n,m,r,  ...
 3                            tol,x,J,Dinv,A,b)
 4  %  LPAlg3 implements LP Algorithn 3 to solve a model
 5  %  Linear Programming Problem.
 6
 7  %  Initialization
 8  fx = c'*x;
```

```
 9  for j=0:10000;
10      %  Step 1: Computation of Search Direction s (= s_j)
11      %  nzero = # components of J having value 0
12      nzerolist = find(J == 0);
13      nzero = length(nzerolist);
14      if nzero ≥ 1
15          % Step 1.1
16          dotprod = abs(c'* Dinv(:,nzerolist) );
17          big = max(dotprod);
18          k = nzerolist(find(big==dotprod,1));
19          gjckj = c'* Dinv(:,k);
20          % get the correct sign for gjckj
21          if gjckj > 0.;
22              s = Dinv(:,k);
23          elseif gjckj ≤ 0.
24              s = − Dinv(:,k);
25          end
26
27      elseif nzero == 0
28          % Step 1.2
29          n12mlist = find(J ≥ 1 & J ≤ m);
30          n12m = length(n12mlist);
31          %  n12m = # components of J having value between
32          %                                  1 and m
33
34
35          if n12m == 0;
36              % We have an optimal solution
37              msg = 'Optimal solution obtained';
38              [dual] = GetDual(Dinv,J,c,n,m,r);
39              return
40          end
41
42          if n12m ≥ 1
43              dotprod = c'* Dinv(:,n12mlist);
44              big = max(dotprod);
45              k = n12mlist(find(big==dotprod,1));
46          end
47
48          if big ≤ tol
49              %  Opimal Solution Obtained
50              msg = 'Optimal solution obtained';
51              [dual] = GetDual(Dinv,J,c,n,m,r);
52              return
53          else
54              s = Dinv(:,k);
55          end
56      end  % end of if nzero ≥ 1
57
58      %  Step 2 of LP  Algorihm 3
59      %  maximum feasible step size (sighat)
```

```
60      inactive = setdiff(1:m,J);
61      sums = A(inactive,:)*s;
62      ind =  find(sums<-tol);
63      if size(ind,1)>0
64          sigtest = (-b(inactive(ind))+A(inactive(ind),:)*x) ...
65              ./sums(ind);
66          sighat = min(sigtest);
67          ell=inactive(ind(find(sigtest==sighat,1))));
68      else
69          sighat = Inf
70      end
71      if sighat == Inf
72          msg = 'unbounded from below';
73      return
74      end
75
76      sigma = sighat;
77
78      %  Step 3: Update
79      x = x - sigma*s;
80      fx = c'*x ;
81      [Dinv] =  Phi(Dinv,A(ell,:)',k,n);
82      J(k) = ell;
83
84 end     % iteration loop for j
85 end
```

Figure 5.13.   LP Algorithm 3: LPAlg3.m.

Discussion of LPAlg3.m.   LP Algorithm 3 is a specialization of QP Algorithm 3 to a linear programming problem (LP). For a QP, the gradient of the objective function is $c + Cx$ whereas for an LP, this gradient is simply c. Furthermore, for the QP Algorithm updating D_j^{-1} is sometimes done with conjugate directions whereas with the LP case updating is always done with gradients of new active constraints.

```
1  %  P1P2.m
2  function [A1,B1,C1,n1,m1,x,J,Dinv,dual,j,alg,fx,msg] =  ...
3                                      P1P2(A,b,C,c,n,m,r)
4  %  This is the Pase 1 - Phase 2 routine.  It formulates
5  %  the Phase 1 problem, solves it with a call to
6  %  LPAlg3 and then solves the given QP it using this
7  %  starting point and QPAlg3
8  %
9  %  zero out d which is necessary if r = 0
10
11 d = zeros(n,1);
12
```

```
13  %  replace equalities with their negatives if rhs < 0
14
15  if r > 0
16      for i=1:r
17          if b(m+i) < 0
18              A(m+i,:) = - A(m+i,:);
19              b(m+i) = - b(m+i);
20          end
21      end
22      for i=1:r
23          d = d - A(m+i,:)';
24      end
25  end     % end of if r > 0
26  n1 = n + 1;
27  m1 = m + r + 1;
28  r1 = 0;
29
30  for i=1:m+r
31      for j=1:n
32          A1(i,j) = A(i,j);
33      end
34  end
35  for i=1:m
36      A1(i,n1) = - 1.;
37  end
38  if r >= 1
39      for i=(m+1):(m+r)
40          A1(i,n1) = 0.;
41      end
42  end
43  for j=1:n
44      A1(m1,j) = 0;
45  end
46  A1(m1,n1) = -1;
47  if r >= 1
48      for j=1:n
49          C1(j) = d(j);
50      end
51  end
52  if r == 0
53      for j=1:n
54          C1(j) = 0.0;
55      end
56  end
57  C1(n1) = 1.;
58  for i= 1:m
59      B1(i) = b(i);
60  end
61  if r >= 1
62      for i = 1:r
63          B1(m+i) = b(m+i);
```

```
64        end
65   end
66   B1(m1) = 0.;

68   r1 = 0;
69   tol = 1.e-6;
70   for i =1:n
71        x(i) = 0.;
72   end
73   big = 0.;
74   for j=1:m
75        if - b(i) > big
76             big = - b(i);
77        end
78   end
79   x(n1) = big;
80   D0inv = eye(n1);
81   for i=1:n1
82        J(i) = 0;
83   end

85   C1 = C1';
86   x = x';
87   B1 = B1';
88   %  We now have the Phase 1 problem data formulated.
89   %  Solve it using LPAlg3

91   [x,J,fx,D0inv,msg,dual] = LPAlg3(C1,n1,m1,r1,tol,x,J,  ...
92                                    D0inv,A1,B1);
93   Dinv = D0inv;
94   for i = 1:n1
95        if J(i) == m1;
96             index = i;
97        end
98   end
99   J(index) = [];
100  D0inv(:,index) = [];
101  D0inv(n1,:) = [];
102  x(n1) = [];
103  %  solve the QP using Alg 3
104  [x,J,fx,Dinv,msg,dual,j,alg] = Alg3(c,C,n,m,r,tol,x,J,  ...
105                                   D0inv,A,b);
106  alg = 'Phase 1 - Phase 2';
```

Figure 5.14.   Phase 1 - Phase 2: P1P2.m.

Discussion of P1P2.m. Lines 15 through 25 replace any equality constraint with its negative if the corresponding rhs is negative. Lines 26 to 28 set the dimensions for the Phase 1 problem. Lines 30 to 46 construct the Phase 1 constraint matrix, A1, from the original constraint matrix A. Lines 47 to 57 construct the Phase 1 objective, C1. Lines 58 to 68 construct the rhs, B1, for the Phase 1 problem. Lines 70 to 79 construct the initial point x for the initial point problem. D_0^{-1} is set to the identity and J is set to all zeros. The Phase 1 data is now complete and the Phase 1 problem is solved by LPAlg3 in line 91.

```
1  %  eg5p3.m
2  %  Illustrates Phase 1 - Phase 2 procedure
3  n = 2; m = 4; r = 0; c = [-10 -4]'; C = [2 0 ; 0 2];
4  A = [ 1 0; 0 -1; -1 0; 0 1]; b = [ 4 -1 -1 3 ]';
5
6  [A1,B1,C1,n1,m1,x,J,Dinv,dual,j,alg,fx,msg] = ...
7                         P1P2(A,b,C,c,n,m,r);
8  checkKKT(A,b,x,n,m,r,dual,c,C,fx,j,msg,alg);
```

Figure 5.15. Example 5.3: eg5p3.m.

```
1  eg5p3
2  Optimization Summary
3  Optimal solution obtained
4  Algorithm: Phase 1 - Phase 2
5  Optimal objective value    -28.00000000
6  Iterations     2
7  Maximum Primal Error       0.00000000
8  Maximum Dual Error         0.00000000
9
10
11   Optimal Primal Solution
12      1    4.00000000
13      2    2.00000000
14
15
16   Optimal Dual Solution
17      1    2.00000000
18      2    0.00000000
19      3    0.00000000
20      4    0.00000000
```

Figure 5.16. Example 5.3: output.

Discussion of eg5p3.m and its output . These two files correspond to Example 5.3. The file eg5p3.m sets up the data for Example 5.3 to be solved by the Phase 1 - Phase 2 procedure and Figure 5.16 shows the output so obtained. Note that the optimal primal solution so obtained is identical to that of Example 5.3 of the text.

```
1  %   eg5p4.m
2  %   Further illustrates the Phase 1 — Phase 2 process
3  n = 3; m = 4; r = 1; c = [0 −2 6]';
4  A = [ 2 1 1 ; 1 0 1; 0 1 1; 1 2 1; 1 1 1];
5  b = [10 5 5 9 6]'; C = [ 2 2 0; 2 3 0; 0 0 4];
6
7  [A1,B1,C1,n1,m1,x,J,Dinv,dual,j,alg,fx,msg] = ...
8                       P1P2(A,b,C,c,n,m,r);
9  checkKKT(A,b,x,n,m,r,dual,c,C,fx,j,msg,alg);
```

Figure 5.17.   Example 5.4: eg5p4.m.

```
1  eg5p4
2  Optimization Summary
3  Optimal solution obtained
4  Algorithm: Phase 1 — Phase 2
5  Optimal objective value   31.00000000
6  Iterations    2
7  Maximum Primal Error      0.00000000
8  Maximum Dual Error        0.00000000
9
10
11   Optimal Primal Solution
12      1    3.00000000
13      2    2.00000000
14      3    1.00000000
15
16
17   Optimal Dual Solution
18      1    0.00000000
19      2    0.00000000
20      3    0.00000000
21      4    0.00000000
22      5  −10.00000000
```

Figure 5.18.   Example 5.4: output.

Discussion of eg5p4.m and its output. This is a second example of the Phase 1 - Phase 2 procedure and corresponds to Example 5.4. The file eg5p4.m (Figure 5.17) sets up the data for Example 5.4 to be solved by the Phase 1 - Phase 2 procedure and Figure 5.18 shows the output so obtained. Note that the optimal primal solutions is identical to that obtained in Example 5.4 of the text.

```
1  %   eg5p5.m
2  %   Illusrtates a hard problem for Algorithm 3; one in which
```

```
3  %  many conjugate directions are constantly destroyed.
4  %  Provides the motivation for Algorihm 4.
5  n = 10; m = 5; tol = 1.e-6; r = 0;
6  C = [9 0 0 0 0 0 0 0 0 0; ...
7        0 8 0 0 0 0 0 0 0 0; ...
8        0 0 7 0 0 0 0 0 0 0; ...
9        0 0 0 5 0 0 0 0 0 0; ...
10       0 0 0 0 4 0 0 0 0 0; ...
11       0 0 0 0 0 3 0 0 0 0; ...
12       0 0 0 0 0 0 1 0 0 0; ...
13       0 0 0 0 0 0 0 1 0 0; ...
14       0 0 0 0 0 0 0 0 1 0; ...
15       0 0 0 0 0 0 0 0 0 1 ];
16 A = [1 1 1 1 1 1 1 1 1 1; ...
17       0 0 0 0 0 0 0 0 0 -1; ...
18       0 0 0 0 0 0 0 1 0 0; ...
19       1 1 1 1 1 1 1 1 1 1; ...
20       0 0 1 1 0 0 0 0 0 0 ];
21 b = [34 -3.6 6.13 9.8  0.06 ]';
22 c = [0 0 0 0 0 0 0 0 0 0 ]';
23 Dinv = eye(n);
24 J = zeros(1,n);
25 x = [ -99 -98 -97 -96 -95 -94 -93 -92 -91 35 ]';
26 checkdata(C)
27 checkconstraints(A,b,x,m,r,n)
28 [x,J,fx,Dinv,msg,dual,j,alg] =  ...
29                               Alg3(c,C,n,m,r,tol,x,J,Dinv,A,b);
30 checkKKT(A,b,x,n,m,r,dual,c,C,fx,j,msg,alg);
```

Figure 5.19.   Example 5.5: eg5p5.m.

```
1  eg5p5
2  Optimization Summary
3  Optimal solution obtained
4  Algorithm: Algorithm 3
5  Optimal objective value    6.48000000
6  Iterations    25
7  Maximum Primal Error       0.00000000
8  Maximum Dual Error         0.00000000
9
10
11   Optimal Primal Solution
12     1     0.00000000
13     2     0.00000000
14     3     0.00000000
15     4    -0.00000000
16     5     0.00000000
17     6     0.00000000
```

```
18      7   -0.00000000
19      8   -0.00000000
20      9    0.00000000
21      10   3.60000000
22

23

24   Optimal Dual Solution
25      1    0.00000000
26      2    3.60000000
27      3    0.00000000
28      4    0.00000000
29      5    0.00000000
```

Figure 5.20.   Example 5.5: output.

Discussion of eg5p5.m and its output. This is a bit larger problem than the previous, having 10 variables and 5 constraints. Algorithm 3 requires 25 iterations to solve it. In Section 5.3, we will formulate a more efficient QP solution algorithm and compare the results for the more efficient algorithm with those for Algorithm 3 solving this same problem. Here, eg5p5.m (Figure 5.19) sets up the problem data for solution by Algorithm 3 and Figure 5.20 shows the results.

```
1  %  Phi2.m
2  function[DhatInv] = Phi2(Dinv,d,nu,J,n)
3  %  Procedure Phi2:  Replaces an existing set of columns which
4  %  are conjugate direcions with a second set for which all
5  %  are orthogonal to a given vector d.
6  y = zeros(n,1);
7  for i=1:n
8      if J(i) == -1
9          y(i) = d' * Dinv(:,i);
10     end
11 end
12 enu = zeros(n,1);
13 enu(nu) = 1;
14 t1 = norm(y) * enu;
15 t2 = norm(t1 - y);
16 if t2 <= 1.e-6
17     mu = -1;
18 else
19     mu = 1;
20 end
21 t3 = (mu*norm(y)*enu - y);
22 t4 = norm(mu*norm(y)*enu - y);
23 t4 = 1./t4;
24 for i = 1:n
25     t3(i) = t4*t3(i);
```

```
26  end
27  t3;
28  w = t3;
29  %  Dinv(:,j) = column j of Dinv
30  p = zeros(n,1);
31  for i=1:n
32      if J(i) == −1
33          p = p + w(i)*Dinv(:,i);
34      end
35  end
36  for i=1:n
37      if J(i) == −1
38          DhatInv(:,i) = Dinv(:,i) − 2*w(i)*p;
39      end
40  end
41
42  for i=1:n
43      if J(i) ≠ −1
44          DhatInv(:,i) = Dinv(:,i);
45      end
46  end
```

Figure 5.21.   Procedure Φ_2: Phi2.m.

```
1  %  eg5p6.m
2  n = 4; C = eye(n); Dinv = eye(n); nu = 1;
3  for i=1:n
4    J(i) = −1;
5    d(i) = 1;
6  end
7  d = d';
8  [DhatInv] = Phi2(Dinv,d,nu,J,n)
```

Figure 5.22.   Example 5.6: eg5p6.m.

```
1  eg5p6
2  DhatInv =
3       0.5000      0.5000      0.5000      0.5000
4       0.5000      0.5000     −0.5000     −0.5000
5       0.5000     −0.5000      0.5000     −0.5000
6       0.5000     −0.5000     −0.5000      0.5000
```

Figure 5.23.   Example 5.6: output.

Discussion of eg5p6.m and its output. This illustrates the computations of Example 5.6 of the text. The given data is stated in eg5p6.m (Figure 5.22). The example requires us to use Procedure Ψ_2 to construct a new set of conjugate directions for which 3 of them are orthogonal to a given vector. The new conjugate directions are shown in Figure 5.23 and are identical to those found in Example 5.6 of the text.

```
1   %  Alg4.m
2   function [x,J,fx,Dinv,msg,dual,j,alg] = ...
3                           Alg4(c,C,n,m,r,tol,x,J,Dinv,A,b)
4   %  Alg4 implements QP Algorithn 4 to solve a model
5   %  Quadratic Programming Problem.
6
7   %  Initialization
8   alg = 'Algorithm 4';
9   fx = c'*x + 0.5*( x'*C*x);
10  g = c + C*x;
11  next = 'Step1';
12  for j=0:10000;
13      if strcmp(next,'Step1')
14          [next,k,s,msg,dual,big] = ...
15                      Step1(J,next,g,Dinv,n,m,r,tol);
16          if strcmp(msg,'Optimal solution obtained')
17              return
18          end
19
20      end
21
22      if strcmp(next,'Step2')
23          [next,sigma,sigtil,sighat,ell,stCs,msg] = ...
24                      Step2(J,m,A,b,C,s,tol,big,x);
25          if strcmp(msg,'Unbounded from below');
26              return
27          end
28      end
29
30      if strcmp(next,'Step3')
31                                          %  Step 3: Update
32          x = x - sigma*s;
33          g = c + C*x;
34          fx = c'*x + 0.5*( x'*C*x);
35          if sigtil ≤ sighat
36              next = 'Step3p1';
37          else
38              next = 'Step3p2';
39          end
40      end
41      if strcmp(next,'Step3p1')
42                                          %  Step 3.1
43          d = stCs^(-.5) * C * s;
44          [Dinv] = Phi1(Dinv,d,k,J,n);   % k is from Step 1
```

```
45          J(k) = -1;
46          next = 'Step1';                        % end of Step 3.1
47       end
48
49    if strcmp(next,'Step3p2')
50                                                 % Step 3.2
51          list = find( J == -1);
52          big = -1.;
53          for i=1:length(list)
54              dotprod = abs( A(ell,:)*Dinv(:,list(i)) );
55              if dotprod > big
56                  big = dotprod;
57              end
58          end
59          if big >= 1.e-6
60              next = 'Step3p3';
61          end
62          if strcmp(next,'Step3p2')
63              [Dinv] =  Phi1(Dinv,A(ell,:)',k,J,n);
64              J(k) = ell;
65              next = 'Step1';
66          end
67       end
68
69    if strcmp(next,'Step3p3')
70                                                 %  Step 3.3
71          big = -1;
72          nu = 0;
73          for i=1:length(list)
74              dotprod = abs( A(ell,:)*Dinv(:,list(i)) ) ;
75              if dotprod > big
76                  big = dotprod;
77                  nu = list(i);
78              end
79          end
80          [Dhatinv] = Phi2(Dinv,A(ell,:)',nu,J,n);
81          [Dinv] = Phi1(Dhatinv,A(ell,:)',nu,J,n);
82          J(nu) = ell;
83          J(k) = 0;
84          next = 'Step1';
85       end
86
87 end    % iteration loop for j
88
89 end
```

Figure 5.24.   Algorithm 4: Alg4.m.

Discussion of Alg4.m. Steps 1 and 2 are identical with those of Algorithm 3. Step 3 begins at line 32 with updates of x, g and fx (lines 32-34) as in

Algorithm 3. If sigtil $\leq$ sighat then next is set to Step3p1 and Step 3.1 is taken. Note that Step 3.1 is identical in both Algorithms 3 and 4. If sigtil $>$ sighat then next is set to Step3p2 and Step 3.2 is taken. When Step 3.2 is taken, lines 51-58 find the biggest inner product, in absolute value, of the gradient of constraint ell with all of the conjugate direction columns of Dinv. If this inner product is greater than some tolerance (line 59), then next is set to Step3p3 (line 60). If this inner product is less than this tolerance, next remains unchanged at Step3p2. If next remains unchanged at Step3p2, line 63 updates Dinv using Procedure Ψ_1 with the gradient of constraint ell being placed in column k. The conjugate direction columns are unchanged because $a_i' c_{ij} = 0$ for all i with $\alpha_{ij} = -1$. Because next is set to Step1, Step 3 is now complete and control passes back to Step1.

Control passes to Step3.3 and if next = Step3p3, this substep is executed as follows. The list of conjugate direction columns of Dinv is still available from Step 3.2. Lines 72-79 calculate the index nu of that conjugate direction column of Dinv which gives the largest inner product (in absolute value) with row ell of A. Lines 80-83 then perform the updating. Procedure Ψ_2 creates Dhatinv from Dinv and row ell of A by modifying the conjugate direction columns of Dinv so that they are orthogonal to row ell of A. The updating continues using Procedure Ψ_1 to construct Dinv from Dhatinv and row ell of A. Next, next is set to Step1 and the iteration is complete.

```
1  %   eg5p7.m
2  n = 3; m = 1; r = 0; tol = 1.e-6; c = [ 0 0 0 ]';
3  C = [1 0 0; 0 1 0; 0 0 1 ]; x = [-2 -2 -2]';
4  J = [ 0 0 0 ]; x = [-2 -2 -2]';
5  Dinv =   [1 0 0; 0 1 0; 0 0 1 ] ;
6  A = [ 3 4 1 ]; b = [ -1 ]';
7  checkdata(C)
8  checkconstraints(A,b,x,m,r,n)
9  [x,J,fx,Dinv,msg,dual,j,alg] = ...
10                      Alg4(c,C,n,m,r,tol,x,J,Dinv,A,b);
11 checkKKT(A,b,x,n,m,r,dual,c,C,fx,j,msg,alg);
```

Figure 5.25.   Example 5.7: eg5p7.m.

```
1  eg5p7
2  Optimization Summary
3  Optimal solution obtained
4  Algorithm: Algorithm 4
5  Optimal objective value     0.01923077
6  Iterations     4
7  Maximum Primal Error        0.00000000
```

```
 8  Maximum Dual Error              0.00000000
 9
10
11    Optimal Primal Solution
12       1   −0.11538462
13       2   −0.15384615
14       3   −0.03846154
15
16
17    Optimal Dual Solution
18       1    0.03846154
```

Figure 5.26.   Example 5.7: output.

Discussion of eg5p7.m and its output. This is Example 5.7 solved by Algorithm 4. The data is set out in eg5p7 (Figure 5.25). The output is displayed in Figure 5.26. As expected, the optimal solution obtained by Algorithm 4 (Figure 5.26) is identical to that obtained in Example 5.7.

```
 1  %   eg5p8.m (and eg5p5.m)
 2  %   This problem was solved by Algorithm 3 as eg5p5.m.
 3  %   For comparison purposes, it is solved here, by Algorithm 4.
 4  n = 10; m = 5; tol = 1.e-6; r = 0; Dinv = eye(n);
 5  c = zeros(n,1); J = zeros(1,n);
 6  b = [34 -3.6 6.13 9.8  0.06 ]';
 7  x = [ -99 -98 -97 -96 -95 -94 -93 -92 -91 35 ]';
 8  C = [9 0 0 0 0 0 0 0 0 0; ...
 9       0 8 0 0 0 0 0 0 0 0; ...
10       0 0 7 0 0 0 0 0 0 0; ...
11       0 0 0 5 0 0 0 0 0 0; ...
12       0 0 0 0 4 0 0 0 0 0; ...
13       0 0 0 0 0 3 0 0 0 0; ...
14       0 0 0 0 0 0 1 0 0 0; ...
15       0 0 0 0 0 0 0 1 0 0; ...
16       0 0 0 0 0 0 0 0 1 0; ...
17       0 0 0 0 0 0 0 0 0 1 ];
18  A = [1 1 1 1 1 1 1 1 1 1; ...
19       0 0 0 0 0 0 0 0 0 -1; ...
20       0 0 0 0 0 0 0 1 0 0; ...
21       1 1 1 1 1 1 1 1 1 1; ...
22       0 0 1 1 0 0 0 0 0 0 ];
23  checkdata(C)
24  checkconstraints(A,b,x,m,r,n)
25  [x,J,fx,Dinv,msg,dual,j,alg] = ...
26                      Alg4(c,C,n,m,r,tol,x,J,Dinv,A,b);
27  checkKKT(A,b,x,n,m,r,dual,c,C,fx,j,msg,alg);
```

Figure 5.27.   Example 5.5 solved by Algorithm 4: eg5p8.m.

```
1   eg5p8
2   Optimization Summary
3   Optimal solution obtained
4   Algorithm: Algorithm 4
5   Optimal objective value    6.48000000
6   Iterations    16
7   Maximum Primal Error       0.00000000
8   Maximum Dual Error         0.00000000
9
10
11    Optimal Primal Solution
12      1     0.00000000
13      2     0.00000000
14      3     0.00000000
15      4     0.00000000
16      5     0.00000000
17      6     0.00000000
18      7     0.00000000
19      8    -0.00000000
20      9    -0.00000000
21     10     3.60000000
22
23
24    Optimal Dual Solution
25      1     0.00000000
26      2     3.60000000
27      3     0.00000000
28      4     0.00000000
29      5     0.00000000
```

Figure 5.28.   Example 5.8 output.

Discussion of eg5p8.m and its output. This is an important example be-cause it compares Algorithms 3 and 4 solving the same problem. Comparison of Figures 5.20 and 5.28 shows that both algorithms determine the same primal and dual optimal solutions as expected. However, Algorithm 3 takes 25 itera-tions while Algorithm 4 requires only 16. This means that Algorithm 4 runs faster by about 50% on this example. The reduction in iterations for Algorithm 4 is a consequence of being able to maintain all but one of the existing conju-gate directions when a new, previously inactive, constraint becomes active. By contrast, when in Algorithm 3 a previously inactive constraint becomes active, all conjugate directions are destroyed and must be rebuilt causing additional iterations to occur. Thus, it is likely that Algorithm 4 will be the most efficient for large problems.

5.6 Exercises

5.1 Use Algorithm 3 to solve each of the following problems using the indicated initial data.

(a)

$$
\begin{aligned}
\text{minimize}:\quad & -\ 4x_1\ -\ 6x_2\ +\ x_1^2\ +\ x_2^2 \\
\text{subject to}:\quad & x_1\ +\ x_2\ \le\ 2, \quad (1) \\
& -\ x_1\ -\ x_2\ \le\ 2, \quad (2) \\
& -\ x_1\ +\ x_2\ \le\ 2, \quad (3) \\
& x_1\ -\ x_2\ \le\ 2. \quad (4)
\end{aligned}
$$

Initial Data: $x_0 = (-2\,,\ 0)'$, $J_0 = \{\,2\,,\,3\,\}$,

$$
D_0^{-1} = \begin{bmatrix} -1/2 & -1/2 \\ -1/2 & 1/2 \end{bmatrix}.
$$

(b)

$$
\begin{aligned}
\text{minimize}:\quad & -\ 2x_1\ -\ 2x_2\ +\ x_1^2\ +\ x_2^2\ -\ x_1x_2 \\
\text{subject to}:\quad & -\ x_1\ +\ x_2\ \le\ -1, \quad (1) \\
& x_1\ +\ x_2\ \le\ 3, \quad (2) \\
& -\ x_2\ \le\ 1. \quad (3)
\end{aligned}
$$

Initial Data: $x_0 = (3\,,\ -1)'$, $J_0 = \{\,0\,,\,3\,\}$,

$$
D_0^{-1} = \begin{bmatrix} 1 & 0 \\ 0 & -1 \end{bmatrix}.
$$

(c)

$$
\begin{aligned}
\text{minimize}:\quad & x_1^2\ +\ x_2^2\ +\ x_3^2\ +\ x_4^2 \\
\text{subject to}:\quad & x_1\ -\ x_2 \ \le\ 1, \quad (1) \\
& x_1\ +\ x_2\ +\ x_3\ +\ x_4\ =\ 8, \quad (2) \\
& x_1\ -\ x_2\ -\ x_3\ -\ x_4\ =\ 0. \quad (3)
\end{aligned}
$$

Initial Data: $x_0 = (4\,,\ 4\,,\ 0\,,\ 0)'$, $J_0 = \{\,2\,,\,3\,,\,0\,,\,0\,\}$,

$$
D_0^{-1} = \begin{bmatrix} 1/2 & 1/2 & 0 & 0 \\ 1/2 & -1/2 & -1 & -1 \\ 0 & 0 & 1 & 0 \\ 0 & 0 & 0 & 1 \end{bmatrix}.
$$

(d)

$$
\begin{aligned}
\text{minimize}:\quad & -\ 3x_1\ -\ 4x_2\ +\ \tfrac{1}{2}(x_1^2\ +\ x_2^2\ -\ 2x_1x_2) \\
\text{subject to}:\quad & -\ 3x_1\ +\ x_2\ \le\ 0, \quad (1) \\
& x_1\ -\ 4x_2\ \le\ 0, \quad (2) \\
& -\ x_1 \ \le\ -1, \quad (3) \\
& -\ x_2\ \le\ -1. \quad (4)
\end{aligned}
$$

Initial Data: $x_0 = (2, 2)'$, $J_0 = \{0, 0\}$,

$$D_0^{-1} = \begin{bmatrix} 1 & 0 \\ 0 & 1 \end{bmatrix}.$$

5.2 (a) Solve the following problem using Algorithm 4 and the indicated initial data.

minimize : $\frac{1}{2}(x_1^2 + x_2^2 + x_3^2 + x_4^2 + x_5^2)$

subject to : $\qquad\qquad\qquad -x_5 \leq -1/2$. (1)

Initial Data: $x_0 = (1, 1, 1, 1, 1)'$, $J_0 = \{0, 0, 0, 0, 0\}$,

$$D_0^{-1} = \begin{bmatrix} 1 & 0 & 0 & 0 & 0 \\ 0 & 1 & 0 & 0 & 0 \\ 0 & 0 & 1 & 0 & 0 \\ 0 & 0 & 0 & 1 & 0 \\ 0 & 0 & 0 & 0 & 1 \end{bmatrix}.$$

(b) Repeat part (a) but with the single constraint replaced with

$$- x_1 - x_2 - x_3 - x_4 - x_5 \leq -1/2. \quad (1)$$

(c) Comment on the updating at iteration 4 for both parts (a) and (b).

5.3 Solve the following problems using Algorithm 4 and the indicated initial data.

(a)

minimize : $\frac{1}{2}(x_1^2 + x_2^2 + x_3^2)$

subject to : $5x_1 + 12x_2 + x_3 \leq -1$. (1)

Initial Data: $x_0 = (-2, -2, -2)'$, $J_0 = \{0, 0, 0\}$,

$$D_0^{-1} = \begin{bmatrix} 1 & 0 & 0 \\ 0 & 1 & 0 \\ 0 & 0 & 1 \end{bmatrix}.$$

(b)

minimize : $- 3x_1 - 4x_2 + \frac{1}{2}(x_1^2 + x_2^2 - 2x_1x_2)$

subject to : $x_1 - x_2 \leq 4$, (1)

$x_1 \qquad\qquad \leq \qquad 6$. (2)

Initial Data: $x_0 = (2, 3)'$, $J_0 = \{0, 0\}$,

$$D_0^{-1} = \begin{bmatrix} 1 & 0 \\ 0 & 1 \end{bmatrix}.$$

5.4 (a) Show that every orthogonal matrix Q is nonsingular and that $Q^{-1} = Q'$.

(b) Given that Q is orthogonal, show that Q' is also orthogonal.

(c) Show that the product of two orthogonal matrices is also orthogonal.

5.5 Let C be an (n , n) positive semidefinite symmetric matrix and let p_1 and p_2 be normalized conjugate directions. Let d be a given n-vector with $p_1' d \neq 0$, let $\alpha_1, \alpha_2, \beta_1, \beta_2$ be scalars, and let

$$\hat{p}_2 = \alpha_1 p_1 + \alpha_2 p_2 ,$$
$$\hat{p}_2 = \beta_1 p_1 + \beta_2 p_2 .$$

Let

$$Q = \begin{bmatrix} \alpha_1 & \beta_1 \\ \alpha_2 & \beta_2 \end{bmatrix}$$

and observe that $(\hat{p}_1 , \hat{p}_2) = (p_1 , p_2)Q$.

(a) Show directly (in terms of $\alpha_1, \alpha_2, \beta_1, \beta_2$) that the requirement that $\hat{p}_1, \hat{p}_2$ be normalized conjugate directions implies that Q is orthogonal.

(b) Suppose that the additional requirement that $\hat{p}_2' d = 0$ is imposed. Show that this completely determines $\alpha_1, \alpha_2, \beta_1, \beta_2$ except for sign and find formulae for $\alpha_1, \alpha_2, \beta_1$, and β_2.

(c) Show directly that $\hat{p}_1$ and $\hat{p}_2$ are the same (up to sign) as those that are obtained from Lemma 5.2 with $\nu = 1$. Assume that $p_2' d \neq 0$.

5.6 (a) Show that every $(2, 2)$ orthogonal matrix Q can be written as $Q = Q_1 Q_2$, where

$$Q_2 = \begin{bmatrix} \cos \theta & \sin \theta \\ -\sin \theta & \cos \theta \end{bmatrix} ,$$

and Q_1 is any of the 8 $(2, 2)$ matrices having two zero elements and the remaining two entries being either $+1$ or -1. State the eight possible values for Q_1.

(b) Let Q be as above and show that the transformation $y = Qx$ represents a rotation of axes through an angle of θ followed by some combination of reversal and interchange of axes. A $(2 , 2)$ orthogonal matrix is sometimes called a **rotation matrix** or a **Givens matrix**.

5.7 (a) Given vectors $p_1, \ldots, p_k$ and d satisfying the hypothesis of Lemma 5.2, show how the results of Exercise 5.5, part (b) can be used repeatedly to construct a set of vectors $\hat{p}_1, \ldots, \hat{p}_k$ satisfying parts (b), (c), and (d) of the lemma. For simplicity, take $\nu = 1$ and assume that $p_1'd \neq 0$.

(b) In terms of n and k, how many additions/subtractions, multiplications/divisions, and square roots are required to implement the procedure of part (a)? How many are required to implement the construction of Lemma 5.2? Which method is numerically more efficient?

5.8 It can be very useful to get some experience as to how large a problem can be solved in a reasonable amount of computer time using Algorithm 3. Using (5.1) as our model problem, we can use random numbers for c, C, A and b as follows. We can use a computer random generator to construct a (k, n) matrix H. Set $C = H'H$. Then C is positive semidefinite (see discussion prior to Theorem 3.4) and has rank no more than k. Then the optimal solution $\hat{x}$ can be chosen to be whatever you like. A can be generated as an (m, n) matrix of random numbers. Now, decide what constraints will be active at the optimal solution and what their multipliers, u, will be (nonnegative, of course) and let K denote the indices of the active constraints. Now define c according to

$$c = -C\hat{x} - \sum_{i \in K} u_i a_i.$$

Finally, set $b_i = a_i \hat{x}$ for all $i \in K$ and $b_i > a_i \hat{x}$ for all $i \notin K$. Tables of execution times can be made with various values of n, k and m.

5.9 Repeat Exercise 5.8 with Algorithm 3 replaced with Algorithm 4. Compare the two algorithms.

Chapter 6

A Dual QP Algorithm

There are three optimality conditions for a QP: primal feasibility, complementary slackness and dual feasibility. In Chapter 5, we developed QP solution algorithms which maintained the first two conditions at each iteration and the sequence of primal objective values is decreasing. Only upon termination is the third optimality (dual feasibility) satisfied. In this chapter, we develop a solution method for a QP which generates a sequence of points satisfying dual feasibility and complementary slackness at each iteration. The corresponding sequence of primal objective function values is increasing. All but the final point is primal infeasible.

6.1   Algorithm 5

We first develop our dual QP algorithm informally. We consider a model problem with a quadratic objective function having a positive semidefinite Hessian and both linear inequality constraints and linear equality constraints.

We consider again the model problem

$$\left.\begin{array}{rlll} \text{minimize}: & c'x & + & \frac{1}{2}x'Cx \\ \text{subject to}: & a_i'x & \leq & b_i, & & i = 1, \ldots, m, \\ & a_i'x & = & b_i, & i = m+1, \ldots, m+r. \end{array}\right\} \quad (6.1)$$

The only assumption we make on the data for (6.1) is that C is positive semidefinite. We assume that (6.1) has been solved with only a subset of the constraints being used. Let x_j be the optimal solution obtained by Algorithm 4. Then there may be many constraints of (6.1) which are violated at x_j.

We assume Algorithm 4 has been used to solve (6.1) with only a subset of the constraints, to obtain an optimal solution x_j with associated final data g_j, J_j and D_j^{-1}. Rename these x_0, g_0, J_0 and D_0^{-1}, respectively. Now suppose we are at a typical iteration j of the dual method.

ALGORITHM 5

Model Problem:

$$\begin{aligned}\text{minimize:} \quad & c'x + \tfrac{1}{2}x'Cx \\ \text{subject to:} \quad & a_i'x \leq b_i, \quad i = 1, \ldots, m, \\ & a_i'x = b_i, \quad i = m+1, \ldots, m+r. \end{aligned}$$

Initialization:

Start with any x_0, $J_0 = \{ \alpha_{10}, \ldots, \alpha_{n0} \}$, and $D_0^{-1} = [c_{10}, \ldots, c_{n0}]$, where $D_0 = [d_1, \ldots, d_n]$ is nonsingular, $d_i = a_{\alpha_{i0}}$ for all i with $1 \leq \alpha_{i0} \leq m+r$, and, with $g_0 = c + Cx_0$, $g_0'c_{i0} \leq 0$ for all i with $1 \leq \alpha_{i0} \leq m$ and $g_0'c_{i0} = 0$ for all i with $\alpha_{i0} \leq 0$. Compute $f(x_0) = c'x_0 + \tfrac{1}{2}x_0'Cx_0$. Set $j = 0$.

Step 1: Computation of a New Active Constraint.

Let $D_j^{-1} = [c_{1j}, \ldots, c_{nj}]$ and $J_j = \{ \alpha_{1j}, \ldots, \alpha_{nj} \}$. If there is at least one i such that $\alpha_{ij} > 0$ and $a_{\alpha_{ij}}'x_j \neq b_{\alpha_{ij}}$, go to Step 1.1. Otherwise, go to Step 1.2.

Step 1.1:

Determine the smallest index $\alpha_{\nu j}$ such that

$$| a_{\alpha_{\nu j}}'x_j - b_{\alpha_{\nu j}} | = \max \{ | a_{\alpha_{ij}}'x_j - b_{\alpha_{ij}} | \, | \, \text{all } \alpha_{ij} \text{ with } 1 \leq \alpha_{ij} \leq m+r \}.$$

Set $l = \alpha_{\nu j}$ and go to Step 2.

Step 1.2:

Determine the smallest index l such that

$$a_l'x_j - b_l = \max\{a_i'x_j - b_i, i = 1, \ldots, m, |a_i'x_j - b_i|, i = m+1, \ldots, m+r\}.$$

If $a_l'x_j - b_l \leq 0$, stop with optimal solution x_j. Otherwise, go to Step 1.3.

Step 1.3:

If there is at least one i with $\alpha_{ij} \leq 0$ and $a_l'c_{ij} \neq 0$, go to Step 1.4. Otherwise, compute

$$\lambda_i = a_l'c_{ij}, \quad \text{for all } i \text{ with } 1 \leq \alpha_{ij} \leq m.$$

If $\lambda_i \leq 0$ for all i, print the message "the problem has no feasible solution" and stop. Otherwise, compute the multipliers

$$v_i = -g_j'c_{ij}, \quad \text{for all } i \text{ with } 1 \leq \alpha_{ij} \leq m \text{ and } \lambda_i > 0.$$

Determine the smallest index ν such that

$$\frac{v_\nu}{\lambda_\nu} = \min \left\{ \frac{v_i}{\lambda_i} \mid \text{all } i \text{ with } 1 \leq \alpha_{ij} \leq m \text{ and } \lambda_i > 0 \right\} .$$

Set $\hat{D}_{j+1}^{-1} = D_j^{-1}$ and go to Step 4.1.

Step 1.4:

If there is at least one i with $\alpha_{ij} = -1$ and $a_l' c_{ij} \neq 0$, go to Step 1.5. Otherwise, determine the smallest index ν with

$$| \, a_l' c_{\nu j} \, | = \max \{ \, | \, a_l' c_{ij} \, | \mid \text{all } i \text{ with } \alpha_{ij} = 0 \, \} .$$

Set $\hat{D}_{j+1}^{-1} = D_j^{-1}$ and go to Step 4.1.

Step 1.5:

Determine the smallest index ν with

$$| \, a_l' c_{\nu j} \, | = \max \{ \, | \, a_l' c_{ij} \, | \mid \text{all } i \text{ with } \alpha_{ij} = -1 \, \} .$$

Set

$$\hat{D}_{j+1}^{-1} = \Phi_2(D_j^{-1} , a_l , \nu , J_j)$$

and go to Step 4.1.

Step 2:   **Computation of Search Direction s_j.**

Set

$$s_j = \begin{cases} c_{\nu j} , & \text{if } a_l' x_j > b_l , \\ -c_{\nu j} , & \text{if } a_l' x_j < b_l . \end{cases}$$

Go to Step 3.

Step 3:   **Computation of Step Size σ_j.**

Set

$$\hat{\sigma}_j = | \, a_l' x_j - b_l \, | .$$

Compute

$$\tau_i = c_{ij}' C s_j , \quad \text{for all } i \text{ with } 0 \leq \alpha_{ij} \leq m .$$

If $\tau_i = 0$ for all i with $\alpha_{ij} = 0$ and $\tau_i \geq 0$ for all i with $1 \leq \alpha_{ij} \leq m$ set $\tilde{\sigma}_j = +\infty$. Otherwise, determine the smallest index k such that

$$\tilde{\sigma}_j = \frac{g_j' c_{kj}}{\tau_k}$$

$$= \min \left\{ \frac{g_j' c_{ij}}{\tau_i} \,\middle|\, \begin{array}{l} \text{all } i \text{ with } \alpha_{ij} = 0 \text{ and } \tau_i \neq 0 \text{ and} \\ \text{all } i \text{ with } 1 \leq \alpha_{ij} \leq m \text{ and } \tau_i < 0 \end{array} \right\} .$$

Set

$$\sigma_j \; = \; \min \, \{ \, \hat{\sigma}_j \, , \, \tilde{\sigma}_j \, \} \, .$$

Go to Step 4.2.

Step 4:   **Update.**

Step 4.1:

Set

$$x_{j+1} \; = \; x_j \, , \quad g_{j+1} \; = \; g_j \, , \quad f(x_{j+1}) \; = \; f(x_j) \, ,$$
$$D_{j+1}^{-1} \; = \; \Phi_1(\hat{D}_{j+1}^{-1} \, , \, a_l \, , \, \nu \, , J_j) \, ,$$

and $J_{j+1} \; = \; \{ \, \alpha_{1j+1}, \ldots, \alpha_{nj+1} \, \}$, where

$$\alpha_{i,j+1} \; = \; \alpha_{ij} \, , \quad i \; = \; 1, \ldots, n \, , \quad i \neq \nu \, ,$$
$$\alpha_{\nu,j+1} \; = \; l \, .$$

Replace j with $j + 1$ and go to Step 1.

Step 4.2:

Set

$$x_{j+1} \; = \; x_j \, - \, \sigma_j s_j \, .$$

Compute $g_{j+1} \; = \; c \, + \, C x_{j+1}$ and $f(x_{j+1}) \; = \; c' x_{j+1} \, + \, \frac{1}{2} x_{j+1}' C x_{j+1}$. If $\sigma_j \; = \; \hat{\sigma}_j$ set $D_{j+1}^{-1} \; = \; D_j^{-1}$ and $J_{j+1} \; = \; J_j$. Replace j with $j + 1$ and go to Step 1. Otherwise, compute $d_j \; = \; (c_{kj}' C c_{kj})^{-1/2} C c_{kj}$, and set $D_{j+1}^{-1} \; = \; \Phi_1(D_j^{-1} \, , \, d_j \, , \, k \, , \, J_j)$ and $J_{j+1} \; = \; \{ \, \alpha_{1j+1}, \ldots, \alpha_{nj+1} \, \}$, where

$$\alpha_{i,j+1} \; = \; \alpha_{ij} \, , \quad i \; = \; 1, \ldots, n \, , \quad i \neq k \, ,$$
$$\alpha_{k,j+1} \; = \; -1 \, .$$

Replace j with $j + 1$ and go to Step 1.

The following theorem validates the most important parts of Algorithm 5. The remainder is left as an exercise (Exercise 6.4).

Theorem 6.1 *Let x_0, x_1, ..., x_j ... be obtained by applying Algorithm 5 to the model problem (6.1). If in every step 4.2, $f(x_{j+1}) > f(x_j)$, then after a finite number of steps, Algorithm 5 terminates with an optimal solution or the information that the problem has no feasible solution.*

Proof:

Justification of Step 1.3   The case of $\lambda_i \leq 0$ for all i will be analyzed below. Suppose $\lambda_i > 0$ for at least one i, compute the multipliers

$$v_i = -g_j' c_{ij}. \text{ All } i \text{ with } 1 \leq i \leq \alpha_{ij} \leq m \text{ and } \lambda_i > 0. \tag{6.2}$$

If a_l replaces column $a_{\alpha_{\nu j}}$ $(1 \leq \alpha_{ij} \leq m)$, then the columns of the revised D_j^{-1} become (Procedure Φ)

$$c_i - \frac{a_l' c_{ij}}{a_l' c_{\nu j}} c_{\nu j}.$$

Then the multipliers for the active inequality constraints are changed to

$$-g_j' c_i + \frac{a_l' c_{ij}}{a_l' c_{\nu j}} g_j' c_{\nu j}.$$

Using the definitions of v_i and λ_i, this gives the new multiplier as

$$v_i - \frac{\lambda_i}{\lambda_\nu} v_\nu.$$

Observing that $\lambda_\nu > 0$, these multipliers will remain nonnegative provided

$$\frac{v_\nu}{\lambda_\nu} \leq \frac{v_i}{\lambda_i}$$

for all i with $1 \leq \alpha_{ij} \leq m$. So, choosing ν such that

$$\frac{v_\nu}{\lambda_\nu} = \min \left\{ \frac{v_i}{\lambda_i} \mid \text{all } i \text{ with } 1 \leq \alpha_{ij} \leq m \text{ and } \lambda_i > 0 \right\}$$

ensures that the new multipliers for the active inequality constraints remain nonnegative.

Justification of Step 3   The multipliers for the active inequality constraints will vary as σ varies in $x_j - \sigma s_j$. Let $\tau_i = c_{ij}' C s_j$, for all i with $1 \leq \alpha_{ij} \leq m$. With the identity $\nabla f(x_j - \sigma s_j) = c + C x_j - \sigma C s_j$, the modified multipliers are

$$\begin{aligned} \nabla f(x_j - \sigma s_j) &= -g_j' c_{ij} + \sigma s_j' C c_{ij}, \\ &= -g_j' c_{ij} + \sigma \tau_i. \end{aligned} \tag{6.3}$$

If $\tau_i \geq 0$, then the multiplier for the inequality constraint α_{ij} is increasing in σ and no restriction is required. If $\tau_i < 0$, we must have

$$\sigma \leq \frac{g_j' c_{ij}}{\tau_i}.$$

This last must hold for all i with $\tau_i < 0$. The largest such value which satisfies this is

$$\min \left\{ \frac{g_j' c_{ij}}{\tau_i} \mid \text{all } i \text{ with } 1 \leq \alpha_{ij} \leq m \text{ and } \tau_i < 0 \right\}. \tag{6.4}$$

There is one further possibility which can be conveniently dealt with here. Suppose there is an i with $\alpha_{ij} = 0$ and $\tau_i \neq 0$. Because $\alpha_{ij} = 0$, we must have $g'_j s_j = 0$ so this situation results in a zero stepsize. This possibility can be accounted for in (6.4). Summarizing, if $\tau_i \geq 0$ for all i with $1 \leq \alpha_{ij} \leq m$ and $\tau_i = 0$ for all i with $\alpha_{ij} = 0$, then there is no upper bound restriction on the optimal step size and we set $\tilde{\sigma}_j = +\infty$. Otherwise we compute k such that

$$
\tilde{\sigma}_j = \frac{g'_j c_{kj}}{\tau_k}
$$
$$
= \min \left\{ \frac{g'_j c_{ij}}{\tau_i} \;\middle|\; \begin{array}{l} \text{all } i \text{ with } \alpha_{ij} = 0 \text{ and } \tau_i \neq 0 \text{ and} \\ \text{all } i \text{ with } 1 \leq \alpha_{ij} \leq m \text{ and } \tau_i < 0 \end{array} \right\}.
$$

Justification of Step 1.3   (The problem has no feasible solution) To get to Step 1.3, we came through Step 1.2. To get to Step 1.2, we must have come though Step 1. Consequently, we know that

$$
a'_{\alpha_{ij}} x_j = b_{\alpha_{ij}} \text{ for all } i \text{ with } \alpha_{ij} \geq 1. \tag{6.5}
$$

If there is at least one i with $\alpha_{ij} \leq 0$ and $a'_l c_{ij} \neq 0$, go to Step 1.4. Otherwise,

$$
a'_l c_{ij} = 0 \text{ for all } i \text{ with } \alpha_{ij} \leq 0. \tag{6.6}
$$

Letting $D'_j = [d_{1j}, d_{2j} \ldots, d_{nj}]$, it follows from Lemma 3.1 that

$$
a_l = \sum_{-1 \leq \alpha_{ij} \leq m+r} (a'_l c_{ij}) d_{ij} \tag{6.7}
$$

so that from (6.6) and (6.7)

$$
a_l = \sum_{1 \leq \alpha_{ij} \leq m} (a'_l c_{ij}) a_{\alpha_{ij}} + \sum_{m+1 \leq \alpha_{ij} \leq m+r} (a'_l c_{ij}) a_{\alpha_{ij}}
$$
$$
= \sum_{1 \leq \alpha_{ij} \leq m} \lambda_i a_{\alpha_{ij}} + \sum_{m+1 \leq \alpha_{ij} \leq m+r} (a'_l c_{ij}) a_{\alpha_{ij}}, \tag{6.8}
$$

where $\lambda_i = a'_l c_{ij}$ for all i with $1 \leq \alpha_{ij} \leq m$.

If $\lambda_i \leq 0$ for all i, the model problem (6.1) has no feasible solution. This is because from (6.6), (6.8) and the fact that $\lambda_i \leq 0$ for all i x_j is optimal for

$$
\begin{array}{rl}
\max & a'_l x \\
\text{subject to}: & a'_{\alpha_{ij}} x \leq b_{\alpha_{ij}}, \text{ all } i \text{ with } 1 \leq \alpha_{ij} \leq m, \\
& a'_{\alpha_{ij}} x = b_{\alpha_{ij}}, \text{ all } i \text{ with } m+1 \leq \alpha_{ij} \leq m+r
\end{array} \right\} \tag{6.9}
$$

and the fact that the feasible region for (6.1) is contained in that for (6.9). $\square$

A Matlab program which implements Algorithm 5 is given as Alg5.m in Figure 6.5, Section 6.2.

Example 6.1 This example is formulated in two parts. First we present a two dimensional QP and then solve it using Algorithm 4. Then we formulate a second QP by adding additional constraints to the first. Then we solve the second problem by using Algorithm 5 for which we show the intermediate computations in detail.

Let

$$f(x) = -157x_1 - 149x_2 + \tfrac{13}{2}x_1^2 + 5x_1x_2 + \tfrac{13}{2}x_2^2.$$

A function similar to this was considered in Example 1.4 and level sets for it were determined by the eigenvector method. The two problems differ only in their linear terms. These linear coefficients do not change the shape of the level sets but only change their centers. For $f(x)$ defined above, it is easy to determine the center as $(9, 8)'$. Consider the following problem.

$$
\begin{aligned}
\text{minimize}: \quad & f(x) \\
\text{subject to}: \quad -\ & 3x_1 + 5x_2 \le 25, \quad (1) \\
& x_1 + 3x_2 \le 29, \quad (2) \\
& 3x_1 + 2x_2 \le 45, \quad (3) \\
-\ & x_1 \qquad\quad \le 0, \quad (4) \\
& -\ x_2 \le 0. \quad (5)
\end{aligned}
$$

The feasible region and level sets for the objective function are shown in Figure 6.1. We solve this problem using Algorithm 4 to find the optimal solution $x_0 = (9.08, 6.64)'$, $f(x_0) = -1290.98$, $J_0 = (2, -1)$ and

$$
D_0^{-1} = \begin{bmatrix} -0.020 & 0.30 \\ 0.34 & -0.1 \end{bmatrix}.
$$

We next consider a modification of the last problem by including two additional constraints $5x_1 + 8x_2 \le 82$ (6) and $x_1 \le 12$ (7). The modified problem is thus

$$
\begin{aligned}
\text{minimize}: \quad & f(x) \\
\text{subject to}: \quad -\ & 3x_1 + 5x_2 \le 25, \quad (1) \\
& x_1 + 3x_2 \le 29, \quad (2) \\
& 3x_1 + 2x_2 \le 45, \quad (3) \\
-\ & x_1 \qquad\quad \le 0, \quad (4) \\
& -\ x_2 \le 0, \quad (5) \\
& 5x_1 + 8x_2 \le 82, \quad (6) \\
& x_1 \qquad\quad \le 12. \quad (7)
\end{aligned}
$$

Here,

$$
c = \begin{bmatrix} -157 \\ -149 \end{bmatrix} \text{ and } C = \begin{bmatrix} 13 & 5 \\ 5 & 13 \end{bmatrix}.
$$

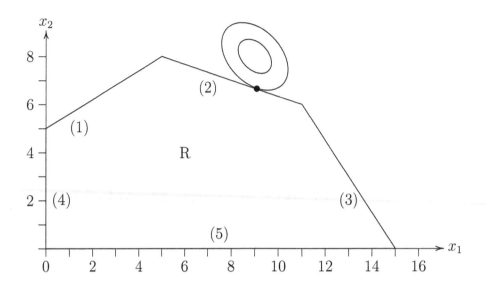

Figure 6.1.   Example 6.1, Original Problem.

We next illustrate the steps of Algorithm 5 by applying it to the modified problem with initial data obtained from applying Algorithm 4 to the unmodified problem.

Initialization:

$$x_0 = \begin{bmatrix} 9.08 \\ 6.64 \end{bmatrix} , \quad J_0 = \{\, 2\,,\, -1\,\} , \quad D_0^{-1} = \begin{bmatrix} -0.02 & 0.30 \\ 0.34 & -0.10 \end{bmatrix} ,$$

$$f(x_0) = -1290.98 , \quad g_0 = \begin{bmatrix} -5.7600 \\ -17.2800 \end{bmatrix} , \quad j = 0 .$$

Iteration 0

Step 1: $a_2' x_0 = b_2$, Go to Step 1.2.

Step 1.2: $a_6' x_0 - b_6 = \min\{-19.04,\ 0,\ -4.48,\ -9.08, -6.64,\ 16.52, \ldots$
$\qquad -2.92\} = 16.52,\ \ l = 6$, Go to Step 1.3.

Step 1.3: $\alpha_{02} = -1 \leq 0$, $a_l' c_{20} = 0.7$, Go to Step 1.4.

Step 1.4: $\alpha_{02} = -1$, $a_l' c_{20} = 0.7$, Go to Step 1.5.

Step 1.5: $|a_l' c_{20}| = \{0.7\} = 0.7$, $\nu = 2$.
$\qquad \hat{D}_1^{-1} = \Psi_2(D_0^{-1}, a_6, 2, J_0) = \begin{bmatrix} -0.0200 & -0.3000 \\ 0.3400 & 0.1000 \end{bmatrix} ,$
$\qquad$ Go to Step 4.1.

Step 4.1: $x_1 = \begin{bmatrix} 9.08 \\ 6.64 \end{bmatrix}$, $g_1 = \begin{bmatrix} -5.7600 \\ -17.2800 \end{bmatrix}$, $f(x_1) = -1290.98$,

$D_1^{-1} = \Psi_1(\hat{D}_1^{-1}, a_2, 2, J_0) = \begin{bmatrix} -1.1429 & 0.4286 \\ 0.7143 & -0.1429 \end{bmatrix}$,

$J_1 = \{2, 6\}$, Go to Step 1.

Iteration 1

Step 1: $a_6' x_1 - b_6 = 16.52$, go to Step 1.1.

Step 1.1: $|a_6' - b_6| = \max\{16.52\} = 16.52$, $l = 6$, $\nu = 2$, Go to Step 2.

Step 2: $s_1 = \begin{bmatrix} 0.4286 \\ -.1429 \end{bmatrix}$, go to Step 3.

Step 3: $\hat{\sigma}_1 = 16.52$ $\tau_1 = -5.3469$, $\tau_2 = 2.0408$
$\tilde{\sigma}_1 = \min\{\frac{-5.7600}{-5.3469}\} = 1.0773$, $k = 1$,
$\sigma_1 = \min\{16.5200, 1.0773\} = 1.0773$. Go to Step 4.2.

Step 4.2: $x_2 = \begin{bmatrix} 8.6183 \\ 6.7939 \end{bmatrix}$, $g_2 = \begin{bmatrix} -10.9924 \\ -17.5878 \end{bmatrix}$, $f(x_2) = -1289.8$

$d_1 = \begin{bmatrix} -2.8713 \\ 0.9086 \end{bmatrix}$,

$D_2^{-1} = \Psi_1(D_1^{-1}, d_1, 1, J_1) = \begin{bmatrix} -0.2908 & 0.0330 \\ 0.1817 & 0.1044 \end{bmatrix}$,

$J_2 = \{-1, 6\}$. Go to Step 1.

Iteration 2

Step 1: $a_6' x_2 - b_6 = 15.4427$,

Step 1.1: $|a_6 x_2 - b_6| = \max\{15.4427\} = 15.4427$, $\nu = 2$, $l = 6$. Go to Step 2.

Step 2: $s_2 = \begin{bmatrix} 0.0330 \\ 0.1044 \end{bmatrix}$. Go to Step 3.

Step 3: $\hat{\sigma}_2 = 15.4427$, $\tau_1 = -$, $\tau_2 = 0.1902$, $\tilde{\sigma}_2 = \infty$,
$\sigma_2 = \min\{15.4427, \infty\} = 15.4427$ Go to Step 4.2.

Step 4.2 $x_3 = \begin{bmatrix} 8.1083 \\ 5.1823 \end{bmatrix}$, $g_3 = \begin{bmatrix} -25.6803 \\ -41.0885 \end{bmatrix}$, $f(x_3) = -1233.2$

$$\sigma_2 = \hat{\sigma}_2 \; ; \; D_3^{-1} = D_2^{-1} = \begin{bmatrix} -0.2908 & 0.0330 \\ 0.1817 & 0.1044 \end{bmatrix},$$
$$J_3 = J_2 = \{-1, 6\}.$$

Iteration 3

Step 1:   Go to Step 1.2.

Step 1.2: $a_6 x_3 - b_6 = \max\{-23.4135, -5.3448, -10.3104, -8.1083,$
$-5.1823, 0, -3.8917\} = 0, l = 6.$

Stop with optimal solution $x_3 = \begin{bmatrix} 8.1083 \\ 5.1823 \end{bmatrix}$

and $f(x_3) = -1233.2.$

The iterates x_0, x_1, x_2, x_3 are shown in Figure 6.2 along with the modified feasible region.

This completes Example 6.1. $\Diamond$

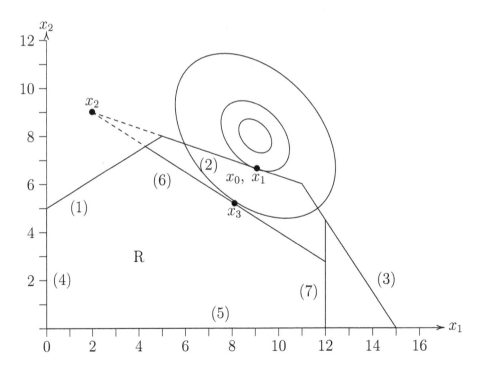

Figure 6.2.   Example 6.1, Modified Constraints.

Example 6.1 is also solved using Alg5.m. The calling program is shown Figure 6.6, Section 6.2 and the output is shown in Figure 6.7.

The Hessian for Example 6.1 is positive definite. Algorithm 5 also includes the possibility of a positive semidefinite Hessian. The following example illustrates this possibility.

Example 6.2 Let $f(x) = -16x_1 - 16x_2 + x_1^2 + 2x_1x_2 + x_2^2$. A similar function is analyzed in Example 1.5. See Figure 1.8. The level sets for the function at hand are pairs of parallel lines centered about the line $x_1 + x_2 = 8$.

We proceed as in Example 6.1, by first solving an unmodified problem using Algorithm 4, then adding additional constraints to create a modified problem. The final data from Algorithm 4 is used as starting data for using Algorithm 5 to solve the modified problem.

The unmodified problem is

$$
\begin{aligned}
\text{minimize}: \quad & f(x) \\
\text{subject to}: \quad -\ x_1\ +\ x_2 \ &\leq\ \ 2, \quad (1) \\
2x_1\ +\ 3x_2 \ &\leq\ 16, \quad (2) \\
x_1 \ &\leq\ \ 5, \quad (3) \\
-\ x_2 \ &\leq\ \ 0, \quad (4) \\
-\ x_1 \ &\leq\ -1. \quad (5)
\end{aligned}
$$

and is shown in Figure 6.3.

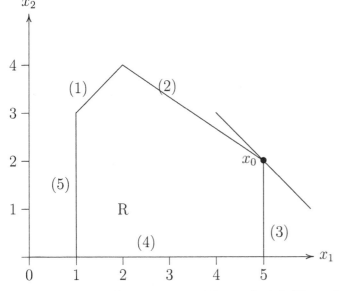

Figure 6.3.   Example 6.2, Unmodified Problem.

By inspection, the optimal solution is $(5, 2)'$. Applying Algorithm 4 to this problem verifies this with $x_0 = (5, 2)'$, $J_0 = \{3, 2\}$, and

$$
D_0^{-1} = \begin{bmatrix} 1.0 & 0.0 \\ -0.6667 & 0.3333 \end{bmatrix}.
$$

We next consider a modification of the last problem by including one additional constraint $-x_2 \leq -3$ (6). The modified problem is thus

$$
\begin{array}{rrrcll}
\text{minimize}: & f(x) & & & & \\
\text{subject to}: & - \quad x_1 & + & x_2 & \leq & 2, \quad (1) \\
& 2x_1 & + & 3x_2 & \leq & 16, \quad (2) \\
& x_1 & & & \leq & 5, \quad (3) \\
& & - & x_2 & \leq & 0, \quad (4) \\
& - \quad x_1 & & & \leq & -1, \quad (5) \\
& & - & x_2 & \leq & -3, \quad (6)
\end{array}
$$

and is shown in Figure 6.4. Notice that the modified feasible region is just the triangle bounded by constraints (1), (2) and (6).

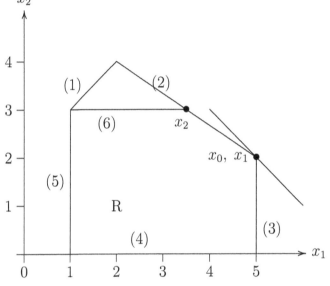

Figure 6.4.   Example 6.2, Modified Problem.

We next illustrate the steps of Algorithm 5 by applying it to the modified problem with initial data obtained from applying Algorithm 4 to the unmodified problem.

Initialization:

$$
x_0 = \begin{bmatrix} 5 \\ 2 \end{bmatrix}, \quad J_0 = \{3,2\}, \quad D_0^{-1} = \begin{bmatrix} 1.0 & 0.0 \\ -2/3 & 1/3 \end{bmatrix},
$$

$$
f(x_0) = -63.0, \quad g_0 = \begin{bmatrix} -2.0 \\ -2.0 \end{bmatrix}, \quad j = 0.
$$

Iteration 0

Step 1: $a_3' x_0 = b_3$, $a_2 x_0 = b_2$ Go to Step 1.2.

Step 1.2: $a_6'x_0 - b_6 = \max\{-5,\ 0,\ 0,\ -2,\ -4,\ 1\} = 1,\ l = 6.$
Go to Step 1.3.

Step 1.3: $\lambda_1 = 0.6667,\ \lambda_2 = -0.3333,\ v_1 = 0.6667,$
$\frac{v_1}{\lambda_1} = \min\{\frac{0.6667}{0.6667}\} = 1,\ \nu = 1.\ \hat{D}_1^{-1} = \begin{bmatrix} 1.0 & 0.0 \\ -2/3 & 1/3 \end{bmatrix}.$
Go to Step 4.1.

Step 4.1: $x_1 = \begin{bmatrix} 5 \\ 2 \end{bmatrix},\ g_1 = \begin{bmatrix} -2.0 \\ -2.0 \end{bmatrix},\ f(x_1) = -63.0,$
$D_1^{-1} = \Psi_1(\hat{D}_0^{-1}, a_6, 1, J_0) = \begin{bmatrix} 1.5000 & 0.5000 \\ -1.0000 & 0.0000 \end{bmatrix},$
$J_1 = \{6,\ 2\}.$ Go to Step 1.

Step 1: $|a_6'x_1 - b_6| = 1.$Go to Step 1.1.

Step 1.1: $|a_6'x_1 - b_6| = 1 = \max\{1,\ 0\},\ \nu = 1,\ l = 6.$ Go to Step 2.

Step 2: $s_1 = \begin{bmatrix} 1.50000 \\ -1.0000 \end{bmatrix}.$ Go to Step 3.

Step 3: $\hat{\sigma}_1 = 1.0,\ \tau_1 = 0.50,\ \tau_2 = 0.50\ \tilde{\sigma}_1 = \infty$
$\sigma_1 = \min\{1.0,\ \infty\} = 1.0.$ Go to Step 4.2.

Step 4.2: $x_2 = \begin{bmatrix} 3.5 \\ 3.0 \end{bmatrix},\ g_1 = \begin{bmatrix} -3.0 \\ -3.0 \end{bmatrix},\ f(x_1) = -61.75,$
$D_2^{-1} = \begin{bmatrix} 1.5000 & 0.5000 \\ -1.0000 & 0.0000 \end{bmatrix},\ J_2 = \{6,\ 2\}.$ Go to Step 1.

Step 1: Go to Step 1.2.

Step 1.2: $a_2'x_2 - b_2 = \max\{-2.5,\ 0.0,\ -1.5,\ -3.0,\ -2.5,\ 0.0\} = 0.0.$
Stop with optimal solution $x_2 = \begin{bmatrix} 3.5 \\ 3.0 \end{bmatrix}$ and
$f(x_2) = -61.75.$

The iterates x_0, x_1, and x_2 are shown in Figure 6.4 along with the modified feasible region.
　This completes Example 6.2.   ◊

　A calling program for Example 6.2 is given in Figure 6.8 and its associated output is shown in Figure 6.9.

6.2   Computer Programs

```
1   %   Alg5.m
2   function [x,J,fx,Dinv,msg,dual,iter,alg] = ...
3                           Alg5(c,C,n,m,r,tol,x,J,Dinv,A,b)
4   %    implements QP Algorithm 5, which is a dual
5   %            QP Algorithm.
6   %  Initialization
7   alg = 'Algorithm 5';
8   fx = c'*x + 0.5*( x'*C*x);
9   g = c + C*x;
10  dfeas = 1;
11  for i=1:n
12     if J(i) ≤ 0
13         if abs(g'*Dinv(:,i)) ≥ tol
14             dfeas = 0;
15          end
16     end
17     if J(i) ≥ 1 & J(i) ≤ m
18         if g'*Dinv(:,i) ≥ tol
19             dfeas = 0;
20         end
21     end
22  end    %end of for loop
23
24  if dfeas == 0;
25     % initial point not dual feasible.
26     msg = 'initial point not dual feasible.'
27     return
28  end
29
30  for iter=0:10000;
31     p11 = 0;
32     for i=1:n
33         if J(i) > 0
34             if abs( A(J(i),:)*x − b(J(i)) ) > tol
35                 p11 =   1;
36             end
37         end
38     end
39     if p11 == 1
40         next = 'Step1p1';
41     else
42         next = 'Step1p2';
43     end
44     if strcmp(next,'Step1p1')
45                                 % Step 1.1:
46         big = − inf;
47         index  = 0;
```

```
48          for i=1:n
49              if J(i) > 0
50                  test = abs( A(J(i),:)*x - b(J(i)) );
51                  if test > big
52                      big = test;
53                      index = i;
54                  end
55              end
56          end     %end of for loop
57          nu = index;
58          ell = J(nu);
59          next = 'Step2';
60      end     % end of Step 1.1
61
62      if strcmp(next,'Step1p2')
63                                      % Step 1.2:
64          big = - inf;
65          ell = 0;
66          for i=1:m+r
67              test = A(i,:)*x - b(i);
68              if i > m;
69                  test = abs(test);
70              end
71              if test > big
72                  big = test;
73                  ell = i;
74              end
75          end     %end of for loop
76          if big <= tol
77              %  Opimal Solution Obtained
78              msg = 'Optimal solution obtained';
79              [dual] = GetDual(Dinv,J,g,n,m,r);
80              return
81          else
82              next = 'Step1p3';
83          end
84      end     % end of Step 1.2
85      if strcmp(next,'Step1p3')
86                                      % Step 1.3
87          p13 = 0;
88          for i=1:n
89              if J(i) <= 0 & abs(A(ell,:)*Dinv(:,i)) >= tol
90                  p13 =   1;
91              end
92          end     %end of for loop
93          if p13 == 1
94              next = 'Step1p4';
95          else
96              nu = 0;
97              minratio = inf;
98              infeas = 1;
```

```
99              for i=1:n
100                 if 1 ≤ J(i) & J(i) ≤ m
101                     lambda = A(ell,:)*Dinv(:,i);
102                     if lambda > 0
103                         infeas = 0;
104                     end
105                     if lambda > 0
106                         vee = − g'*Dinv(:,i);
107                         ratio = vee/lambda;
108                         if ratio < minratio
109                             minratio = ratio;
110                             nu = i;
111                         end
112                     end
113                 end
114             end      %end of for loop
115             if infeas == 1;
116                 msg = 'No feasible solution'
117             return;
118             end
119             Dhatinv = Dinv;
120             next = 'Step4p1';
121         end             %end of if−else loop
122     end     % end of Step 1.3
123
124     if strcmp(next,'Step1p4')
125                                     %Step 1.4
126         p14 = 0;
127         for i=1:n
128             if J(i) == −1 & abs(A(ell,:)*Dinv(:,i)) ≥ tol
129                 p14 =  1;
130             end
131         end    %end of for loop
132         if p14 == 1
133             next = 'Step1p5';
134         else
135         % continue with Step 1.4
136             nu = 0;
137             big = −inf;
138             for i=1:n
139                 if J(i) == 0
140                     testbig = abs(A(ell,:)*Dinv(:,i));
141                     if testbig > big
142                         big = testbig;
143                         nu = i;
144                     end
145                 end
146             end      %end of for loop
147             Dhatinv = Dinv;
148             next = 'Step4p1';
149         end    %end of if−else loop
```

```
150      end     % end of Step 1.4
151
152      if strcmp(next,'Step1p5')
153                                  %Step 1.5
154          nu = 0;
155          big = -inf;
156          for i=1:n
157              if J(i) == -1
158                  testbig = abs(A(ell,:)*Dinv(:,i));
159                  if testbig > big
160                      big = testbig;
161                      nu = i;
162                  end
163              end
164          end     %end of for loop
165          [Dhatinv] = Phi2(Dinv,A(ell,:)',nu,J,n);
166          next = 'Step4p1';
167      end     % end of Step 1.5
168
169      if strcmp(next,'Step2')
170                                  %Step 2
171          if A(ell,:)*x > b(ell)
172              s = Dinv(:,nu);
173          else
174              s = - Dinv(:,nu);
175          end
176          next = 'Step3';
177      end     % end of Step 2
178
179      if strcmp(next,'Step3');
180                                  %Step3
181          t1 =  ( A(ell,:)*x - b(ell) );
182          t2 = A(ell,:)*s;
183          sighat = abs(t1/t2);
184          kay = 0;
185          ratio = inf;
186          Csj = C*s;
187          for i=1:n
188              if J(i) == 0
189                  tau = Dinv(:,i)'*Csj;
190                  if abs(tau) >= tol
191                      test = g' * Dinv(:,i)/tau;
192                      if test <= ratio
193                          ratio = test;
194                          kay = i;
195                      end
196                  end
197              ratio
198              kay
199              elseif J(i) >= 1 & J(i) <= m
200                  tau = Dinv(:,i)'*Csj;
```

```
201              vee = g' * Dinv(:,i);
202              if tau < -tol
203                  test = g' * Dinv(:,i)/tau;
204                  if test ≤ ratio
205                      ratio = test;
206                      kay = i;
207                  end
208              end
209          end
210      end  % end for i=1:n
211      sigtil = ratio;
212      sigma = min(sighat,sigtil);
213      next = 'Step4p2';
214  end     % end of Step 3
215
216
217  if strcmp(next,'Step4p1')
218                              %Step 4p1
219      % no updates required for x, g or fx
220      Dinv = Phil(Dhatinv,A(ell,:)',nu,J,n);
221      J(nu) = ell;
222      next = 'End';
223  end     % end of Step 4.1
224
225  if strcmp(next,'Step4p2')
226                              % Step 4p2 :
227      x = x - sigma*s;
228      g = c + C*x;
229      fx = c'*x + 0.5*( x'*C*x);
230      if sigma == sighat
231          %sigma = sighat: nothing to do
232          next = 'End';
233      else          %  sigma = sigtil
234          d = (Dinv(:,kay)'*C*Dinv(:,kay)).^(-0.5) ...
235                              *C*Dinv(:,kay);
236          Dinv = Phil(Dinv,d,kay,J,n);
237          J(kay) = -1;
238          next = 'End';
239      end
240  end     % end of Step 4.2
241  if strcmp(next,'End')
242  end
243
244  end   %end of for j=0:10000;
```

Figure 6.5.   Algorithm 5: A Dual Method, Alg5.m.

Discussion of Alg5.m. The initial point "x" given to Algorithm 5 is assumed to be dual feasible. Lines 10–28 check this assumption. In line 10 "dfeas" is set to 1. Any dual infeasibility detected results in "dfeas" being set to 0. If "dfeas" is found to be 0 in line 24, an error message is issued and ALG5 returns to the calling program on line 27.

Step 1 begins on line 31 initializing the flag "p11" to zero. The "for" loop (lines 32–38) looks for a constraint whose gradient is a column of D'_j and that primal constraint is either inactive or violated. In this case, the flag "p11" is set to 1. If "p11" is one, control passes to Step1p1 by setting "next" = 'Step1p1'. If "p11" is zero control passes to Step1p2 by setting "next" = 'Step1p2'.

If Step1p1 is taken, the maximum of $|a_{\alpha_{ij}} - b_{\alpha_{ij}}|$ is computed for inequality and equality constraints in J_j. The index for which the maximum occurs is ν and Step1p1 continues by setting $l = \alpha_{ij}$ and control passes to Step2.

If Step1p2 is taken, the max of all the inequality constraints, $a_i x_j - b_i$ and $|a_i x_j - b_i|$ for all the equality constraints is computed and has index l. If $a_l x_j - b_l \leq 0$, then x_j is optimal and we stop. Otherwise control transfers to Step1p3.

In Step1p3, a check is first made to see if there is an index i with $\alpha_{ij} \leq 0$ and $a'_l c_{ij} \neq 0$. If so, control passes to Step1p4. Otherwise, compute $\lambda_i = a'_l c_{ij}$ for all inequality constraints i. If all such λ_i are nonnegative, the problem has no feasible solution. Otherwise, compute the multipliers $v_i = -g'_j c_{ij}$ for all active constraints i and $\lambda_i > 0$. Now determine ν so that v_ν / λ_ν is the smallest among all v_i / λ_i over all i for which v_i is defined. Set $\hat{D}^{-1}_{j+1} = D^{-1}_j$ set "next" = 'Step4p1' which will transfer control to Step 4.1.

In Step1.4 a check is made to see if there is at least one i with $\alpha_{ij} = -1$ and $a'_l c_{ij} \neq 0$ and if so control transfers to Step 1.5 with "next" = 'Step1p5'. Otherwise determine ν so that $a'_l c_{\nu j}$ is the largest of $|a'_l c_{ij}|$ over all i with $\alpha_{ij} = 0$. Set $\hat{D}^{-1}_{j+1} = D^{-1}_j$ and "next" = 'Step4p1' which will transfer control to Step 4.1.

In Step 1.5, the index ν is computed such that $|a'_l c_{n_j}|$ is the largest among all such quantities for which column i is a conjugate direction column of D^{-1}_j. This inner product is known to be non zero because Step 1.5 was reached from the first part of Step 1.4. Finally, procedure Φ_2 is used with arguments D^{-1}_j, a_l, ν and J_j to produce $\hat{D}^{-1}_j$ and control transfers to Step 4.1 using "next" = 'Step4p1'.

Step 2 sets the search direction s_j to be $c_{\nu j}$ if $a'_l x_j > b_l$ and $-c_{\nu j}$ if $a'_l x_j < b_l$. Control now transfers to Step 3 by setting "next" = 'Step3'.

In Step 3, the maximum feasible step size $\hat{\sigma}_j$ is computed as $\hat{\sigma}_j = |a_l x_j - b_l|$.

Next we compute $\tau_i = c'_{ij} C s_j$ for all i with $0 \leq \alpha_{ij} \leq m$ The computations the proceed exactly as in Step 3 of the algorithm to determine

$$\tilde{\sigma}_j = \frac{g'_j c_{kj}}{\tau_k}.$$

We now compute $\sigma_j = \min\{\hat{\sigma}_j, \tilde{\sigma}_j\}$ and control transfers to Step 4.2 by setting "next" = 'Step4p2'.

In Step 4.1, x_{j+1}, g_{j+1} and f_{j+1}, are unchanged from their previous values. D_{j+1}^{-1} is updated using $\hat{D}_{j+1}^{-1}$, a_l, ν and J_j. The index set J_j is updated to J_{j+1} by replacing the ν-th element with l. Replace j with $j+1$ and go to Step 1 by setting "next" = 'Step1'.

In Step 4.2, x_j is updated according to $x_{j+1} = x_j - \sigma_j s_j$ and g_{j+1} as well as $f(x_{j+1})$ are updated accordingly. If $\sigma_j = \hat{\sigma}_j$, D_{j+1}^{-1} and J_{j+1} remain at their previous values. Replace j with $j+1$ and go to Step 1 by setting "next" = 'Step1'. Otherwise, compute $d_j = (c'_{kj} C c_{kj})^{-\frac{1}{2}} C c_{kj}$, $D_{j+1}^{-1} = \Phi_1(D_j^{-1}, d_j, k, J_{j+1}$ where J_{j+1} is obtained from J_j by replacing the k-th element with -1. Replace j with $j+1$ and go to Step 1 by setting "next" = 'Step1'.

```
1  %   eg6p1.m
2  %   Modified eg6p1 solution using Alg5
3  n = 2;   m = 7;   r = 0;   tol = 1.e-6;
4  c = [ -157 -149 ]';
5  C = [13 5; 5 13];
6  A = [-3 5; 1 3; 3 2; -1 0; 0 -1; 5 8; 1 0 ];
7  b = [25 29 45 0 0 82 12 ]';
8  x = [9.0800000 6.6400000]';
9  Dinv = [-0.0200 0.3000000 ; 0.340000 -0.10000];
10 J = [2 -1];
11 checkdata(C);
12 [x,J,fx,Dinv,msg,dual,j,alg] = ...
13                     Alg5(c,C,n,m,r,tol,x,J,Dinv,A,b);
14 checkKKT(A,b,x,n,m,r,dual,c,C,fx,j,msg,alg);
```

Figure 6.6.　　Example 6.1 (modified): eg6p1.m.

```
1  eg6p1
2  Optimization Summary
3  Optimal solution obtained
4  Algorithm: Algorithm 5
5  Optimal objective value    -1233.16314399
6  Iterations    3
7  Maximum Primal Error       0.00000000
8  Maximum Dual Error         0.00000000
9
10
11   Optimal Primal Solution
12      1    8.10832232
13      2    5.18229855
14
15
16   Optimal Dual Solution
```

```
17    1    0.00000000
18    2    0.00000000
19    3    0.00000000
20    4    0.00000000
21    5    0.00000000
22    6    5.13606341
23    7    0.00000000
```

Figure 6.7. Example 6.1 (modified), output: eg6p1(a).m.

```
1   %   eg6p2.m
2   %   Modified eg6p2 soln using Alg5
3   n = 2;  m = 6;  r = 0;   tol = 1.e−6;
4   c = [ −16 −16 ]';
5   C = [2 2; 2 2];
6   A = [−1 1; 2 3; 1 0; 0  −1; −1 0 ; 0. −1 ];
7   b = [2 16 5 0 −1 −3 ]';
8   x = [5 2]';
9   Dinv = [1 0; −0.666666667  0.333333333];
10  J = [3 2];
11  checkdata(C);
12  [x,J,fx,Dinv,msg,dual,j,alg] = ...
13                        Alg5(c,C,n,m,r,tol,x,J,Dinv,A,b);
14  checkKKT(A,b,x,n,m,r,dual,c,C,fx,j,msg,alg);
```

Figure 6.8. Example 6.2 (modified): eg6p2.m.

```
1   eg6p2
2   Optimization Summary
3   Optimal solution obtained
4   Algorithm: Algorithm 5
5   Optimal objective value    −61.75000000
6   Iterations    2
7   Maximum Primal Error      0.00000000
8   Maximum Dual Error        0.00000001
9
10
11    Optimal Primal Solution
12      1    3.50000000
13      2    3.00000000
14
15
16    Optimal Dual Solution
17      1    0.00000000
18      2    1.50000000
19      3    0.00000000
```

```
20      4      0.00000000
21      5      0.00000000
22      6      1.50000000
```

Figure 6.9.   Example 6.2 (modified), output: eg6p2(a).m.

6.3   Exercises

6.1 Consider the problem of Exercise 4.5. Find the unconstrained minimum of the objective function for this problem and call it x_0. Solve the problem using Algorithm 5 with the starting point x_0.

6.2 Consider the problem of Exercise 5.8. Find the unconstrained minimum for the objective function of this problem. Use this as a starting point and solve the problem using Algorithm 5.

6.3 Consider the problem of Exercise 4.5. Find the unconstrained minimum for the objective function of this problem. Use this as a starting point and solve the problem using Algorithm 5.

6.4 Complete the proof of Theorem 6.1.

Chapter 7

General QP and Parametric QP Algorithms

Algorithms 1 through 5 have all used conjugate directions to perform part of the quadratic minimization. This is one way this can be done but it is not the only way. The critical thing is that certain linear equations must be solved and the distinguishing feature of a particular quadratic programming algorithm is the method that it uses to solve the linear equations. In Section 7.1, we present a general quadratic programming algorithm. It is general in the sense that it leaves unspecified the method for solving the relevant linear equations. The "simplex method for quadratic programming" will be derived in Chapter 8 as a special case of this general algorithm. The model problem for this method has nonnegativity constraints and its updating relies on the assumption that many variables will be zero at intermediate iterations.

It has been shown in Best [1] that the methods of Best and Ritter (Algorithm 4 here) [5], Fletcher [9], Gill and Murray [10] and the "simplex method for quadratic programming" are equivalent in the sense that they all generate the same sequence of points when applied to a common problem with the same initial point and ties are resolved in a common way. Thus, most quadratic programming algorithms differ only in the way that they solve the linear equations. The only assumption made is that C is positive semidefinite.

In Section 7.2, we give the parametric extension of the general quadratic programming algorithm. It will be used in Chapter 8 to obtain the analogous parametric extension of the "simplex method for quadratic programming".

Section 7.3 will provide updating formulae for a symmetric matrix. The update finds the inverse of the modified matrix when a row and column are added to the matrix (Procedure Ψ_1), when a row and column are deleted from the matrix (Procedure Ψ_2) and when a row and column are exchanged (Procedure Ψ_3). These updates are used to make the general QP algorithm and its parametric extension into implementable algorithms.

251

Throughout this chapter, we will assume that the objective function for the QP is convex.

7.1   A General QP Algorithm: Algorithm 6

Consider the model problem

$$\left.\begin{array}{rl} \text{minimize}: & c'x + \frac{1}{2}x'Cx \\ \text{subject to}: & a_i'x \leq b_i, \quad i = 1, \ldots, m, \\ & a_i x = b_i, \quad i = m+1, \ldots, m+r. \end{array}\right\} \tag{7.1}$$

Assume that C is positive semidefinite and that $a_{m+1}, \ldots, a_{m+r}$ are linearly independent. Let $f(x) = c'x + \frac{1}{2}x'Cx$. We now formulate an algorithm for the solution of (7.1) which constructs a sequence of points x_0, x_1, x_2, $\ldots$ according to $x_{j+1} = x_j - \sigma_j s_j$, $j = 0, 1, \ldots$ and after finitely many steps either determines an optimal solution for (7.1) or determines that (7.1) is unbounded from below. As with Algorithm 4, we will call s_j and σ_j the search direction and step size, respectively, at iteration j. For any $x \in R$, let

$$I(x) = \{ i \mid a_i'x = b_i, \ 1 \leq i \leq m + r \}$$

so that $I(x)$ denotes the (unordered) set of indices of constraints active at x. By definition, each of $m + 1, \ldots, m + r$ is in $I(x)$. At iteration j, let K_j denote an ordered subset of $I(x_j)$ such that a_i, all $i \in K_j$, are linearly independent. We refer to K_j as the <u>active set</u> at iteration j. K_j is analogous to J_j used in Algorithms 1 through 6.

Suppose first that x_j is not a quasistationary point. We choose s_j such that s_j is an optimal solution for

$$\min \{ f(x_j - s_j) \mid a_i'(x_j - s_j) = b_i, \text{ all } i \in K_j \}, \tag{7.2}$$

provided that (7.2) has an optimal solution. Let A_j' denote the matrix whose columns are those a_i with $i \in K_j$ and ordered according to K_j. From Taylor's series (3.6) the objective function for (7.2) may be written

$$f(x_j - s_j) = f(x_j) - g_j's_j + \frac{1}{2}s_j'Cs_j,$$

where $g_j = g(x_j)$. The constraints become

$$A_j s_j = 0.$$

Omitting the constant term $f(x_j)$ in the objective function, (7.2) is equivalent to

$$\min \{ -g_j's_j + \frac{1}{2}s_j'Cs_j \mid A_j s_j = 0 \}. \tag{7.3}$$

According to Theorem 3.11, s_j is optimal for (7.3) if and only if there is a vector v_j which with s_j satisfies

$$Cs_j + A'_j v_j = g_j,$$
$$A_j s_j = 0.$$

These can be written in partitioned matrix form as

$$\begin{bmatrix} C & A'_j \\ A_j & 0 \end{bmatrix} \begin{bmatrix} s_j \\ v_j \end{bmatrix} = \begin{bmatrix} g_j \\ 0 \end{bmatrix}. \tag{7.4}$$

Let

$$H_j = \begin{bmatrix} C & A'_j \\ A_j & 0 \end{bmatrix}$$

denote the coefficient matrix for (7.4). If H_j is nonsingular, we call s_j in the solution of (7.4) the <u>Newton direction</u> at x_j.[1] Having obtained s_j from (7.4), we compute the maximum feasible step size $\hat{\sigma}_j$ for s_j and set $\sigma_j = \min\{1, \hat{\sigma}_j\}$. If $\sigma_j = 1$, then x_{j+1} is a quasistationary point for (7.1). If $\sigma_j < 1$, then some constraint $l \notin K_j$ becomes active at x_{j+1} and we augment K_j with l to obtain the new active set K_{j+1}. If there are ρ_j constraints in the active set at iteration j, then in at most $n - \rho_j$ such iterations either a quasistationary point will be determined or there will be n constraints in the active set at iteration $j + n - \rho_j$. In the latter case, the corresponding point is an extreme point. Since an extreme point is a special case of a quasistationary point, it follows that a quasistationary point must be obtained in at most $n - \rho_j$ iterations.

Now suppose that $\sigma_j = 1$ so that x_{j+1} is a quasistationary point. The multipliers for the constraints active at x_{j+1} can be obtained from v_j in the solution of (7.4) as follows. By definition, $g_{j+1} = c + Cx_{j+1}$, so that from the first partition of (7.4),

$$g_{j+1} = c + Cx_{j+1} = A'_j v_j.$$

That is,

$$-g_{j+1} = A'_j(-v_j), \tag{7.5}$$

and the multipliers for the active constraints may be obtained by negating the components of v_j. The $(m + r)$-dimensional vector of multipliers u_{j+1} can be obtained by setting the components associated with inactive constraints to

---

[1]Newton's method for nonlinear optimization uses a search direction which minimizes a quadratic approximation to the objective function.

zero and using K_j to set the remaining to the appropriate component of $-v_j$. Let k be such that

$$(u_{j+1})_k = \min \{ (u_{j+1})_i \mid i = 1, \ldots, m \} . \tag{7.6}$$

If $(u_{j+1})_k \geq 0$, then x_{j+1} is an optimal solution. If $(u_{j+1})_k < 0$, we proceed by dropping constraint k from the active set. That is, we obtain K_{j+1} from K_j by deleting k. If H_{j+1} is nonsingular, we continue by determining the new Newton direction s_{j+1} from (7.4) with j replaced by $j + 1$. It remains to consider the case that H_{j+1} is singular. Suppose that H_{j+1} is singular. By definition of singularity, there is a nonzero vector, which when multiplied by H_{j+1}, gives zero. Let this vector be $(s'_{j+1} , v'_{j+1})'$. Then

$$H_{j+1} \begin{bmatrix} s_{j+1} \\ v_{j+1} \end{bmatrix} = \begin{bmatrix} C & A'_{j+1} \\ A_{j+1} & 0 \end{bmatrix} \begin{bmatrix} s_{j+1} \\ v_{j+1} \end{bmatrix} = \begin{bmatrix} Cs_{j+1} + A'_{j+1}v_{j+1} \\ A_{j+1}s_{j+1} \end{bmatrix} = 0,$$

so that

$$Cs_{j+1} + A'_{j+1}v_{j+1} = 0 \quad \text{and} \tag{7.7}$$
$$A_{j+1}s_{j+1} = 0. \tag{7.8}$$

Multiplying (7.7) on the left with s'_{j+1} and using (7.8) gives

$$s'_{j+1}Cs_{j+1} + (A_{j+1}s_{j+1})'v_{j+1} = s'_{j+1}Cs_{j+1} = 0.$$

Since C is positive semidefinite, this implies (Exercise 3.7(a)) that $Cs_{j+1} = 0$. With (7.7), this implies that $A'_{j+1}v_{j+1} = 0$. But since A_{j+1} has full row rank, it must be the case that $v_{j+1} = 0$. Therefore, H_{j+1} singular implies that there is an $s_{j+1} \neq 0$ such that

$$Cs_{j+1} = 0 \quad \text{and} \quad A_{j+1}s_{j+1} = 0. \tag{7.9}$$

Furthermore, let k denote the index of the constraint which was dropped from the active set at the previous iteration. Then

$$A'_j = \begin{bmatrix} A'_{j+1} , & a_k \end{bmatrix}. \tag{7.10}$$

If $a'_k s_{j+1} = 0$, then from (7.9) $A_j s_{j+1} = 0$ and $Cs_{j+1} = 0$. This would imply that

$$H_j \begin{bmatrix} s_{j+1} \\ 0 \end{bmatrix} = 0 ,$$

which in turn implies that H_j is singular. This is a contradiction and establishes that $a'_k s_{j+1} \neq 0$. By replacing s_{j+1} with $(a'_k s_{j+1})^{-1}s_{j+1}$, we may now assume that

$$a'_k s_{j+1} = 1 . \tag{7.11}$$

Finally, from (7.5), (7.6), (7.9), (7.10), and (7.11)

$$g'_{j+1}s_{j+1} = -(u_{j+1})_k a'_k s_{j+1} = -(u_{j+1})_k ,$$

so that $g'_{j+1}s_{j+1} > 0$. But then from Taylor's series (3.6) and (7.9)

$$
\begin{aligned}
f(x_{j+1} - \sigma s_{j+1}) &= f(x_{j+1}) - \sigma g'_{j+1}s_{j+1} + \tfrac{1}{2}\sigma^2 s'_{j+1}Cs_{j+1} \\
&= f(x_{j+1}) - \sigma g'_{j+1}s_{j+1} \\
&\to -\infty \text{ as } \sigma \to +\infty.
\end{aligned}
$$

From (7.9), $x_{j+1} - \sigma s_{j+1}$ is feasible for

$$\min \{ -g'_{j+1}s_{j+1} + \tfrac{1}{2}s'_{j+1}Cs_{j+1} \mid A_{j+1}s_{j+1} = 0 \} \tag{7.12}$$

for all $\sigma \geq 0$. This implies that deleting constraint k from the active set results in the next direction finding subproblem (7.12) being unbounded from below. Let $\hat{\sigma}_{j+1}$ denote the maximum feasible step size for s_{j+1}. If $\hat{\sigma}_{j+1} = +\infty$, then (7.1) is unbounded from below. If $\hat{\sigma}_{j+1} < +\infty$, let l denote the index of the restricting constraint. Setting $\sigma_{j+1} = \hat{\sigma}_{j+1}$ results in $f(x_{j+2}) < f(x_{j+1})$ (provided that $\hat{\sigma}_{j+1} > 0$). Furthermore, augmenting K_{j+1} with l to give K_{j+2} results in H_{j+2} being nonsingular as will be shown in Lemma 7.1(c). Thus there can never be two consecutive instances of H_j being singular. Throughout the algorithm, we set $\gamma_j = 1$ if H_j is nonsingular and $\gamma_j = 0$ if H_j is singular.

In order to initiate the algorithm, we require a feasible point x_0 and $K_0 \subseteq I(x_0)$ satisfying

Assumption 7.1.
  (i) a_i, all $i \in K_0$ are linearly independent,
  (ii) $s'Cs > 0$ for all $s \neq 0$ with $A_0 s = 0$.

As a consequence of Assumption 7.1, we will show that H_0 is nonsingular. Note that Assumption 7.1(ii) would be satisfied if x_0 were a nondegenerate extreme point. Also, Assumption 7.1(ii) would be satisfied if C is positive definite.

A detailed statement of the algorithm follows.

ALGORITHM 6

Model Problem:

$$
\begin{aligned}
\text{minimize:} \quad & c'x + \tfrac{1}{2}x'Cx \\
\text{subject to:} \quad & a'_i x \leq b_i, \quad i = 1,\ldots,m, \\
& a'_i x = b_i, \quad i = m+1,\ldots,m+r.
\end{aligned}
$$

Initialization:
Start with any feasible point x_0 and active set $K_0 \subseteq I(x_0)$ such that Assumption 7.1 is satisfied. Compute $f(x_0) = c'x_0 + \frac{1}{2}x_0'Cx_0$, $g_0 = c + Cx_0$, and set $j = 0$.

Step 1:   **Computation of Search Direction s_j.**
Let

$$H_j = \begin{bmatrix} C & A_j' \\ A_j & 0 \end{bmatrix} .$$

If H_j is nonsingular, go to Step 1.1. Otherwise, go to Step 1.2.

Step 1.1:
Compute the Newton direction s_j and multipliers v_j from the solution of the linear equations

$$H_j \begin{bmatrix} s_j \\ v_j \end{bmatrix} = \begin{bmatrix} g_j \\ 0 \end{bmatrix} .$$

Set $\gamma_j = 0$ and go to Step 2.

Step 1.2:
Compute s_j such that $A_j s_j = 0$, $C s_j = 0$, and $a_k' s_j = 1$. Set $\gamma_j = 1$ and go to Step 2.

Step 2:   **Computation of Step Size σ_j.**
Set $\tilde{\sigma}_j = 1$ if $\gamma_j = 0$ and $\tilde{\sigma}_j = +\infty$ if $\gamma_j = 1$. If $a_i' s_j \geq 0$ for $i = 1, \ldots, m$, set $\hat{\sigma}_j = +\infty$. Otherwise, compute the smallest index l and $\hat{\sigma}_j$ such that

$$\hat{\sigma}_j = \frac{a_l' x_j - b_l}{a_l' s_j} = \min \left\{ \frac{a_i' x_j - b_i}{a_i' s_j} \;\middle|\; \text{all } i \notin K_j \text{ with } a_i' s_j < 0 \right\} .$$

If $\tilde{\sigma}_j = \hat{\sigma}_j = +\infty$, print the message "the objective function is unbounded from below" and stop. Otherwise, set $\sigma_j = \min \{ \tilde{\sigma}_j, \hat{\sigma}_j \}$ and go to Step 3.

Step 3:   **Update.**
Set $x_{j+1} = x_j - \sigma_j s_j$, $g_{j+1} = c + C x_{j+1}$, and $f(x_{j+1}) = c'x_{j+1} + \frac{1}{2}x_{j+1}'Cx_{j+1}$. If $\sigma_j = \hat{\sigma}_j$, go to Step 3.1. Otherwise, go to Step 3.2.

Step 3.1:
Set $K_{j+1} = K_j + l$, form A_{j+1} and H_{j+1}, replace j with $j + 1$ and go to Step 1.1.

Step 3.2:
Compute the multiplier vector u_{j+1} from $-v_j$ and K_j and compute k such that

$$(u_{j+1})_k = \min \{ (u_{j+1})_i \mid i = 1, \ldots, m \} .$$

If $(u_{j+1})_k \geq 0$, then stop with optimal solution x_{j+1}. Otherwise, set $K_{j+1} = K_j - k$, form A_{j+1} and H_{j+1}, replace j with $j + 1$ and go to Step 1.

We illustrate Algorithm 6 by applying it to the following example. The problem of the example was used in Example 5.1 to illustrate the steps of Algorithm 3.

Example 7.1

$$
\begin{array}{lrcll}
\text{minimize:} & - 10x_1 - 4x_2 + x_1^2 + x_2^2 & & & \\
\text{subject to:} & x_1 & \leq & 4, & (1) \\
& - x_2 & \leq & -1, & (2) \\
& - x_1 & \leq & -1, & (3) \\
& x_2 & \leq & 3. & (4)
\end{array}
$$

Here

$$
c = \begin{bmatrix} -10 \\ -4 \end{bmatrix} \text{ and } C = \begin{bmatrix} 2 & 0 \\ 0 & 2 \end{bmatrix}.
$$

Initialization:

$$
x_0 = \begin{bmatrix} 1 \\ 3 \end{bmatrix}, \quad K_0 = \{ 3 , 4 \}, \quad A_0 = \begin{bmatrix} a_3' \\ a_4' \end{bmatrix} = \begin{bmatrix} -1 & 0 \\ 0 & 1 \end{bmatrix},
$$

$$
H_0 = \begin{bmatrix} 2 & 0 & -1 & 0 \\ 0 & 2 & 0 & 1 \\ -1 & 0 & 0 & 0 \\ 0 & 1 & 0 & 0 \end{bmatrix}, \quad f(x_0) = -12, \quad g_0 = \begin{bmatrix} -8 \\ 2 \end{bmatrix}, \quad j = 0.
$$

Iteration 0

Step 1: H_0 is nonsingular (x_0 is a nondegenerate extreme point).

Transfer to Step 1.1.

Step 1.1: Solving $H_0 \begin{bmatrix} s_0 \\ v_0 \end{bmatrix} = \begin{bmatrix} g_0 \\ 0 \end{bmatrix}$ gives $s_0 = \begin{bmatrix} 0 \\ 0 \end{bmatrix}$ and

$$
v_0 = \begin{bmatrix} 8 \\ 2 \end{bmatrix}.
$$

Set $\gamma_0 = 0$.

Step 2: $\tilde{\sigma}_0 = 1$,

$$\hat{\sigma}_0 = +\infty,$$

$$\sigma_0 = \min \{ 1 , +\infty \} = 1 .$$

Step 3: $x_1 = \begin{bmatrix} 1 \\ 3 \end{bmatrix}$, $g_1 = \begin{bmatrix} -8 \\ 2 \end{bmatrix}$, $f(x_1) = -12 .$

Transfer to Step 3.2.

Step 3.2: $u_1 = (0 , 0 , -8 , -2)' ,$

$(u_1)_3 = \min \{ 0 , 0 , -8 , -2 \} = -8 , \quad k = 3 ,$

$$K_1 = \{ 4 \} , \quad A_1 = [\, 0 \ \ 1 \,] , \quad H_1 = \begin{bmatrix} 2 & 0 & 0 \\ 0 & 2 & 1 \\ 0 & 1 & 0 \end{bmatrix} ,$$

$j = 1 .$

Iteration 1

Step 1: H_1 is nonsingular. Transfer to Step 1.1.

Step 1.1: Solving $H_1 \begin{bmatrix} s_1 \\ v_1 \end{bmatrix} = \begin{bmatrix} g_1 \\ 0 \end{bmatrix}$ gives $s_1 = \begin{bmatrix} -4 \\ 0 \end{bmatrix}$ and

$v_1 = [\, 2 \,] .$

Set $\gamma_1 = 0 .$

Step 2: $\tilde{\sigma}_1 = 1 ,$

$$\hat{\sigma}_1 = \min \left\{ \frac{-3}{-4} , - , - , - \right\} = \frac{3}{4} , \quad l = 1 ,$$

$$\sigma_1 = \min \left\{ 1 , \frac{3}{4} \right\} = \frac{3}{4} .$$

Step 3: $x_2 = \begin{bmatrix} 1 \\ 3 \end{bmatrix} - \frac{3}{4} \begin{bmatrix} -4 \\ 0 \end{bmatrix} = \begin{bmatrix} 4 \\ 3 \end{bmatrix} , \quad g_2 = \begin{bmatrix} -2 \\ 2 \end{bmatrix} ,$

$f(x_2) = -27 .$ Transfer to Step 3.1.

Step 3.1: $K_2 = \{ 4 , 1 \} , \quad A_2 = \begin{bmatrix} 0 & 1 \\ 1 & 0 \end{bmatrix} ,$

$$H_2 = \begin{bmatrix} 2 & 0 & 0 & 1 \\ 0 & 2 & 1 & 0 \\ 0 & 1 & 0 & 0 \\ 0 & 0 & 1 & 0 \end{bmatrix} , \quad j = 2 .$$

Iteration 2

Step 1: H_2 is nonsingular. Transfer to Step 1.1.

Step 1.1:   Solving $H_2 \begin{bmatrix} s_2 \\ v_2 \end{bmatrix} = \begin{bmatrix} g_2 \\ 0 \end{bmatrix}$ gives $s_2 = \begin{bmatrix} 0 \\ 0 \end{bmatrix}$ and

$v_2 = \begin{bmatrix} 2 \\ -2 \end{bmatrix}$.

Set $\gamma_2 = 0$.

Step 2: $\tilde{\sigma}_2 = 1$,

$\hat{\sigma}_2 = +\infty$,

$\sigma_2 = \min\{1, +\infty\} = 1$.

Step 3: $x_3 = \begin{bmatrix} 4 \\ 3 \end{bmatrix}$, $g_3 = \begin{bmatrix} -2 \\ 2 \end{bmatrix}$, $f(x_3) = -27$.

Transfer to Step 3.2.

Step 3.2: $u_3 = (2, 0, 0, -2)'$,

$(u_3)_4 = \min\{2, 0, 0, -2\} = -2$, $k = 4$,

$K_3 = \{1\}$, $A_3 = \begin{bmatrix} 1 & 0 \end{bmatrix}$, $H_3 = \begin{bmatrix} 2 & 0 & 1 \\ 0 & 2 & 0 \\ 1 & 0 & 0 \end{bmatrix}$,

$j = 3$.

Iteration 3

Step 1: H_3 is nonsingular. Transfer to Step 1.1.

Step 1.1:   Solving $H_3 \begin{bmatrix} s_3 \\ v_3 \end{bmatrix} = \begin{bmatrix} g_3 \\ 0 \end{bmatrix}$ gives $s_3 = \begin{bmatrix} 0 \\ 1 \end{bmatrix}$ and

$v_3 = \begin{bmatrix} -2 \end{bmatrix}$.

Set $\gamma_3 = 0$.

Step 2: $\tilde{\sigma}_3 = 1$,

$\hat{\sigma}_3 = \min\left\{ -, \dfrac{-2}{-1}, -, - \right\} = 2$, $l = 2$,

$$\sigma_3 = \min \{ 1, 2 \} = 1.$$

Step 3: $x_4 = \begin{bmatrix} 4 \\ 3 \end{bmatrix} - \begin{bmatrix} 0 \\ 1 \end{bmatrix} = \begin{bmatrix} 4 \\ 2 \end{bmatrix}, \quad g_4 = \begin{bmatrix} -2 \\ 0 \end{bmatrix},$

$f(x_4) = -28$. Transfer to Step 3.2.

Step 3.2: $u_4 = (2, 0, 0, 0)'$,

$(u_4)_2 = \min \{ 2, 0, 0, 0 \} = 0, \quad k = 2.$

$(u_4)_2 \geq 0$; stop with optimal solution $x_4 = (4, 2)'$.

This completes Example 7.1.   ◇

The progress of Algorithm 6 for the problem of Example 7.1 is shown in Figure 7.1. The same problem was solved using Algorithm 3 in Example 5.1 and the progress of that algorithm was shown in Figure 5.2. Other than duplicating x_0 and x_1, and, x_2 and x_3, Algorithm 6 produced exactly the same sequence of points as did Algorithm 3.

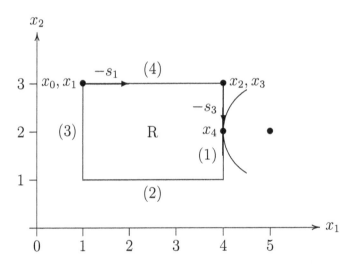

Figure 7.1.   Progress of Algorithm 6 in solving Example 7.1.

Because $s_0 = 0$ and $s_2 = 0$ in Example 7.1, these iterations do very little other than compute the multipliers in Step 3.2 and then modify the active set so that positive progress can be made in the next iteration. In general, $s_j = 0$ implies that x_j is a quasistationary point. In Example 7.1, both x_0 and x_2 are extreme points. We could modify Algorithm 6 so that when $s_j = 0$ in Step 1.1, control transfers to Step 3.2. Indeed, if A_j has n rows, then x_j is an extreme point and we need not even solve for s_j. Computationally, it would be useful to do such things. However, doing so would complicate the algorithm and subsequent discussion of it. For example, we would sometimes have two s_j's; the first being 0 and the second being the "real" s_j. Consequently, we avoid additional tests in favor of a "cleanly" stated algorithm.

The linear equations in Example 7.1 are sufficiently simple to be solved by hand. In general, of course, this will not always be the case. However, the equations in Step 1.1 of Algorithm 6 may be solved in a variety of ways. One way is to maintain H_j^{-1} explicitly (or at least the symmetric part of it) and develop updating formulae for H_{j+1}^{-1} in terms of H_j^{-1} and the modified portion of H_j. Such an approach must also account for the possibility that H_{j+1} is singular and determine the appropriate search direction for Step 1.2. We shall develop this approach after presenting Lemma 7.1 which shows that singularity of the coefficient matrix H_j cannot occur for two consecutive iterations.

Alternatively, the linear equations may be solved using some factorization of H_j and advantage taken of the fact that H_j and H_{j+1} differ in only a single row and column.

A third approach is to employ one or more matrices, typically of dimension less that H_j, to solve the equations. Indeed, this is the approach used by most methods, including Algorithm 4 and the "simplex method for quadratic programming" (Section 8.1) as well as several other methods.

Let C be as in (7.1) and let A be any $(k \, , \, n)$ matrix. Define

$$H(A) \; = \; \begin{bmatrix} C & A' \\ A & 0 \end{bmatrix} .$$

Then $H_j \; = \; H(A_j)$ is the coefficient matrix used at the jth iteration of Algorithm 6. The critical analytic properties of $H(A)$ are established in the following lemma.

Lemma 7.1 *The following hold.*

(a) Suppose that A has full row rank. Then $H(A)$ is nonsingular if and only if $s'Cs > 0$ for all $s \neq 0$ with $As = 0$.

(b) Suppose that A_0 has full row rank and that $H(A_0)$ is nonsingular. Let $A_1' \; = \; [\, A_0' \, , \, d \,]$, where d is any vector such that there is an s_0 with $A_0 s_0 \; = \; 0$ and $d's_0 \neq 0$. Then A_1 has full row rank and $H(A_1)$ is nonsingular.

(c) Suppose that A_0 has full row rank, $A_0' \; = \; [\, A_1' \, , \, d \,]$, and that $H(A_0)$ is nonsingular. Then $H(A_1)$ is singular if and only if there exists a unique n-vector s_1 such that $A_1 s_1 \; = \; 0$, $Cs_1 \; = \; 0$, and $d's_1 \; = \; 1$. Furthermore, if $H(A_1)$ is singular, then A_2 has full row rank and $H(A_2)$ is nonsingular, where $A_2' \; = \; [\, A_1' \, , \, e \,]$ and e is any n-vector with $e's_1 \neq 0$ and s_1 is the uniquely determined n-vector of the previous statement.

Proof: (a) Suppose first that $H(A)$ is nonsingular. Let s be nonzero and let $As = 0$. Because C is positive semidefinite, $s'Cs \geq 0$, and furthermore, $s'Cs = 0$ implies that $Cs = 0$ (Exercise 3.7(a)). But then $s'Cs = 0$ implies that

$$\begin{bmatrix} C & A' \\ A & 0 \end{bmatrix} \begin{bmatrix} s \\ 0 \end{bmatrix} = \begin{bmatrix} 0 \\ 0 \end{bmatrix}$$

which in turn implies that $H(A)$ is singular. The assumption that $s'Cs = 0$ leads to a contradiction and is therefore false. Thus $s'Cs > 0$ which establishes the forward implication of part (a).

Conversely, suppose that $s'Cs > 0$ for all nonzero s with $As = 0$. If $H(A)$ is singular, there exist s and v, at least one of which is nonzero, such that

$$\begin{bmatrix} C & A' \\ A & 0 \end{bmatrix} \begin{bmatrix} s \\ v \end{bmatrix} = 0 .$$

Writing the two partitions separately gives

$$Cs + A'v = 0 , \tag{7.13}$$

$$As = 0 . \tag{7.14}$$

Multiplying (7.13) on the left with s' and using (7.14) gives

$$s'Cs + (As)'v = s'Cs = 0 .$$

But then from hypothesis and (7.14), $s = 0$. From (7.13), this implies that $A'v = 0$ and $v \neq 0$. Because by assumption A has full row rank, this is a contradiction. The assumption that $H(A)$ is singular leads to a contradiction and is therefore false. Thus $H(A)$ is indeed nonsingular which completes the proof of part (a).

(b) If A_1 does not have full row rank, then there is a vector u and a scalar θ such that $(u' , \theta)' \neq 0$ and

$$A_1' \begin{bmatrix} u \\ \theta \end{bmatrix} = 0.$$

That is,

$$A_0'u + \theta d = 0. \tag{7.15}$$

Since A_0 has full row rank, $\theta \neq 0$. Furthermore, taking the inner product of both sides of (7.15) with s_0 gives

$$u'A_0 s_0 + \theta d' s_0 = 0 .$$

Because $A_0 s_0 = 0$, this implies that $\theta d' s_0 = 0$. But both θ and $d' s_0$ are nonzero. This contradiction establishes that indeed, A_1 has full row rank.

Finally, since A_0 has full row rank and $H(A_0)$ is nonsingular, part (a) asserts that $s'Cs > 0$ for all $s \neq 0$ with $A_0 s = 0$. Thus it is also true that $s'Cs > 0$ for all $s \neq 0$ with $A_1 s = 0$ and from part (a) this implies that $H(A_1)$ is nonsingular.

(c) From part (a), Exercise 3.7(a) and the assumption that $H(A_0)$ is nonsingular, it follows that $H(A_1)$ is singular if and only if there is a nonzero s with $Cs = 0$, $A_1 s = 0$, and $d's \neq 0$. Letting $s_1 = (d's)^{-1}s$, it remains to show that s_1 is uniquely determined. But this is established by observing that s_1 satisfies

$$
\begin{bmatrix} C & A_1' & d \\ A_1 & 0 & 0 \\ d' & 0' & 0 \end{bmatrix} \begin{bmatrix} s_1 \\ 0 \\ 0 \end{bmatrix} = \begin{bmatrix} 0 \\ 0 \\ 1 \end{bmatrix}
$$

and because the coefficient matrix $H(A_0)$ is nonsingular, s_1 must be unique. Finally, suppose that $H(A_1)$ is singular. Because $A_1 s_1 = 0$ and $e's \neq 0$, it follows that A_2 has full row rank. If $H(A_2)$ is singular, then from part (a) there is an $s \neq 0$ with $Cs = 0$ and $A_1 s = 0$, and $e's = 0$. From above, this s must be parallel to s_1. But then $e's = 0$ is in contradiction to $e's \neq 0$. Thus $H(A_2)$ is indeed nonsingular and the proof of Lemma 7.1 is complete. $\square$

Lemma 7.1 can be used to analyze Algorithm 6 as follows. Suppose that we begin with x_0, A_0, and H_0 such that H_0 is nonsingular. If, for example, x_0 is a nondegenerate extreme point so that A_0 is square and nonsingular then Lemma 7.1 asserts that H_0 is nonsingular. Indeed, in this case a formula can be given for H_0^{-1} in terms of A_0^{-1} and C (Exercise 7.1). Alternatively, if C is positive definite and x_0 is any nondegenerate feasible point, then it follows from Lemma 7.1(a) that H_0 is nonsingular.

Now suppose that at iteration j, H_j is nonsingular so that s_j is constructed from Step 1.1. Also suppose that $\sigma_j = \hat{\sigma}_j$ and let l be the index of the new active constraint. Since $A_j s_j = 0$ and $a_l' s_j < 0$, it follows from Lemma 7.1(b) that

$$
A_{j+1} = \begin{bmatrix} A_j \\ a_l' \end{bmatrix}
$$

has full row rank and that H_{j+1} is nonsingular. Thus adding a constraint to the active set results in a nonsingular coefficient matrix.

Next suppose that H_j is nonsingular and that $\sigma_j = \tilde{\sigma}_j = 1 < \hat{\sigma}_j$. Suppose also that in Step 3.2 $(u_{j+1})_k < 0$ so that termination with an optimal solution does not occur. Then H_{j+1} is obtained from H_j by deleting the row and column containing

$$\begin{bmatrix} a_k \\ 0 \end{bmatrix},$$

or equivalently,

$$A'_j = \begin{bmatrix} A'_{j+1} , & a_k \end{bmatrix}.^2$$

According to Lemma 7.1(c), H_{j+1} is singular if and only if there exists a unique n-vector s_{j+1} satisfying

$$A_{j+1}s_{j+1} = 0, \quad Cs_{j+1} = 0, \quad \text{and} \quad a'_k s_{j+1} = 1.$$

This uniquely determined s_{j+1} is precisely the search direction required by Step 1.2 of Algorithm 6.

We continue by further considering the case that H_{j+1} is singular. If $\hat{\sigma}_j = +\infty$, the model problem is unbounded from below. Assume that $\hat{\sigma}_j < +\infty$ and let l be the index of the new active constraint. Algorithm 6 continues by setting

$$A'_{j+2} = \begin{bmatrix} A'_{j+1} , & a_l \end{bmatrix}.$$

Lemma 7.1(c) asserts that H_{j+2} is nonsingular. Pictorially we have

$$H_j = \begin{bmatrix} C & A'_{j+1} & a_k \\ A_{j+1} & 0 & 0 \\ a'_k & 0' & 0 \end{bmatrix} \quad \text{and} \quad H_{j+2} = \begin{bmatrix} C & A'_{j+1} & a_l \\ A_{j+1} & 0 & 0 \\ a'_l & 0' & 0 \end{bmatrix}.$$

In this case H_{j+2} is obtained from H_j by replacing the row and column associated with a_k, with those associated with a_l.

In summary, beginning with nonsingular H_j dropping a constraint can result in H_{j+1} being singular but then H_{j+2} must be nonsingular. Adding a constraint always results in H_{j+1} being nonsingular.

The termination properties of Algorithm 6 are easily demonstrated as follows.

Theorem 7.1 *Let Algorithm 6 be applied to the model problem*

$$\begin{aligned} minimize: \quad & c'x + \tfrac{1}{2}x'Cx \\ subject\ to: \quad & a'_i x = b_i, \quad i = 1, \ldots, m, \\ & a'_i x = b_i, \quad i = m+1, \ldots, m+r, \end{aligned}$$

beginning with a feasible point x_0 for which Assumption 7.1 is satisfied and let $x_1, x_2, \ldots, x_j, \ldots$ be the sequence so obtained. If each quasistationary point

---

[2]For notational convenience, we assume that a_k is the last column of A'_j.

determined by Algorithm 6 is nondegenerate, then Algorithm 6 terminates in a finite number of steps with either an optimal solution x_j or the information that the problem is unbounded from below. In the former case, Algorithm 6 terminates with a vector u_j such that (x_j , u_j) is an optimal solution for the dual problem.

Proof: By construction, the sequence of objective function values $\{ f(x_j) \}$ is monotone decreasing. During any iteration j either the number of constraints in the active set increases by exactly one, or a quasistationary point is obtained. From Lemma 7.1 parts (b) and (c), A_j has full row rank at every iteration j. Therefore a quasistationary point must be obtained in at most n iterations. If x_j is such a quasistationary point, by assumption it is nondegenerate so that $f(x_{j+1}) < f(x_j)$. Thus no quasistationary set can ever be repeated. Since there are only finitely many quasistationary sets, termination in a finite number of steps must occur. If termination occurs with "objective function is unbounded from below" then by construction that is indeed the case. If termination occurs with "optimal solution x_j," then from Theorem 4.9 (Strong Duality Theorem) that is indeed the case and (x_j , u_j) is an optimal solution for the dual problem. $\square$

7.2   A General PQP Algorithm: Algorithm 7

Consider the model problem

$$\left.\begin{array}{ll} \text{minimize}: & (c + tq)'x + \frac{1}{2}x'Cx \\ \text{subject to}: & a_i'x \leq b_i + tp_i, \qquad i = 1, \ldots, m, \\ & a_ix = b_i + tp_i, \qquad i = m+1, \ldots, m+r, \end{array}\right\} \qquad (7.16)$$

where c, q, $a_1, \ldots, a_{m+r}$ are n-vectors, C is an (n , n) symmetric positive semidefinite matrix, and $b_1, \ldots, b_{m+r}$, $p_1, \ldots, p_{m+r}$ are scalars. We develop an algorithm to solve (7.16) for all values of the parameter t satisfying $\underline{t} \equiv t_0 \leq t \leq \bar{t}$, where $\underline{t}$ and $\bar{t}$ are the given initial and final values, respectively, of the parameter. The algorithm determines finitely many critical parameter values $t_1 , t_2, \ldots, t_\nu$ satisfying

$$\underline{t} \equiv t_0 < t_1 < t_2 < \cdots < t_{\nu-1} < t_\nu \leq \bar{t}.$$

The optimal solution and associated multipliers are obtained as linear functions of t for all t between consecutive critical parameter values. The algorithm terminates when either $t_\nu = +\infty$, t_ν is finite and either (7.16) has no feasible solution for $t > t_\nu$ or (7.16) is unbounded from below for $t > t_\nu$. The relevant reason for termination is also made explicit.

We begin the solution of (7.16) by setting $t_0 = t$ and solving (7.16) using Algorithm 7 with $t = t_0$. Let $x_0(t_0)$ be the optimal solution so obtained and let K_0, A_0, and

$$H_0 \equiv H(A_0) = \begin{bmatrix} C & A_0' \\ A_0 & 0 \end{bmatrix}$$

be the associated index set of constraints active at $x_0(t_0)$, $K_0 \subseteq \{1, 2, \ldots, m + r\}$, matrix whose rows are the a_i', all $i \in K_0$, and coefficient matrix, respectively. Let b_0 and p_0 be the vectors whose components are those b_i and p_i associated with rows of A_0, respectively. Because C is positive semidefinite, the optimality conditions are both necessary and sufficient for optimality (Theorem 4.5). Thus there is a vector of multipliers $v_0(t_0)$ whose components are associated with constraints of (7.16) which are active at $x_0(t_0)$. The components of $v_0(t_0)$ associated with active inequality constraints are necessarily nonnegative. We require

Assumption 7.2.
   (i) A_0 has full row rank,
   (ii) $s'Cs > 0$ for all $s \neq 0$ with $A_0 s = 0$,
   (iii) Each component of $v_0(t_0)$ associated with an active inequality constraint is strictly positive.

Assumptions 7.2 (i) and (ii), together with Lemma 7.1(a), imply that H_0 is nonsingular. The optimality conditions then assert that $x_0(t_0)$ and $v_0(t_0)$ are the uniquely determined solution of the linear equations

$$H_0 \begin{bmatrix} x_0(t_0) \\ v_0(t_0) \end{bmatrix} = \begin{bmatrix} -c \\ b_0 \end{bmatrix} + t_0 \begin{bmatrix} -q \\ p_0 \end{bmatrix}.$$

The full; i.e., $(m + r)$-dimensional, vector of multipliers $u_0(t_0)$ is obtained from K_0 and $v_0(t_0)$ by assigning zero to those components of $u_0(t_0)$ associated with constraints inactive at $x_0(t_0)$ and the appropriately indexed component of $v_0(t_0)$, otherwise. The first m components of $u_0(t_0)$ correspond to the m inequality constraints while that last r components correspond to the r equality constraints.

Now suppose that t is increased from t_0. Let $x_0(t)$ and $v_0(t)$ denote the optimal solution and associated active constraint multipliers as functions of the parameter t. Provided that there are no changes in the active set, $x_0(t)$ and $v_0(t)$ are the uniquely determined solution of the linear equations

$$H_0 \begin{bmatrix} x_0(t) \\ v_0(t) \end{bmatrix} = \begin{bmatrix} -c \\ b_0 \end{bmatrix} + t \begin{bmatrix} -q \\ p_0 \end{bmatrix}. \tag{7.17}$$

The solution can conveniently be obtained by solving the two sets of linear equations:

$$H_0 \begin{bmatrix} h_{10} \\ g_{10} \end{bmatrix} = \begin{bmatrix} -c \\ b_0 \end{bmatrix}, \quad H_0 \begin{bmatrix} h_{20} \\ g_{20} \end{bmatrix} = \begin{bmatrix} -q \\ p_0 \end{bmatrix},$$

which both have coefficient matrix H_0. Having solved these for h_{10}, h_{20}, g_{10}, and g_{20} (the second subscript refers to the first critical interval $\{ t_0 , t_1 \}$, t_1 being defined below) the solution of (7.17) is

$$x_0(t) = h_{10} + th_{20} \quad \text{and} \tag{7.18}$$
$$v_0(t) = g_{10} + tg_{20} .$$

The full vector of multipliers $u_0(t)$ may then be obtained from K_0 and $v_0(t)$ as described above. We write $u_0(t)$ as

$$u_0(t) = w_{10} + tw_{20} . \tag{7.19}$$

The optimal solution for (7.16) and the associated multiplier vector are given by (7.18) and (7.19), respectively, for all $t > t_0$ for which $x_0(t)$ is feasible for (7.16) and $(u_0(t))_i \geq 0$, $i = 1, \ldots, m$. The first restriction implies that $t \leq \hat{t}_1$, where

$$\hat{t}_1 = \min \left\{ \frac{b_i - a_i' h_{10}}{a_i' h_{20} - p_i} \mid \text{all } i = 1, \ldots, m \text{ with } a_i' h_{20} > p_i \right\}$$
$$= \frac{b_l - a_l' h_{10}}{a_l' h_{20} - p_l}. \tag{7.20}$$

The second restriction implies that $t \leq \tilde{t}_1$, where

$$\tilde{t}_1 = \min \left\{ \frac{-(w_{10})_i}{(w_{20})_i} \mid \text{all } i = 1, \ldots, m \text{ with } (w_{20})_i < 0 \right\}$$
$$= \frac{-(w_{10})_k}{(w_{20})_k} . \tag{7.21}$$

We have used $(w_{10})_i$ to denote the ith component of w_{10}. We use the convention $\hat{t}_1 = +\infty$ to mean that $a_i' h_{20} \leq p_i$ for $i = 1, \ldots, m$. In this case, l is undefined in (7.20). However, l is not required for further calculations and thus no difficulty occurs. The analogous convention is used when $(w_{20})_i \geq 0$ for $i = 1, \ldots, m$. Setting $t_1 = \min \{ \hat{t}_1 , \tilde{t}_1 \}$, it follows that the optimal solution is given by (7.18) that the multipliers are given by (7.19) for all t satisfying $t_0 \leq t \leq t_1$. Furthermore Assumption 7.2 (iii) implies that $\tilde{t}_1 > t_0$, and the inclusion of all indices of constraints active at $x_0(t_0)$ in K_0 implies that $\hat{t}_1 > t_0$. Therefore, $t_1 > t_0$.

For $t > \hat{t}_1$, $x_0(t)$ violates constraint l and for $t > \tilde{t}_1$, $(u_0(t))_k$ becomes negative. This suggests that the analysis be continued by setting $K_1 = K_0 + l$ if $t_1 = \hat{t}_1$ and $K_1 = K_0 - k$ if $t_1 = \tilde{t}_1$. A_1 would then be obtained from A_0 by either adding the new row a'_l or deleting the existing row a'_k. Similarly, b_1 and p_1 would be obtained from b_0 and p_0 by either adding the components b_l and p_l, respectively, or deleting the components b_k and p_k, respectively. $H_1 = H(A_1)$ would be obtained accordingly. The optimal solution for the next interval would then be $x_1(t) = h_{11} + t h_{21}$ with multiplier vector $u_1(t) = w_{11} + t w_{21}$ for all t with $t_1 \leq t \leq t_2$, where h_{11}, h_{21}, w_{11}, and w_{21} are obtained from the linear equations

$$H_1 \begin{bmatrix} h_{11} \\ g_{11} \end{bmatrix} = \begin{bmatrix} -c \\ b_1 \end{bmatrix}, \quad H_1 \begin{bmatrix} h_{21} \\ g_{21} \end{bmatrix} = \begin{bmatrix} -q \\ p_1 \end{bmatrix}, \quad (7.22)$$

$t_2 = \min \{ \hat{t}_2 , \tilde{t}_2 , \bar{t} \}$, and $\hat{t}_2$ and $\tilde{t}_2$ are computed from formulae analogous to those for $\hat{t}_1$ and $\tilde{t}_1$, respectively. We will justify this intuitive approach in two cases (Cases 1.1 and 2.1). In two other cases (Cases 1.2.1 and 2.2.1) we will show that K_1 must be obtained from K_0 by exchanging the indices of two constraints. Nonetheless, all quantities for the next parametric interval are to be found using (7.22) and are completely specified in terms of K_1. In the two remaining cases, we will show that for $t > t_1$ either (7.16) has no feasible solution (Case 1.2.2) or (7.16) is unbounded from below (Case 2.2.2).

We make the assumption that $\hat{t}_1 \neq \tilde{t}_1$ and that no ties occur in the computation of $\hat{t}_1$ and $\tilde{t}_1$; i.e., the indices l and k of (7.20) and (7.21) are uniquely determined.

Case 1: $t_1 = \hat{t}_1$

The analysis continues according to whether or not a_l is linearly dependent on a_i, all $i \in K_0$. The relevant possibility may be determined by solving the linear equations

$$H_0 \begin{bmatrix} s_0 \\ \hat{\xi}_0 \end{bmatrix} = \begin{bmatrix} a_l \\ 0 \end{bmatrix}. \quad (7.23)$$

If the solution for (7.23) has $s_0 = 0$, then a_l is indeed linearly dependent on the a_i, all $i \in K_0$ and the components of $\hat{\xi}_0$ are the coefficients of the linear combination. There is one component of $\hat{\xi}_0$ for each active constraint. From $\hat{\xi}_0$ it is useful to construct an $(m + r)$-vector ξ_0 for which the ith component is zero if constraint i is not in the active set and is the corresponding component of $\hat{\xi}_0$ if constraint i is in the active set. Thus if row i of A_0 is a'_k then $(\xi_0)_k = (\hat{\xi}_0)_i$. If the solution has $s \neq 0$, then because H_0 is nonsingular, we may conclude that a_l and a_i, all $i \in K_0$, are linearly independent. We note that solving (7.23) involves little additional work since H_0 is presumed to be available in a suitably factored form in order to obtain h_{10} and h_{20}.

Case 1.1: a_l and a_i, all $i \in K_0$, are linearly independent

We set $K_1 = K_0 + l$, form A_1, b_1, p_1, and $H_1 = H(A_1)$ accordingly. We set $x_1(t) = h_{11} + th_{21}$ and $u_1(t) = w_{11} + tw_{21}$, where h_{11}, h_{21}, w_{11}, and w_{21} are obtained from the solution of (7.22). Observe that $u_0(t_1) = u_1(t_1)$ and consequently $(u_1(t_1))_l = 0$. In order to show that $t_2 > t_1$, we must show that $(u_1(t))_l$ is strictly increasing in t. However, it is straightforward to show from the defining equations for $x_0(t)$ and $x_1(t)$ that for $t > t_1$,

$$(u_1(t))_l = \frac{(x_1(t) - x_0(t))'C(x_1(t) - x_0(t))}{a_l'x_0(t) - (b_l + tp_l)} .$$

Now $A_0(x_1(t) - x_0(t)) = 0$, so from Assumption 7.2 (ii) the numerator of the above expression is strictly positive for $t > t_1$. By definition of $\hat{t}_1$, constraint l is violated for $t > t_1$. Consequently,

$$(u_1(t))_l > 0 \text{ for } t > t_1 .$$

Case 1.2: a_l **is linearly dependent on** a_i, **all** $i \in K_0$
By construction of ξ_0 from $\hat{\xi}_0$,

$$a_l = \sum_{i \in K_0} (\xi_0)_i a_i . \tag{7.24}$$

There are two subcases to be considered.

Case 1.2.1:   **There is at least one** $(\xi_0)_i > 0$, $i \in K_0$, **and** $1 \le i \le m$
The hypothesis of Case 1.2.1 implies that the multipliers for those constraints active at $x_0(t_1)$ are not uniquely determined. The defining equations for $x_0(t_1)$ assert that

$$-c - t_1q - Cx_0(t_1) = \sum_{i \in K_0} (u_0(t_1))_i a_i . \tag{7.25}$$

Let λ be any nonnegative scalar. Multiplying (7.24) by λ and subtracting the result from (7.25) gives

$$-c - t_1q - Cx_0(t_1) = \lambda a_l + \sum_{i \in K_0} ((u_0(t_1))_i - \lambda(\xi_0)_i) a_i .$$

Thus λ and the coefficients of the a_i are also valid multipliers for the active constraints for all λ for which these coefficients remain nonnegative. The largest such value of λ is

$$\lambda^* = \min \left\{ \frac{(u_0(t_1))_i}{(\xi_0)_i} \mid \text{all } i \in K_0, \ 1 \le i \le m, \text{ with } (\xi_0)_i > 0 \right\}$$
$$= \frac{(u_0(t_1))_k}{(\xi_0)_k} ,$$

for which the multiplier associated with constraint k is reduced to zero.

We set $K_1 = K_0 + l - k$ and obtain the quantities giving the optimal solution and multipliers for the next parametric interval from the solution of (7.22).

The multipliers for the new active set at $t = t_1$ are

$$
\begin{aligned}
(u_1(t_1))_i &= (u_0(t_1))_i - \lambda^*(\xi_0)_i , \quad \text{all } i \in K_0 , \\
(u_1(t_1))_l &= \lambda^* , \quad \text{and} \\
(u_1(t_1))_k &= 0 .
\end{aligned}
$$

Constraint k, which is active at $x_1(t_1)$, has been deleted from the active set. In order to show that $t_2 > t_1$, we must show that constraint k is inactive at $x_1(t)$ for $t > t_1$. We do this as follows. By definition of t_1, constraint l is violated for $t > t_1$. Because $A_0 x_0(t) = b_0 + t p_0$ for all t, it follows that

$$
b_l + t p_l < \sum_{i \in K_0} (\xi_0)_i (b_i + t p_i) , \quad \text{for } t > t_1 . \tag{7.26}
$$

By definition of K_1, we have for all t that

$$
\begin{aligned}
a'_i x_1(t) &= b_i + t p_i , \quad \text{all } i \in K_0 , \quad i \neq k , \quad \text{and} \\
a'_l x_1(t) &= b_l + t p_l .
\end{aligned}
$$

Therefore

$$
b_l + t p_l = \sum_{i \in K_0} (\xi_0)_i (b_i + t p_i) + (\xi_0)_k (a'_k x_1(t) - b_k - t p_k) .
$$

From (7.26) and the fact that $(\xi_0)_k > 0$, this implies that

$$
a'_k x_1(t) < b_k + t p_k , \quad \text{for } t > t_1 ,
$$

as required.

Case 1.2.2: $(\xi_0)_i \leq 0$ **for all** $i \in K_0$ **and** $1 \leq i \leq m$

In this case, we claim that there is no feasible solution for (7.16) for $t > t_1$. Suppose, to the contrary, that for some $t > t_1$ there is an x satisfying the constraints of (7.16). In particular, x must satisfy

$$
a'_i x \leq b_i + t p_i , \quad \text{all } i \in K_0 .
$$

Multiplying each such inequality by $(\xi_0)_i \leq 0$ and adding gives

$$
a'_l x \geq \sum_{i \in K_0} (\xi_0)_i (b_i + t p_i) .
$$

But then from (7.26)

$$a_l'x > b_l + tp_l \; ;$$

i.e., constraint l is violated. The assumption that (7.16) possesses a feasible solution leads to a contradiction and is therefore false.

Case 2: $t_1 = \tilde{t}_1$

Consider the effect of deleting constraint k from the active set. Let $\tilde{A}_1$ be obtained from A_0 by deleting the row containing a_k'. The analysis continues according to whether $H(\tilde{A}_1)$ is nonsingular or singular. The relevant possibility can be determined by solving the linear equations

$$H_0 \begin{bmatrix} s_0 \\ \xi_0 \end{bmatrix} = \begin{bmatrix} 0 \\ -e_\rho \end{bmatrix}, \tag{7.27}$$

where e_ρ is a unit vector with one in the same row as is a_k' is A_0. It follows from Lemma 7.1(b) that $H(A_1)$ is nonsingular if and only if $\xi_0 \neq 0$. As in Case 1, we note that solving (7.27) requires little additional work since H_0 is presumed to be available in a suitably factored form.

Case 2.1: $\xi_0 \neq 0$

We set $K_1 = K_0 - k$. Observe that $x_1(t_1) = x_0(t_1)$ and consequently constraint k is active at $x_1(t_1)$. In order to show that $t_2 > t_1$, we must show that k becomes inactive at $x_1(t)$ for $t > t_1$. As in Case 1.1, it is straightforward to show from the defining equations for $x_0(t)$ and $x_1(t)$ that for $t > t_1$,

$$(a_k'x_1(t) - (b_k + tp_k)) = \frac{(x_1(t) - x_0(t))'C(x_1(t) - x_0(t))}{(u_0(t))_k}.$$

Now $A_1(x_1(t) - x_0(t)) = 0$ and H_1 is nonsingular so that from Lemma 7.1(a), the numerator of the above expression is strictly positive for $t > t_1$. By definition of $\tilde{t}_1$, $(u_0(t))_k < 0$ for $t > t_1$. Consequently,

$$a_k'x_1(t) < b_k + tp_k , \quad \text{for } t > t_1 .$$

Case 2.2: $\xi_0 = 0$

By definition, s_0 satisfies

$$Cs_0 = 0 , \quad a_k's_0 = -1 , \quad \tilde{A}_1 s_0 = 0 .$$

Thus all points $x_0(t_1) + \sigma s_0$ are alternate optimal solutions for (7.16) when $t = t_1$, for all σ such that $x_0(t_1) + \sigma s_0$ is feasible. The largest such value of σ is

$$\sigma_0 = \min \left\{ \frac{b_i + t_1 p_i - a_i'x_0(t_1)}{a_i's_0} \;\middle|\; \text{all } i \text{ with } a_i's_0 > 0 \right\}$$

$$= \frac{b_l + t_1 p_l - a_l'x_0(t_1)}{a_l's_0},$$

where we adopt the convention that $\sigma_0 = +\infty$ means that $a_i' s_0 \leq 0$ for $i = 1, \ldots, m$. There are two subcases to be considered.

Case 2.2.1: $\sigma_0 < +\infty$

We set $K_1 = K_0 + l - k$. Observe that $x_1(t_1) = x_0(t_1) + \sigma_0 s_0$, $u_1(t_1) = u_0(t_1)$, and constraint k, which was active at $x_0(t_1)$, has become inactive at $x_1(t_1)$. Also, constraint l, which was inactive at $x_0(t_1)$, has become active at $x_1(t_1)$. Since $(u_1(t_1))_l = 0$, we must show that $u_1(t_1)$ is a strictly increasing function of t in order to establish that $t_2 > t_1$. We do this by first observing that from defining equations for $x(t_0)$ and the definition of s_0,

$$(u_0(t))_k = (c + tq)' s_0 .$$

By definition of t_1 we have

$$(c + tq)' s_0 < 0 , \quad \text{for } t > t_1 .$$

The defining equations for $x_1(t)$ and the definition of s_0 imply that

$$(u_1(t))_l = \frac{-(c + tq)' s_0}{a_l' s_0} ,$$

and thus by definition of σ_0 we have that $(u_1(t))_l$ is a positive, strictly increasing function of t for $t > t_1$.

Case 2.2.2: $\sigma_0 = +\infty$

In this case, $x_0(t_1) + \sigma s_0$ is feasible for all $\sigma \geq 0$. As in Case 2.2.1, we have

$$(u_0(t))_k = (c + tq)' s_0 ,$$

for $t > t_1$. Let $f_t(x)$ denote the objective function for (7.16) evaluated at x for specified t. Because $C s_0 = 0$, Taylor's series gives

$$f_t(x_0(t) + \sigma s_0) = f_t(x_0(t)) + \sigma(c + tq)' s_0$$
$$\to -\infty \text{ as } \sigma \to +\infty,$$

demonstrating that (7.16) is unbounded from below for $t > t_1$.

We note that for each of the four cases (1.1, 1.2.1, 2.1, 2.2.1), it follows from Lemma 7.1 that $H_1 = H(A_1)$ is nonsingular.

Based on the analysis of the previous section, we now formulate the algorithm for the solution of (7.16).

ALGORITHM   7

Model Problem:

$$\text{minimize:} \quad (c + tq)' x + \tfrac{1}{2} x' C x$$

$$\text{subject to:} \qquad a_i'x \leq b_i + tp_i, \qquad i = 1, \ldots, m,$$
$$a_i'x = b_i + tp_i, \qquad i = m + 1, \ldots, m + r,$$

for $\underline{t} \leq t \leq \bar{t}$.

Initialization:
Set $t_0 = \underline{t}$. Use Algorithm 6 to determine an optimal solution $x_0(t_0)$ for the problem for $t = t_0$. Let K_0 denote the set of indices of active constraints. Set $j = 0$.

Step 1: **Computation of Optimal Solution, Multipliers and End of Interval**

Let

$$H_j = \begin{bmatrix} C & A_j' \\ A_j & 0 \end{bmatrix}.$$

Solve the linear equations

$$H_j \begin{bmatrix} h_{1j} \\ g_{1j} \end{bmatrix} = \begin{bmatrix} -c \\ b_j \end{bmatrix}, \quad H_j \begin{bmatrix} h_{2j} \\ g_{2j} \end{bmatrix} = \begin{bmatrix} -q \\ p_j \end{bmatrix},$$

for h_{1j}, h_{2j}, g_{1j}, and g_{2j}. Form w_{1j} and w_{2j} from g_{1j} and g_{2j}, respectively. If $a_i'h_{2j} - p_i \leq 0$ for $i = 1, \ldots, m$, set $\hat{t}_{j+1} = +\infty$. Otherwise, compute l and $\hat{t}_{j+1}$ such that

$$\hat{t}_{j+1} = \frac{b_l - a_l'h_{1j}}{a_l'h_{2j} - p_l}$$
$$= \min \left\{ \frac{b_i - a_i'h_{1j}}{a_i'h_{2j} - p_i} \mid \text{all } i \text{ such that } a_i'h_{2j} > p_i \right\}.$$

If $(w_{2j})_i \geq 0$ for $i = 1, \ldots, m$, set $\hat{t}_{j+1} = +\infty$. Otherwise, compute k and $\tilde{t}_{j+1}$ such that

$$\tilde{t}_{j+1} = \frac{-(w_{1j})_k}{(w_{2j})_k}$$
$$= \min \left\{ \frac{-(w_{1j})_i}{(w_{2j})_i} \mid \text{all } i \text{ such that } (w_{2j})_i < 0 \right\}.$$

Set $t_{j+1} = \min \{ \hat{t}_{j+1}, \tilde{t}_{j+1}, \bar{t} \}$. Print "the optimal solution is $x_j(t) = h_{1j} + th_{2j}$ with multiplier vector $u_j(t) = w_{1j} + tw_{2j}$ for all t with $t_j \leq t \leq t_{j+1}$." If $t_{j+1} = \bar{t}$, stop. Otherwise, go to Step 2.

Step 2: **Computation of K_{j+1}.**
If $t_{j+1} = +\infty$, then stop. If $t_{j+1} = \hat{t}_{j+1}$ then go to Step 2.1, and otherwise, go to Step 2.2.

Step 2.1:   **New Active Constraint.**
Solve the linear equations

$$H_j \begin{bmatrix} s_j \\ \hat{\xi}_j \end{bmatrix} = \begin{bmatrix} a_l \\ 0 \end{bmatrix},$$

for s_j and $\hat{\xi}_j$. For $i = 1, \ldots, m + r$, set $(\xi_j)_i = (\hat{\xi}_j)_k$ if row k of A_j is a'_i and zero otherwise. If $s_j \neq 0$, then go to Step 2.1.1 and if $s_j = 0$, then go to Step 2.1.2.

Step 2.1.1:
Set $K_{j+1} = K_j + l$, form A_{j+1} and H_{j+1}, replace j with $j + 1$ and go to Step 1.

Step 2.1.2:
If $(\xi_j)_i \leq 0$ for $i = 1, \ldots, m$, print "there is no feasible solution for $t > t_{j+1}$" and stop. Otherwise, compute ρ such that

$$\frac{(g_{1j})_\rho + t_{j+1}(g_{2j})_\rho}{(\xi_j)_\rho} = \min\left\{ \frac{(g_{1j})_i + t_{j+1}(g_{2j})_i}{(\xi_j)_i} \;\middle|\; \begin{array}{l} \text{all } i \in K_j, \ 1 \leq i \leq m, \\ \text{with } (\xi_j)_i > 0 \end{array} \right\}$$

and let k be the index of the constraint associated with row ρ of A_j. Set $K_{j+1} = K_j + l - k$, form A_{j+1} and H_{j+1}, replace j with $j + 1$ and go to Step 1.

Step 2.2:   **Deletion of a Constraint.**
Solve the linear equations

$$H_j \begin{bmatrix} s_j \\ \xi_j \end{bmatrix} = \begin{bmatrix} 0 \\ -e_\rho \end{bmatrix}$$

for s_j and ξ_j, where e_ρ is a vector of zeros except for component ρ which has value unity, and ρ is the index of the row of A_j associated with constraint k. If $\xi_j \neq 0$ then go to Step 2.2.1 and if $\xi_j = 0$ then go to Step 2.2.2.

Step 2.2.1:
Set $K_{j+1} = K_j - k$, form A_{j+1} and H_{j+1}, replace j with $j + 1$ and go to Step 1.

Step 2.2.2:
If $a'_i s_j \leq 0$ for $i = 1, \ldots, m$, print "the objective function is unbounded from below for $t > t_{j+1}$" and stop. Otherwise, compute the smallest index l and σ_j according to

$$\sigma_j = \frac{b_l + t_{j+1}p_l - a'_l x_j(t_{j+1})}{a'_l s_j}$$

$$= \min\left\{ \frac{b_i + t_{j+1}p_i - a'_i x_j(t_{j+1})}{a'_i s_j} \;\middle|\; \text{all } i \text{ such that } a'_i s_j > 0 \right\}.$$

Set $K_{j+1} = K_j + l - k$, form A_{j+1} and H_{j+1}, replace j with $j + 1$ and go to Step 1.

The steps of Algorithm 7 are illustrated in the following two examples.

Example 7.2

$$
\begin{aligned}
\text{minimize:} \quad & tx_1 - (1 - t)x_2 + \tfrac{1}{2}(4x_1^2 + x_2^2 + 4x_1x_2) \\
\text{subject to:} \quad & - x_1 && \leq \ 0, && (1) \\
& - 2x_1 - x_2 && \leq -2, && (2) \\
& x_2 && \leq \ 3, && (3)
\end{aligned}
$$

for all t with $t \geq 0$.
Here

$$
c = \begin{bmatrix} 0 \\ -1 \end{bmatrix}, \quad
q = \begin{bmatrix} 1 \\ 1 \end{bmatrix}, \quad
C = \begin{bmatrix} 4 & 2 \\ 2 & 1 \end{bmatrix}, \quad
b = \begin{bmatrix} 0 \\ -2 \\ 3 \end{bmatrix},
$$

$$
p = \begin{bmatrix} 0 \\ 0 \\ 0 \end{bmatrix}, \quad \underline{t} = 0, \ \text{and} \ \bar{t} = +\infty.
$$

Initialization:

$$
t_0 = 0, \quad x_0 = \begin{bmatrix} 0 \\ 2 \end{bmatrix}, \quad K_0 = \{1, 2\},
$$

$$
A_0 = \begin{bmatrix} a_1' \\ a_2' \end{bmatrix} = \begin{bmatrix} -1 & 0 \\ -2 & -1 \end{bmatrix}, \quad
H_0 = \begin{bmatrix} 4 & 2 & -1 & -2 \\ 2 & 1 & 0 & -1 \\ -1 & 0 & 0 & 0 \\ -2 & -1 & 0 & 0 \end{bmatrix},
$$

$j = 0$.

Iteration 0

Step 1: $b_0 = (0, -2)'$, $p_0 = (0, 0)'$,

Solving $H_0 \begin{bmatrix} h_{10} \\ g_{10} \end{bmatrix} = \begin{bmatrix} -c \\ b_0 \end{bmatrix}$ and $H_0 \begin{bmatrix} h_{20} \\ g_{20} \end{bmatrix} = \begin{bmatrix} -q \\ p_0 \end{bmatrix}$

gives $h_{10} = \begin{bmatrix} 0 \\ 2 \end{bmatrix}$, $h_{20} = \begin{bmatrix} 0 \\ 0 \end{bmatrix}$,

$$g_{10} = \begin{bmatrix} 2 \\ 1 \end{bmatrix} , \quad g_{20} = \begin{bmatrix} -1 \\ 1 \end{bmatrix} ,$$

$$u_0(t) = \begin{bmatrix} 2 \\ 1 \\ 0 \end{bmatrix} + t \begin{bmatrix} -1 \\ 1 \\ 0 \end{bmatrix} ,$$

$$\hat{t}_1 = +\infty ,$$

$$\tilde{t}_1 = \min \left\{ \frac{-2}{-1} , - , - \right\} = 2 , \quad k = 1 ,$$

$$t_1 = \min \{ +\infty , 2 , +\infty \} = 2 .$$

The optimal solution is $x_0(t) = \begin{bmatrix} 0 \\ 2 \end{bmatrix} + t \begin{bmatrix} 0 \\ 0 \end{bmatrix}$

with multipliers $u_0(t) = \begin{bmatrix} 2 \\ 1 \\ 0 \end{bmatrix} + t \begin{bmatrix} -1 \\ 1 \\ 0 \end{bmatrix}$ for all t with

$0 \le t \le 2 .$

Step 2:   Transfer to Step 2.2.

Step 2.2: $\rho = 1, \quad e_\rho = (-1 , 0)' .$

Solving $H_0 \begin{bmatrix} s_0 \\ \xi_0 \end{bmatrix} = \begin{bmatrix} 0 \\ -e_\rho \end{bmatrix}$ gives $s_0 = \begin{bmatrix} 1 \\ -2 \end{bmatrix}$ and

$\xi_0 = 0 .$ Transfer to Step 2.2.2.

Step 2.2.2: $a_i' s_0 \le 0, \quad i = 1 , 2 , 3 ;$ stop, the objective function is unbounded from below for $t > t_1 = 2 .$

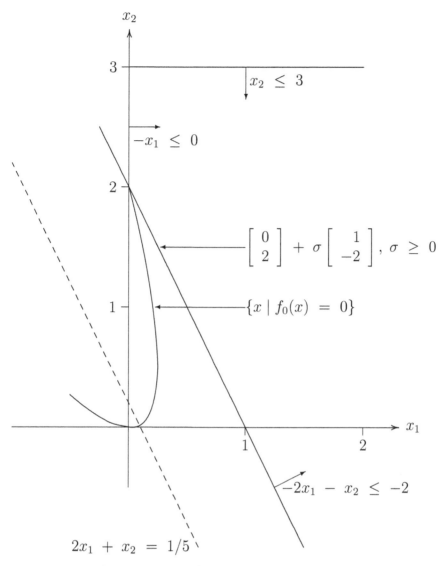

Figure 7.2.   Geometry of Example 7.2.

The geometry of Example 7.2 is illustrated in Figure 7.2. Letting $f_t(x)$ denote the objective function for $0 \leq t < 2$ the level sets of $f_t(x)$ are parabolas, symmetric about the line $2x_1 + x_2 = 1/5 - 3t/5$. The parabolas are open, up and to the left. They become "thinner" as t approaches 2. For $t = 2$, the level sets become pairs of parallel lines, symmetric about the line $2x_1 + x_2 = -1$ and all points

$$\begin{bmatrix} 0 \\ 2 \end{bmatrix} + \sigma \begin{bmatrix} 1 \\ -2 \end{bmatrix}, \quad \sigma \geq 0,$$

are alternate optima.

For $t > 2$, the level sets $f_t(x)$ are again parabolas opening down and to the right and the objective function is unbounded from below.

Example 7.3

$$\text{minimize:} \quad tx_1 - (1 - t)x_2 + \tfrac{1}{2}(4x_1^2 + x_2^2 + 4x_1 x_2)$$

$$\text{subject to:} \quad -\ x_1 \qquad\qquad \le\ \ 0, \qquad (1)$$

$$-\ 2x_1 - x_2 \ \le\ -2, \qquad (2)$$

$$4x_1 + x_2 \ \le\ \ \ 6 - t, \qquad (3)$$

for all t with $t \ge 0$.

This example was obtained from Example 7.5 by replacing the constraint $x_2 \le 3$ with $4x_1 + x_2 \le 6 - t$. Here

$$c = \begin{bmatrix} 0 \\ -1 \end{bmatrix}, \quad q = \begin{bmatrix} 1 \\ 1 \end{bmatrix}, \quad C = \begin{bmatrix} 4 & 2 \\ 2 & 1 \end{bmatrix}, \quad b = \begin{bmatrix} 0 \\ -2 \\ 6 \end{bmatrix},$$

$$p = \begin{bmatrix} 0 \\ 0 \\ -1 \end{bmatrix}, \quad \underline{t} = 0, \quad \text{and } \bar{t} = +\infty .$$

Initialization:

$$t_0 = 0, \quad x_0 = \begin{bmatrix} 0 \\ 2 \end{bmatrix}, \quad K_0 = \{1, 2\},$$

$$A_0 = \begin{bmatrix} a_1' \\ a_2' \end{bmatrix} = \begin{bmatrix} -1 & 0 \\ -2 & -1 \end{bmatrix}, \quad H_0 = \begin{bmatrix} 4 & 2 & -1 & -2 \\ 2 & 1 & 0 & -1 \\ -1 & 0 & 0 & 0 \\ -2 & -1 & 0 & 0 \end{bmatrix}, \quad j = 0.$$

Iteration 0

Step 1: $b_0 = (0, -2)', \quad p_0 = (0, 0)',$

Solving $H_0 \begin{bmatrix} h_{10} \\ g_{10} \end{bmatrix} = \begin{bmatrix} -c \\ b_0 \end{bmatrix}$ and $H_0 \begin{bmatrix} h_{20} \\ g_{20} \end{bmatrix} = \begin{bmatrix} -q \\ p_0 \end{bmatrix}$

gives $h_{10} = \begin{bmatrix} 0 \\ 2 \end{bmatrix}, \quad h_{20} = \begin{bmatrix} 0 \\ 0 \end{bmatrix},$

$$g_{10} = \begin{bmatrix} 2 \\ 1 \end{bmatrix}, \quad g_{20} = \begin{bmatrix} -1 \\ 1 \end{bmatrix},$$

$$u_0(t_0) = \begin{bmatrix} 2 \\ 1 \\ 0 \end{bmatrix} + t \begin{bmatrix} -1 \\ 1 \\ 0 \end{bmatrix},$$

$$\hat{t}_1 = \min\left\{ -, -, \frac{4}{1} \right\} = 4, \ l = 3,$$

$$\tilde{t}_1 = \min\left\{ \frac{-2}{-1}, -, - \right\} = 2, \ k = 1,$$

$$t_1 = \min\{4, 2, +\infty\} = 2.$$

The optimal solution is $x_0(t) = \begin{bmatrix} 0 \\ 2 \end{bmatrix} + t \begin{bmatrix} 0 \\ 0 \end{bmatrix}$

with multipliers $u_0(t) = \begin{bmatrix} 2 \\ 1 \\ 0 \end{bmatrix} + t \begin{bmatrix} -1 \\ 1 \\ 0 \end{bmatrix}$ for all t with

$0 \le t \le 2$.

Step 2: Transfer to Step 2.2.

Step 2.2: $\rho = 1, \ e_\rho = (-1, 0)'$.

Solving $H_0 \begin{bmatrix} s_0 \\ \xi_0 \end{bmatrix} = \begin{bmatrix} 0 \\ -e_\rho \end{bmatrix}$ gives $s_0 = \begin{bmatrix} 1 \\ -2 \end{bmatrix}$ and

$\xi_0 = 0$. Transfer to Step 2.2.2.

Step 2.2.2: $\sigma_0 = \min\left\{ -, -, \frac{2}{2} \right\} = 1, \ l = 3,$

$$K_1 = \{2, 3\}, \ A_1 = \begin{bmatrix} -2 & -1 \\ 4 & 1 \end{bmatrix},$$

$$H_1 \leftarrow \begin{bmatrix} 4 & 2 & -2 & 4 \\ 2 & 1 & -1 & 1 \\ -2 & -1 & 0 & 0 \\ 4 & 1 & 0 & 0 \end{bmatrix}, \ j = 1.$$

Iteration 1

Step 1: $b_1 = (-2, 6)', \ p_1 = (0, -1)',$

Solving $H_1 \begin{bmatrix} h_{11} \\ g_{11} \end{bmatrix} = \begin{bmatrix} -c \\ b_1 \end{bmatrix}$ and $H_1 \begin{bmatrix} h_{21} \\ g_{21} \end{bmatrix} = \begin{bmatrix} -q \\ p_1 \end{bmatrix}$

gives $h_{11} = \begin{bmatrix} 2 \\ -2 \end{bmatrix}, \ h_{21} = \begin{bmatrix} -1/2 \\ 1 \end{bmatrix},$

$g_{11} = \begin{bmatrix} 0 \\ -1 \end{bmatrix}, \ g_{21} = \begin{bmatrix} 3/2 \\ 1/2 \end{bmatrix},$

$$u_1(t) = \begin{bmatrix} 0 \\ 0 \\ -1 \end{bmatrix} + t \begin{bmatrix} 0 \\ 3/2 \\ 1/2 \end{bmatrix},$$

$$\hat{t}_2 = \min \left\{ \frac{2}{1/2}, -, - \right\} = 4, \; l = 1,$$

$$\tilde{t}_2 = +\infty,$$

$$t_2 = \min \{ 4, +\infty, +\infty \} = 4.$$

The optimal solution is $x_1(t) = \begin{bmatrix} 2 \\ -2 \end{bmatrix} + t \begin{bmatrix} -1/2 \\ 1 \end{bmatrix}$

with multipliers $u_1(t) = \begin{bmatrix} 0 \\ 1 \\ -1 \end{bmatrix} + t \begin{bmatrix} 0 \\ 3/2 \\ 1/2 \end{bmatrix}$ for all t with

$2 \leq t \leq 4$.

Step 2:   Transfer to Step 2.1.

Step 2.1: $l = 1$,

Solving $H_1 \begin{bmatrix} s_1 \\ \hat{\xi}_1 \end{bmatrix} = \begin{bmatrix} a_l \\ 0 \end{bmatrix}$ gives $s_1 = \begin{bmatrix} 0 \\ 0 \end{bmatrix}$,

$\hat{\xi}_1 = \begin{bmatrix} -1/2 \\ -1/2 \end{bmatrix}$, and $\xi_1 = \begin{bmatrix} 0 \\ -1/2 \\ -1/2 \end{bmatrix}$.

Transfer to Step 2.1.2.

Step 2.1.2: $(\xi_1)_i \leq 0, \; i = 1, 2, 3$; stop, there is no feasible solution
for $t > t_2 = 4$.

$\diamond$

The geometry of Example 7.3 differs from that of Example 7.2 as follows.
For $0 \leq t \leq 2$, the optimal solutions coincide. At $t = 2$, movement to
an alternate optima in Example 7.3 results in constraint (3) becoming active.
The optimal solution then lies at the intersection of constraints (2) and (3) as
t is increased to 4. At $t = 4$, all three constraints are active and the feasible
region consists of the single point $(0, 2)'$. Increasing t beyond 4 moves the
line $4x_1 + x_2 = 6 - t$ up and to the left, and the feasible region is null.
    The properties of Algorithm 7 are established in

Theorem 7.2 *Let Assumption 7.2 be satisfied and let $x_j(t)$, $u_j(t)$, t_j, $j =$ 1, ... be obtained by applying Algorithm 7 to the model problem*

$$
\left.
\begin{aligned}
\textit{minimize:} \quad & (c + tq)'x + \tfrac{1}{2}x'Cx \\
\textit{subject to:} \quad & a_i'x \leq b_i + tp_i, \quad i = 1,\ldots,m, \\
& a_i'x = b_i + tp_i, \quad i = m+1,\ldots,m+r,
\end{aligned}
\right\}
\tag{7.28}
$$

for t with $\underline{t} \leq t \leq \bar{t}$. Assume that for each iteration j, the indices k, l, and ρ are uniquely determined and that $\hat{t}_j \neq \tilde{t}_j$. Then Algorithm 7 terminates after $\nu < +\infty$ iterations, and for $j = 1,\ldots,\nu - 1$, $\tilde{x}_j$ is optimal for (7.28) with multiplier vector $u_j(t)$ for all t with $t_j \leq t \leq t_{j+1}$ and either $t_\nu = +\infty$, $t_\nu < +\infty$ and (7.28) has no feasible solution for $t > t_\nu$, or, $t_\nu < +\infty$ and (7.28) is unbounded from below for $t > t_\nu$.

Proof: Other than termination after finitely many steps, the theorem follows directly from Cases 1.2, 1.2.1, 1.2.2, 2.2, 2.2.1 and 2.2.2 discussed in the development of Algorithm 7. Finite termination is established by showing that no active set is ever repeated, as follows. Suppose to the contrary that there are iterations γ, δ with $\delta > \gamma$ and $K_\gamma = K_\delta$. By construction both $x_\gamma(t)$ and $x_\delta(t)$ are optimal solutions for

$$
\min \left\{ (c + tq)'x + \tfrac{1}{2}x'Cx \mid a_i'x \leq b_i + tp_i , \text{ all } i \in K_\gamma \right\},
$$

so that necessarily $t_\gamma = t_\delta$ and $t_{\gamma+1} = t_{\delta+1}$. But by construction, each $t_{j+1} > t_j$. Thus $\gamma = \delta$ and this contradiction establishes that indeed each active set is uniquely determined. Termination after finitely many steps now follows from the fact that there are only finitely many distinct subsets of $\{1,\ldots,m\}$. $\qquad\square$

7.3 Symmetric Matrix Updates

The general algorithms of the previous two sections, Algorithms 6 and 7, are general in that they leave unspecified the method by which key linear equations are to be solved. If a specialized QP algorithm is being formulated to solve a particular class of problems which have some structure, then it may be advantageous to formulate special methods to solve linear equations having a coefficient matrix of the form

$$
H = \begin{bmatrix} C & A' \\ A & 0 \end{bmatrix}.
$$

   In this section we develop three types of updates for the modification of a symmetric coefficient matrix: addition of a row and column (Procedure

Ψ_1), deletion of a row and column (Procedure Ψ_2), and exchange of a row and column (Procedure Ψ_3). The first two of these updates can be viewed as special cases of the Bordered Inverse Lemma [13]. The third is obtained from the Sherman–Morrison–Woodbury formula [15] and [17].

The simplest version of the the Bordered Inverse Lemma concerns the addition/deletion/exchange of the last row and column. We shall present this simplest version first and then generalize it so that the changed row and column can lie anywhere in the matrix.

Lemma 7.2 *(Bordered Inverse Lemma) Let A be an (n, n) nonsingular matrix. Let*

$$H = \begin{bmatrix} A & u \\ v' & \alpha \end{bmatrix}.$$

(a) Let $\theta = (\alpha - v'A^{-1}u)$. Then H is nonsingular if and only if $\theta \neq 0$. If $\theta = 0$, then $w = \begin{bmatrix} -A^{-1}u \\ 1 \end{bmatrix}$ is in the null space of H.

(b) Assume $\theta \neq 0$. Let H^{-1} be partitioned as

$$H^{-1} = \begin{bmatrix} B & p \\ q' & \beta \end{bmatrix},$$

where B is an (n, n) matrix, u, v, p, and q are n-vectors, and α and β are scalars. Then the partitions of H^{-1} are given by

$$\begin{aligned} p &= -\beta A^{-1}u, & q' &= -\beta v'A^{-1}, \\ B &= A^{-1} + \beta A^{-1}uv'A^{-1}, & \beta &= (\alpha - v'A^{-1}u)^{-1}, \end{aligned}$$

(c) Furthermore,

$$A^{-1} = B - \frac{1}{\beta}pq'.$$

Proof: (a) H is nonsingular if and only if

$$H \begin{bmatrix} x \\ \lambda \end{bmatrix} = \begin{bmatrix} A & u \\ v' & \alpha \end{bmatrix} \begin{bmatrix} x \\ \lambda \end{bmatrix} = \begin{bmatrix} Ax + u\lambda \\ v'x + \alpha\lambda \end{bmatrix} = \begin{bmatrix} 0 \\ 0 \end{bmatrix}$$

implies that $x = 0$ and that $\lambda = 0$. We have $x = -A^{-1}u\lambda$ and $-v'A^{-1}u\lambda + \alpha\lambda = 0$ or $(\alpha - v'A^{-1}u)\lambda = 0$. This implies that $\lambda = 0$ and hence $x = 0$ if and only if $\alpha - v'A^{-1}u \neq 0$.

To show that when $\theta = 0$ w is in the null space of H, we calculate

$$Hw = \begin{bmatrix} -AA^{-1}u + u \\ -v'A^{-1}u + \alpha \end{bmatrix} = \begin{bmatrix} 0 \\ 0 \end{bmatrix}.$$

(b) Because

$$HH^{-1} = \begin{bmatrix} A & u \\ v' & \alpha \end{bmatrix} \begin{bmatrix} B & p \\ q' & \beta \end{bmatrix} = \begin{bmatrix} I_{(n,n)} & 0 \\ 0' & 1 \end{bmatrix},$$

we have

$$AB + uq' = I_{(n,n)}, \tag{7.16}$$

$$Ap + u\beta = 0, \tag{7.17}$$

$$v'B + \alpha q' = 0', \tag{7.18}$$

$$v'p + \alpha\beta = 1. \tag{7.19}$$

Multiplying both sides of (7.17) on the left by A^{-1} and rearranging gives

$$p = -\beta A^{-1} u. \tag{7.20}$$

Multiplying both sides of (7.16) on the left by A^{-1} gives

$$B = A^{-1} - A^{-1} uq'. \tag{7.21}$$

Combining (7.19) and (7.20) gives

$$v'p + \alpha\beta = \alpha\beta - v'A^{-1}u\beta = 1 \tag{7.22}$$

Upon simplification this becomes

$$\beta = (\alpha - v'A^{-1}u)^{-1}. \tag{7.23}$$

From (7.18) we have

$$v'B + \alpha q' = v'A^{-1} + (\alpha - v'A^{-1}u)q' = 0. \tag{7.24}$$

Using β in (7.23) and the second equality in (7.24) gives

$$q' = -\beta v'A^{-1}. \tag{7.25}$$

(c) Using (7.21) in (7.25) gives

$$B = A^{-1} + \beta A^{-1} uv'A^{-1}. \tag{7.26}$$

$$\square$$

Lemma 7.2 can be utilized in several ways. Suppose that A^{-1}, u, v, and α are known. Then H^{-1} can be easily calculated in terms of an (n, n) matrix and the new row and column which define it. Further, suppose that H^{-1} is known and we wish to compute A^{-1}. This can be done using Lemma 7.2(c); that is, the inverse of an (n, n) submatrix of an $(n+1, n+1)$ matrix can easily be computed in terms of the partitions of the inverse of the larger matrix.

We illustrate the Bordered Inverse Lemma (Lemma 7.2) in

Example 7.4

$$A = \begin{bmatrix} 1 & 2 \\ 0 & 1 \end{bmatrix} \text{ and } A^{-1} = \begin{bmatrix} 1 & -2 \\ 0 & 1 \end{bmatrix}$$

are given. Let

$$H = \begin{bmatrix} 1 & 2 & 1 \\ 0 & 1 & 4 \\ 3 & 2 & -14 \end{bmatrix} .$$

We compute H^{-1} using the Bordered Inverse Lemma (Lemma 7.2) as follows. Here

$$v = \begin{bmatrix} 3 \\ 2 \end{bmatrix} , \quad u = \begin{bmatrix} 1 \\ 4 \end{bmatrix} , \quad \text{and } \alpha = -14 .$$

We calculate

$$A^{-1}u = \begin{bmatrix} -7 \\ 4 \end{bmatrix} , \quad v'A^{-1} = (3 , -4) ,$$

$$\beta = (-14 - (-13))^{-1} = -1 ,$$

$$p = \begin{bmatrix} -7 \\ 4 \end{bmatrix} , \quad q' = (3 , -4) ,$$

$$B = \begin{bmatrix} 1 & -2 \\ 0 & 1 \end{bmatrix} + (-1) \begin{bmatrix} -7 \\ 4 \end{bmatrix} (3 , -4)$$

$$= \begin{bmatrix} 22 & -30 \\ -12 & 17 \end{bmatrix} .$$

Thus

$$H^{-1} = \begin{bmatrix} 22 & -30 & -7 \\ -12 & 17 & 4 \\ 3 & -4 & -1 \end{bmatrix} ,$$

and this completes Example 7.4. $\Diamond$

In order to solve the linear equations required by Algorithm 6 by maintaining H_j^{-1} explicitly (when it exists), we require three types of updates:

addition of a row and column (Procedure Ψ_1), deletion of a row and column (Procedure Ψ_2), and exchange of a row and column (Procedure Ψ_3). In the case of the first two updates, if the modified matrix is singular, a logical variable "singular" will be returned with the value "true" and a nonzero vector in the null space of the modified matrix will also be returned. If the modified matrix is nonsingular, "singular" will be returned with the value "false" and the procedure will give the inverse of the modified matrix.

In order to formulate the first two of these three procedures, we need a generalization of Lemma 7.2 (Bordered Inverse Lemma). Lemma 7.2 formulates the inverse of a matrix with a new row and column in terms of the inverse of the unmodified matrix inverse and the data defining the new row and column. The new row and column are assumed to be in the *last* row and column. This is too restrictive for our purposes. We next formulate a version of Lemma 7.2 which allows the new row and column to be in any position.

Let H_0 be symmetric, nonsingular, and have dimension (p, p). Let H_0 be partitioned as

$$H_0 = \begin{bmatrix} (H_0)_{11} & (H_0)_{12} \\ (H_0)_{21} & (H_0)_{22} \end{bmatrix}, \tag{7.27}$$

where $(H_0)_{11}$ has dimension (q, q) and $(H_0)_{21}$, $(H_0)_{12}$, and $(H_0)_{22}$ have dimensions $(p - q, q)$, $(q, p - q)$, and $(p - q, p - q)$, respectively. Let d_1 and d_2 be given q- and $(p - q)$-vectors, respectively, and let γ be a scalar. Define

$$H_1 = \begin{bmatrix} (H_0)_{11} & d_1 & (H_0)_{12} \\ d_1' & \gamma & d_2' \\ (H_0)_{21} & d_2 & (H_0)_{22} \end{bmatrix}, \tag{7.28}$$

so that H_1 is of dimension $(p + 1, p + 1)$ and is obtained from H_0 by inserting the new row and column (d_1', γ, d_2') and $(d_1', \gamma, d_2')'$, respectively, immediately after row and column q of H_0. Finally, let

$$\theta \equiv \gamma - (d_1', d_2') H_0^{-1} \begin{bmatrix} d_1 \\ d_2 \end{bmatrix}. \tag{7.29}$$

We will show that if $\theta \neq 0$ then H_1 is nonsingular. In this case, we partition H_1^{-1} in a similar manner to H_1 as follows.

$$H_1^{-1} = \begin{bmatrix} B_{11} & e_1 & B_{12} \\ e_1' & \beta & e_2' \\ B_{21} & e_2 & B_{22} \end{bmatrix}. \tag{7.30}$$

We can now state and prove the Generalized Bordered Inverse Lemma.

Lemma 7.3 *(Generalized Bordered Inverse Lemma) Let H_0, H_1, θ and H_1^{-1} (when it exists) be as in (7.27), (7.28), (7.29) and (7.30), respectively. Then*

(a) *$\theta = 0$ if and only if H_1 is singular. Furthermore, if H_1 is singular, then defining*

$$\begin{bmatrix} \tilde{d}_1 \\ \tilde{d}_2 \end{bmatrix} = H_0^{-1} \begin{bmatrix} d_1 \\ d_2 \end{bmatrix} \quad \text{and} \quad w = \begin{bmatrix} \tilde{d}_1 \\ -1 \\ \tilde{d}_2 \end{bmatrix},$$

it follows that w is in the null space of H_1.

(b) *If $\theta \neq 0$, the partitions of H_1^{-1} given by (7.30) can be expressed in term of H_0^{-1}, d_1, γ and d_2 as follows.*

$$\beta = \frac{1}{\theta},$$

$$B \equiv \begin{bmatrix} B_{11} & B_{12} \\ B_{21} & B_{22} \end{bmatrix} = H_0^{-1} + \beta H_0^{-1} \begin{bmatrix} d_1 \\ d_2 \end{bmatrix} (d_1', d_2') H_0^{-1},$$

$$e = -\beta H_0^{-1} \begin{bmatrix} d_1 \\ d_2 \end{bmatrix} \equiv \begin{bmatrix} e_1 \\ e_2 \end{bmatrix}.$$

(c)

$$H_0^{-1} = \begin{bmatrix} B_{11} & B_{12} \\ B_{21} & B_{22} \end{bmatrix} - \frac{1}{\beta} \begin{bmatrix} e_1 \\ e_2 \end{bmatrix} (e_1', e_2'),$$

where B, β, and e are as in part (b).

Proof:

(a) H_1 is nonsingular if and only if

$$H_1 \begin{bmatrix} x_1 \\ \lambda \\ x_2 \end{bmatrix} , = \begin{bmatrix} (H_0)_{11}x_1 + d_1\lambda + (H_0)_{12}x_2 \\ d_1'x_1 + \gamma\lambda + d_2'x_2 \\ (H_0)_{21}x_1 + d_2\lambda + (H_0)_{22}x_2 \end{bmatrix} = \begin{bmatrix} 0 \\ 0 \\ 0 \end{bmatrix} , \quad (7.31)$$

implies $x_1 = 0$, $\lambda = 0$ and $x_2 = 0$. Suppose (7.31) is satisfied. Letting $x = (x_1', x_2')'$ and $d = (d_1', d_2')'$, it follows from the first and third partitions of (7.31) that $H_0x - \lambda d = 0$ and thus

$$x = \lambda H_0^{-1}d. \quad (7.32)$$

Using (7.32) in the second partition of (7.31) gives

$$\lambda(-d'H_0^{-1}d + \gamma) = 0.$$

If $\theta \neq 0$, then $\lambda = 0$ and from (7.32) it follows that $x = 0$. Thus we have shown that $\theta \neq 0$ implies H_1 is nonsingular.

Next we show that if $\theta = 0$, that w is in the null space of H_1. Assume $\theta = 0$. By definition of H_1 and w, and letting the result of forming $H_1 w$ be denoted by $s = (s_1', s_2, s_3')'$, we have

$$H_1 w = \begin{bmatrix} (H_0)_{11}\tilde{d}_1 - d_1 + (H_0)_{12}\tilde{d}_2 \\ d_1'\tilde{d}_1 - \gamma + d_2'\tilde{d}_2 \\ (H_0)_{21}\tilde{d}_1 - d_2 + (H_0)_{22}\tilde{d}_2 \end{bmatrix} = \begin{bmatrix} s_1 \\ s_2 \\ s_3 \end{bmatrix}. \tag{7.33}$$

From the first and third partitions of the above we have

$$\begin{bmatrix} s_1 \\ s_3 \end{bmatrix} = H_0 \begin{bmatrix} \tilde{d}_1 \\ \tilde{d}_2 \end{bmatrix} - \begin{bmatrix} d_1 \\ d_2 \end{bmatrix}.$$

But by definition,

$$\begin{bmatrix} \tilde{d}_1 \\ \tilde{d}_2 \end{bmatrix} = H_0^{-1} \begin{bmatrix} d_1 \\ d_2 \end{bmatrix}.$$

Combining these last two gives

$$\begin{bmatrix} s_1 \\ s_3 \end{bmatrix} = \begin{bmatrix} 0 \\ 0 \end{bmatrix}.$$

Furthermore, from the second partition of (7.33), $\theta = 0$ implies $s_2 = 0$. We have thus shown that $\theta = 0$ implies $H_1 w = 0$ and that H_1 is singular.

(b) Now we assume that $\theta \neq 0$ and validate the formulae giving H_1^{-1}. Note that by definition

$$H_1 H_1^{-1} = \begin{bmatrix} (H_0)_{11} & d_1 & (H_0)_{12} \\ d_1' & \gamma & d_2' \\ (H_0)_{21} & d_2 & (H_0)_{22} \end{bmatrix} \begin{bmatrix} B_{11} & e_1 & B_{12} \\ e_1' & \beta & e_2' \\ B_{21} & e_2 & B_{22} \end{bmatrix}.$$

This last must be equal to the $(p+1, p+1)$ identity matrix. Thus

$$(H_0)_{11} B_{11} + d_1 e_1' + (H_0)_{12} B_{21} = I, \tag{7.34}$$

$$d_1' B_{11} + \gamma e_1' + d_2' B_{21} = 0, \tag{7.35}$$

$$(H_0)_{21} B_{11} + d_2 e_1' + (H_0)_{22} B_{21} = 0, \tag{7.36}$$

$$d_1' e_1 + \gamma \beta + d_2' e_2 = 1, \tag{7.37}$$

$$(H_0)_{21}e_1 + d_2\beta + (H_0)_{22}e_2 = 0, \tag{7.38}$$

$$(H_0)_{21}B_{12} + d_2e_2' + (H_0)_{22}B_{22} = I, \tag{7.39}$$

$$(H_0)_{11}e_1 + d_1\beta + (H_0)_{12}e_2 = 0, \tag{7.40}$$

$$(H_0)_{11}B_{12} + d_1e_2' + (H_0)_{12}B_{22} = 0, \tag{7.41}$$

$$d_1'B_{12} + \gamma e_2' + d_2'B_{22} = 0. \tag{7.42}$$

Combining (7.34) and (7.39), we get

$$\begin{bmatrix} (H_0)_{11} & (H_0)_{12} \\ (H_0)_{21} & (H_0)_{22} \end{bmatrix} \begin{bmatrix} B_{11} & B_{12} \\ B_{21} & B_{22} \end{bmatrix} = I - \begin{bmatrix} d_1e_1' \\ d_2e_2' \end{bmatrix}. \tag{7.43}$$

The matrix on the left hand-side of (7.40) is H_0 so that (7.40) becomes

$$B = \begin{bmatrix} B_{11} & B_{12} \\ B_{21} & B_{22} \end{bmatrix} = H_0^{-1} - H_0^{-1} \begin{bmatrix} d_1e_1' \\ d_2e_2' \end{bmatrix}. \tag{7.44}$$

Putting (7.35) and (7.42) in matrix format gives (because B is symmetric)

$$e = -\frac{1}{\gamma}Bd. \tag{7.45}$$

Equations (7.38) and (7.40) give

$$\begin{bmatrix} e_1 \\ e_2 \end{bmatrix} = -\beta H_0^{-1} \begin{bmatrix} d_1 \\ d_2 \end{bmatrix}. \tag{7.46}$$

It follows from (7.46) that

$$\begin{bmatrix} e_1 \\ e_2 \end{bmatrix}' = -\beta[d_1', d_2']H_0^{-1}.$$

Substituting this last into (7.44) gives

$$B = \begin{bmatrix} B_{11} & B_{12} \\ B_{21} & B_{22} \end{bmatrix} = H_0^{-1} - \beta H_0^{-1} \begin{bmatrix} d_1 \\ d_2 \end{bmatrix} [d_1', d_2']H_0^{-1}. \tag{7.47}$$

From (7.37),

$$[d_1', d_2'] \begin{bmatrix} e_1 \\ e_2 \end{bmatrix} + \gamma\beta = 1,$$

so that from (7.46) we have

$$\beta = \left[\gamma - [d_1', d_2']H_0^{-1} \begin{bmatrix} d_1 \\ d_2 \end{bmatrix} \right]^{-1} = \theta^{-1}. \tag{7.48}$$

Equations (7.46), (7.47) and (7.48) now complete the verification of part (b).

(c) From part (b),

$$H_0^{-1} \begin{bmatrix} d_1 \\ d_2 \end{bmatrix} = -\frac{1}{\beta} \begin{bmatrix} e_1 \\ e_2 \end{bmatrix}.$$

Substituting this in the expression for B in part (b), (7.47), and solving for H_0^{-1} gives

$$H_0^{-1} = \begin{bmatrix} B_{11} & B_{12} \\ B_{21} & B_{22} \end{bmatrix} - \frac{1}{\beta} \begin{bmatrix} e_1 \\ e_2 \end{bmatrix} (e_1', \ e_2'),$$

as required.

□

We next use Lemma 7.3 to formulate procedures Ψ_1 (add a row and column), Ψ_2 (delete a row and column) and then find the inverse of the new matrix in terms of the inverse of the original matrix. We first formulate

Procedure Ψ_1

Let H_0 be symmetric, nonsingular, and have dimension (p, p). Let H_0 be partitioned as

$$H_0 = \begin{bmatrix} (H_0)_{11} & (H_0)_{12} \\ (H_0)_{21} & (H_0)_{22} \end{bmatrix},$$

where $(H_0)_{11}$ has dimension (q, q) and $(H_0)_{21}$, $(H_0)_{12}$, and $(H_0)_{22}$ have dimensions $(p - q, q)$, $(q, p - q)$, and $(p - q, p - q)$, respectively. Let d_1 and d_2 be given q- and $(p - q)$-vectors, respectively, and let γ be a scalar. Define

$$H_1 = \begin{bmatrix} (H_0)_{11} & d_1 & (H_0)_{12} \\ d_1' & \gamma & d_2' \\ (H_0)_{21} & d_2 & (H_0)_{22} \end{bmatrix}$$

so that H_1 is of dimension $(p + 1, p + 1)$ and is obtained from H_0 by inserting the new row and column $(d_1', \ \gamma, \ d_2')'$ immediately after row and column q of H_0. We assume that H_0^{-1} is known and we wish to find H_1^{-1} in terms of H_0^{-1} and the other quantities defining H_1. We allow $0 \le q \le p$.

When Procedure Ψ_1 is invoked as

$$[\, H_1^{-1}, \text{singular}, w \,] \leftarrow \Psi_1(H_0^{-1}, d_1, \gamma, d_2, p, q),$$

the following results. If singular is returned as "false" then H_1 is nonsingular, its inverse is returned in H_1^{-1} and w is null. If singular is returned as "true," then H_1 is singular and w is in the null space of H_1.

Let

$$\theta \equiv \gamma - (d_1', \ d_2') H_0^{-1} \begin{bmatrix} d_1 \\ d_2 \end{bmatrix}.$$

If $\theta = 0$, then H_1 is singular. Furthermore, defining

$$\begin{bmatrix} \tilde{d}_1 \\ \tilde{d}_2 \end{bmatrix} = H_0^{-1} \begin{bmatrix} d_1 \\ d_2 \end{bmatrix}$$

and

$$w = \begin{bmatrix} \tilde{d}_1 \\ -1 \\ \tilde{d}_2 \end{bmatrix}$$

it follows that w is in the null space of H_1.

If $\theta \neq 0$, then H_1 is nonsingular and one way to compute H_1^{-1} is as follows. Let H_1^{-1} be partitioned similar to H_1 so that

$$H_1^{-1} = \begin{bmatrix} B_{11} & e_1 & B_{12} \\ e_1' & \beta & e_2' \\ B_{21} & e_2 & B_{22} \end{bmatrix}.$$

Then the components of H_1^{-1} are given by

$$\beta = \frac{1}{\theta},$$

$$B \equiv \begin{bmatrix} B_{11} & B_{12} \\ B_{21} & B_{22} \end{bmatrix} = H_0^{-1} + \beta H_0^{-1} \begin{bmatrix} d_1 \\ d_2 \end{bmatrix} (d_1', \ d_2') H_0^{-1},$$

$$e = -\beta H_0^{-1} \begin{bmatrix} d_1 \\ d_2 \end{bmatrix} \equiv \begin{bmatrix} e_1 \\ e_2 \end{bmatrix}.$$

The validity of Procedure Ψ_1 follows directly from Lemma 7.3 (a) and (b). $\triangle$

Procedure Ψ_1 allows the possibility $q = 0$ and $q = p$. In either of these cases, some quantities used in this procedure are null. Some computer programs might have difficulty with interpreting these null quantities. Table 7.1 summarizes the results for Procedure Ψ_1 for the cases $q = 0$ and $p = q$.

Proc Ψ_1 $q = 0$	$H_1 = \begin{bmatrix} \gamma & d_2' \\ d_2 & (H_0)_{22} \end{bmatrix}$, $H_1^{-1} = \begin{bmatrix} \beta & e_2' \\ e_2 & B_{22} \end{bmatrix}$, $(H_0)_{11}$, $(H_0)_{12}$, $(H_0)_{21}$ and d_1 are all null, $(H_0)_{22}$ is (p, p), $(H_0)_{22} = H_0$, $\theta = \gamma - d_2' H_0^{-1} d_2$. If $\theta \neq 0$, then H_1 is nonsingular and is given as above with $\beta = \theta^{-1}$, $B_{22} = H_0^{-1} + \beta H_0^{-1} d_2 d_2' H_0^{-1}$, $e_2 = -\beta H_0^{-1} d_2$. If $\theta = 0$, then H_1 is singular. Set $\tilde{d}_2 = H_0^{-1} d_2$ and define $w = (-1, \tilde{d}_2')'$. Then w is in the null space of H_1.
Proc Ψ_1 $q = p$	$H_1 = \begin{bmatrix} (H_0)_{11} & d_1 \\ d_1' & \gamma \end{bmatrix}$, $H_1^{-1} = \begin{bmatrix} B_{11} & e_1 \\ e_1' & \beta \end{bmatrix}$, $(H_0)_{12}$, $(H_0)_{21}$, $(H_0)_{22}$ and d_2 are all null, $(H_0)_{11}$ is (p, p), $(H_0)_{11} = H_0$, $\theta = \gamma - d_1' H_0^{-1} d_1$. If $\theta \neq 0$ then H_1 is nonsingular and is given as above with $\beta = \theta^{-1}$, $B_{11} = H_0^{-1} + \beta H_0^{-1} d_1 d_1' H_0^{-1}$, $e_1 = -\beta H_0^{-1} d_1$. If $\theta = 0$, then H_1^{-1} is singular. Set $\tilde{d}_1 = H_0^{-1} d_1$ and define $w = (\tilde{d}_1', -1)'$. Then w is in the null space of H_1.

Table 7.1   Procedure $\Psi_1 : q = 0$ and $q = p$

A computer program which implements Ψ_1 is given in Figure 7.3.

Example 7.5
With

$$H_0 = \begin{bmatrix} 1 & 0 \\ 0 & 1 \end{bmatrix} \quad \text{and} \quad H_0^{-1} = \begin{bmatrix} 1 & 0 \\ 0 & 1 \end{bmatrix},$$

use Procedure Ψ_1 to determine H_1^{-1} where

$$H_1 = \begin{bmatrix} 1 & 0 & 1 \\ 0 & 1 & 1 \\ 1 & 1 & 0 \end{bmatrix}.$$

Here

$$H_1 = \begin{bmatrix} H_0 & & 1 \\ & & 1 \\ 1 & 1 & 0 \end{bmatrix},$$

so $p = 2, q = 2, d_1 = (1, 1)', \gamma = 0$, and $d_2 = $ null . Since $q = p$, we use the second entry of Table 7.1. Invoking Procedure Ψ_1 as

$$[\, H_1^{-1}, \text{ singular },\ w\,] \leftarrow \Psi_1 \left(H_0^{-1},\ \text{null},0\,,\ \begin{bmatrix} 1 \\ 1 \end{bmatrix},\ 2\,,0 \right),$$

the intermediate computations for Ψ_1 proceed as follows.

$$\theta = 0 - (1\,,\ 1) H_0^{-1} \begin{bmatrix} 1 \\ 1 \end{bmatrix} = -2,$$

$\theta \neq 0$ and singular is assigned "false,"

$$\beta = \frac{-1}{2},$$

$$B = \begin{bmatrix} 1 & 0 \\ 0 & 1 \end{bmatrix} + (\frac{-1}{2}) \begin{bmatrix} 1 \\ 1 \end{bmatrix} (1,1)$$

$$= \begin{bmatrix} 1 & 0 \\ 0 & 1 \end{bmatrix} + \begin{bmatrix} -1/2 & -1/2 \\ -1/2 & -1/2 \end{bmatrix} = \begin{bmatrix} 1/2 & -1/2 \\ -1/2 & 1/2 \end{bmatrix},$$

$$e = \frac{1}{2} \begin{bmatrix} 1 \\ 1 \end{bmatrix} = \begin{bmatrix} 1/2 \\ 1/2 \end{bmatrix},$$

which gives

$$H_1^{-1} = \begin{bmatrix} 1/2 & -1/2 & 1/2 \\ -1/2 & 1/2 & 1/2 \\ 1/2 & 1/2 & -1/2 \end{bmatrix},$$

singular = "false" and w is null.   ◊

The following procedure finds the inverse of a matrix after the original matrix has a row and column deleted. The inverse of the new matrix is formulated in terms of the inverse of the original matrix and the changed data.

Procedure Ψ_2

Let H_0 be a (p, p) symmetric nonsingular matrix partitioned as

$$H_0 = \begin{bmatrix} (H_1)_{11} & d_1 & (H_1)_{12} \\ d_1' & \gamma & d_2' \\ (H_1)_{21} & d_2 & (H_1)_{22} \end{bmatrix}, \tag{7.49}$$

where $(H_1)_{11}$, $(H_1)_{21}$, $(H_1)_{12}$ and $(H_1)_{22}$ have dimensions $(q - 1\ , q - 1)$, $(p - q, q - 1)$, $(q - 1, p - q)$, and $(p - q, p - q)$, respectively. In addition, d_1 and d_2 are vectors of dimension $(q - 1)$ and $(p - q)$, respectively. Finally, γ is a scalar.

Define

$$H_1 = \begin{bmatrix} (H_1)_{11} & (H_1)_{12} \\ (H_1)_{21} & (H_1)_{22} \end{bmatrix},$$

so that H_1 is obtained from H_0 by deleting the qth row and column. We assume that H_0^{-1} is known and we wish to find H_1^{-1} in terms of H_0^{-1} and the other quantities defining H_1 We allow $0 \le q \le p$.

Let H_0^{-1} be partitioned similar to H_0 with

$$H_0^{-1} = \begin{bmatrix} B_{11} & e_1 & B_{12} \\ e_1' & \beta & e_2' \\ B_{21} & e_2 & B_{22} \end{bmatrix}. \tag{7.50}$$

When Procedure Ψ_2 is invoked as

$$[\, H_1^{-1}, \text{singular}, w \,] \leftarrow \Psi_2(H_0^{-1}, p, q)$$

the following results. If singular is returned as "false," then H_1 is nonsingular, its inverse is returned in H_1^{-1} and w is null. If singular is returned as "true," then H_1 is singular and w is in the null space of H_1.

If $\beta = 0$, then H_1 is singular and defining $w = (e_1' \, , \, e_2')'$ it follows that w is in the null space of H_1. We verify this as follows. Multiply the first row partition of H_0 in (7.49) with the second column partition of H_0^{-1} in (7.50). The result must be 0 since $H_0 H_0^{-1} = I$. Thus

$$(H_1)_{11}e_1 + (H_1)_{12}e_2 = 0.$$

Next, multiply the third row partition of H_0 in (7.49) with the second column partition of H_0^{-1} in (7.50). The result must be 0 since $H_0 H_0^{-1} = I$. Thus

$$(H_1)_{21}e_1 + (H_1)_{22}e_2 = 0.$$

Combining these last two equations gives

$$H_1 \begin{bmatrix} e_1 \\ e_2 \end{bmatrix} = 0.$$

Therefore H_1 is singular and w is in its null space.

If $\beta \ne 0$, then H_1 is nonsingular and one way to compute H_1^{-1} is

$$H_1^{-1} = \begin{bmatrix} B_{11} & B_{12} \\ B_{21} & B_{22} \end{bmatrix} - \frac{1}{\beta} \begin{bmatrix} e_1 \\ e_2 \end{bmatrix} (e_1' \, , \, e_2'),$$

where, θ is defined in (7.29) and $\beta = \theta^{-1}$. For the case of $\beta \ne 0$, the validity of Procedure Ψ_2 including the nonsingularity of H_1 follows immediately from

Lemma 7.3(c).   △

Procedure Ψ_2 allows the possibility $q = 1$ and $q = p$. In either of these cases, some quantities used in this procedure are null. Some computer programs might have difficulty with interpreting these null quantities. Table 7.2 summarizes the results for Procedure Ψ_1 for the cases $q = 1$ and $q = p$.

Proc Ψ_2	$H_0 = \begin{bmatrix} \gamma & d_2' \\ d_2 & H_1 \end{bmatrix}$ $H_0^{-1} = \begin{bmatrix} \beta & e_2' \\ e_2 & B_{22} \end{bmatrix}$,
$q = 1$	$(H_0)_{11}$, $(H_0)_{21}$, $(H_0)_{12}$, d_1 and e_1 are all null, H_1 is $(p-1, p-1)$, $H_1 = (H_0)_{22}$. If $\beta \neq 0$, then H_1 is nonsingular and is given as above with $\quad H_1^{-1} = B_{22} - \beta^{-1} e_2 e_2'$. If $\beta = 0$ then H_1 is singular. Define $w = e_2$. Then w is in $\quad$ the null space of H_1.
Proc Ψ_2	$H_0 = \begin{bmatrix} H_1 & d_1 \\ d_1' & \gamma \end{bmatrix}$ and $H_0^{-1} = \begin{bmatrix} B_{11} & e_1 \\ e_1' & \beta \end{bmatrix}$,
$q = p$	$(H_0)_{22}$, $(H_0)_{21}$, $(H_0)_{12}$, d_2 and e_2 are all null, H_1 is $(p-1, p-1)$, $H_1 = (H_0)_{11}$. If $\beta \neq 0$, then H_1 is nonsingular and $\quad H_1^{-1} = B_{11} - \beta^{-1} e_1 e_1'$. If $\beta = 0$, then H_1 is singular. Define $w = e_1$. Then w is in $\quad$ the null space of H_1.

Table 7.2   Procedure Ψ_2 : $q = 1$, $q = p$

A computer program which implements Ψ_2 is given in Figure 7.4.

Example 7.6
With
$$H_0^{-1} = \begin{bmatrix} 1/2 & -1/2 & 1/2 \\ -1/2 & 1/2 & 1/2 \\ 1/2 & 1/2 & -1/2 \end{bmatrix},$$
use Procedure Ψ_2 to determine H_1^{-1}, where H_1 is obtained from H_0 by deleting the third row and column of H_0.

Because $q = p = 3$, we use the second entry of Table 7.2 for Procedure Ψ_2. Invoking Procedure Ψ_2 as

$$[\, H_1^{-1}, \text{singular}, w \,] \leftarrow \Psi_2(H_0^{-1}, 3, 3),$$

the intermediate calculations are: $\beta = -1/2 \neq 0$ so H_1 is nonsingular and

$$H_1^{-1} = \begin{bmatrix} 1/2 & -1/2 \\ -1/2 & 1/2 \end{bmatrix} - (-2) \begin{bmatrix} 1/2 \\ 1/2 \end{bmatrix} (1/2\, ,\ 1/2)$$

$$= \begin{bmatrix} 1 & 0 \\ 0 & 1 \end{bmatrix}.$$

$\Diamond$

Note that Example 7.6 is the reverse of Example 7.5 in the sense that Example 7.6 deletes the last row and column of a $(3, 3)$ matrix whereas Example 7.5 augments the same $(2, 2)$ matrix with the identical row and column.

The following procedure finds the inverse of a matrix after the original matrix has a row and column replaced by a new row and column. The inverse of the new matrix is formulated in terms of the inverse of the original matrix and the changed data. It is difficult to formulate this procedure in terms of the Bordered Inverse Lemma. Rather, it is more natural to derive it from the Sherman–Morrison-Woodbury formula [15] and [17].

Let A be an $(n\,,\ n)$ nonsingular matrix and let u and v be $(n\,,\ k)$ matrices with $1 \leq k \leq n$ and rank $u = $ rank $v = k$. Then $A + uv'$ can be thought of as a rank k modification of A. The Sherman–Morrison–Woodbury formula expresses $(A + uv')^{-1}$ in terms of A^{-1}, u, and v as follows.

Lemma 7.4 *(Sherman–Morrison–Woodbury formula)*

(a) $(A + uv')$ *is nonsingular if and only if* $(I + v'A^{-1}u)$ *is nonsingular.*

(b) *If* $(A + uv')$ *is nonsingular, then*

$$(A + uv')^{-1} = A^{-1} - A^{-1}u(I + v'A^{-1}u)^{-1}v'A^{-1}\,.$$

Proof:

The proof of part (a) is left as an exercise (Exercise 7.8). Part (b) is verified by observing that

$$(A + uv')(A^{-1} - A^{-1}u(I + v'A^{-1}u)^{-1}v'A^{-1})$$
$$= I - u(I + v'A^{-1}u)^{-1}v'A^{-1} + uv'A^{-1} - uv'A^{-1}u(I + v'A^{-1}u)^{-1}v'A^{-1}$$
$$= I + uv'A^{-1} - u((I + v'A^{-1}u)^{-1} + v'A^{-1}u(I + v'A^{-1}u)^{-1})v'A^{-1}$$
$$= I + uv'A^{-1} - u((I + v'A^{-1}u)(I + v'A^{-1}u)^{-1})v'A^{-1}$$
$$= I + uv'A^{-1} - uv'A^{-1}$$
$$= I .$$

$\square$

Note that $(I + v'A^{-1}u)$ has dimension (k , k) so use of the Sherman–Morrison–Woodbury formula for a rank k modification requires the inversion of a (k , k) matrix.

The best known version of the Sherman–Morrison–Woodbury formula corresponds to $k = 1$. In this case, $(I + v'A^{-1}u)$ is simply a scalar and provided it is nonzero,

$$(A + uv')^{-1} = A^{-1} - \frac{1}{1 + v'A^{-1}u}A^{-1}uv'A^{-1} .$$

The case of $k = 1$ is generally referred to as the Sherman–Morrison formula. The case of $k > 1$ is referred to as the Sherman–Morrison–Woodbury formula.

In our case, we shall use Lemma 7.4 with $k = 2$. To see how it is used, let d be any n–vector and let $\tilde{d}$ be from d by replacing its q–th element d_q by $d_q/2$. Further, let e_q denote the q–th unit vector. Define

$$u = [\, e_q,\ d\,] \quad \text{and} \quad v = [\, d,\ e_q\,].$$

Then

$$uv' = [\, e_q, \tilde{d}\,] \begin{bmatrix} \tilde{d}' \\ e_q' \end{bmatrix} = e_q\tilde{d} + \tilde{d}e_q'.$$

This shows that uv' is an (n, n) matrix of zeros but with the q–th row and column replaced by d' and d, respectively and element (q, q) is $2d_q$.

Procedure Ψ_3

Let H_0 be a (p, p) symmetric nonsingular matrix with q–th row and column d_0' and d_0, respectively. Let H_0^{-1} denote the inverse of H_0. Suppose H_1 is obtained from H_0 by replacing the q–th row and column with d_1' and d_1, respectively. Define the vector d according to

$$d_i = ((d_1)_i - (d_0)_i)_i\ i = 1, p,\ i \neq q \quad \text{and} \quad d_q = ((d_1)_q - (d_0)_q)/2.$$

Then

(a) $H_1 = H_0 + uv'$ with $u = [\, e_q, \ d\,]$ and $v = [\, d, \ e_q\,]$.

(b) $H_1^{-1} = H_0^{-1} - H_0^{-1}u(I + v'H_0^{-1}u)^{-1}v'H_0^{-1}$.

When we perform computations using parts (a) and (b) of Procedure Ψ_3, we use the notation

$$H_1^{-1} \leftarrow \Psi_3(H_0^{-1}, \ d, \ p, \ q\,).$$

A computer program which implements Ψ_3 is given in Figure 7.5.

Note that in Procedure Ψ_3, no provision is made for H_1 being singular. Although in general it could be singular it cannot be singular in the situations in which it is used here. See Lemma 7.1(b) and Lemma 8.11(b).

The computations of Procedure Ψ_3 are illustrated in

Example 7.7
With

$$H_0 = \begin{bmatrix} 1 & 0 & 1 \\ 0 & 1 & 1 \\ 1 & 1 & 0 \end{bmatrix} \quad \text{and } H_0^{-1} = \begin{bmatrix} 1/2 & -1/2 & 1/2 \\ -1/2 & 1/2 & 1/2 \\ 1/2 & 1/2 & -1/2 \end{bmatrix},$$

use Procedure Ψ_3 to determine H_1^{-1} where

$$H_1 = \begin{bmatrix} 1 & 0 & 1 \\ 0 & 1 & 0 \\ 1 & 0 & 0 \end{bmatrix}.$$

Here $q = 2$, $p = 3$ and H_1 is obtained from H_0 by replacing row and column 2 of H_0 with $(0, 1, 0)$ and $(0, 1, 0)'$, respectively. Here, $d = (0-0, (1-1)/2, 0-1)' = (0, 0, -1)'$. We need to evaluate

$$H_1^{-1} \leftarrow \Psi_3(H_0^{-1}, \ d, \ , 3, \ 2)\,.$$

We do this by calculating

$$(I + v'H_0^{-1}u) = \begin{bmatrix} 0.5 & -0.5 \\ 0.5 & 0.5 \end{bmatrix},$$

and

$$(I + v'H_0^{-1}u)^{-1} = \begin{bmatrix} 1. & 1. \\ -1. & 1. \end{bmatrix}.$$

Furthermore, we calculate

$$H_0^{-1}u = \begin{bmatrix} -0.5 & -0.5 \\ 0.5 & -0.5 \\ 0.5 & 0.5 \end{bmatrix}$$

and

$$v'H_0^{-1} = \begin{bmatrix} -0.5 & -0.5 & 0.5 \\ -0.5 & 0.5 & 0.5. \end{bmatrix}.$$

Assembling these results gives

$$H_1^{-1} = \begin{bmatrix} 0 & 0 & 1 \\ 0 & 1 & 0 \\ 1 & 0 & -1 \end{bmatrix}.$$

$\Diamond$

We complete this by discussing how Procedures Ψ_1, Ψ_2 and Ψ_3 can be used to make Algorithm 6 into an implementable algorithm. We call this Algorithm 6a and the Matlab program which implements it we call Alg6a.m. Alg6a.m is shown in Figure 7.6. In Alg6a, Hinv plays the role of H_j^{-1} in Algorithm 6 and is initialized in lines 11-13 of Figure 7.6. It is then used to determine s_j and v_j in lines 27-30. It is then updated in lines 27-30 when Step 3.1 applies with H_j^{-1} being nonsingular. If singularity would occur, Alg6a uses Procedure Ψ_3 to do the relevant update in line 102.

When a constraint is to be dropped, Alg6a does this in lines 125-126 by using Procedure Ψ_2.

Algorithm 7 can be formulated into an implementable algorithm which uses just Procedures Ψ_1, Ψ_2 and Ψ_3 in a similar way. See Exercise 7.4.

7.4   Computer Programs

```
1  %     Psi1.m
2  function [ H1inv,singular,w ] = Psi1 ( H0inv,d1,gamma,d2,p,q )
3
4  %Psi1 implements Procedure Psi 1 of Chapter 7, text.
5  %A matrix with a known inverse (H0inv) is augmented with a
6  %new row and column (d1,gamma,d2) and Procedure Psi1 finds
7  %inverse (H1inv) of the new matrix.  If the new matrix is
8  %singular, Psi1 returns a vector w in its null space.
9  %
10 d = vertcat(d1,d2);
11 tol = 1.e-6;
12 if q ≥  1 & q < p
13     theta = gamma - d' * H0inv * d;
14     if abs(theta) ≤ tol                    % H1 is singular
15         singular = 'true';
16         dtilde = H0inv * d;
```

```
17              dtilde1 = dtilde(1:q);    dtilde2 = dtilde(q+1:p);
18              w = vertcat(d1,-1,d2);    % w is in null space of H1
19              H1inv = [];
20              return
21
22       else                                     % H1 is nonsingular
23              singular = 'false';
24              w = [];
25              beta = 1./theta;
26              temp = H0inv * d;
27              B = H0inv + beta * (temp * temp');
28              e = - beta * temp;  e1 = e(1:q);  e2 = e(q+1:p);
29              B11 = B(1:q,1:q);         B12 = B(1:q,q+1:p);
30              B21 = B(q+1:p,1:q);       B22 = B(q+1:p,q+1:p);
31              H1inv = vertcat([B11,e1,B12],[e1',beta,e2'], ...
32                                          [B21,e2,B22]);
33              return
34       end
35   end
36
37   if q == 0
38       theta = gamma - d2'*H0inv*d2;
39       if abs (theta) < tol
40              singular = 'true';                    % H1 is singular
41              d2tilde = H0inv*d2;
42              w = vertcat(-1,d2tilde);
43              H1inv = [];
44       else                                     % H1 is nonsingular
45              singular = 'false';
46              w = [];
47              beta = 1./theta;
48              B22 = H0inv + beta*H0inv*d2*d2'*H0inv;
49              e2 = - beta*H0inv*d2;
50              H1inv = vertcat([beta,e2'],[e2,B22]);
51              return
52       end
53   end
54
55   if q == p
56       theta = gamma - d1'*H0inv* d1;
57       if abs(theta) < tol
58              singular = 'true';                    % H1 is singular
59              d1tilde = H0inv*d1;
60              w = vertcat(d1tilde,-1);
61              H1inv = [];
62              return
63       else                                     % H1 is nonsingular
64              singular = 'false';                   % H1 is nonsingular
65              w = [];
66              beta = 1./theta;
67              B11 = H0inv + beta*H0inv*d1*d1'*H0inv;
```

```
68          e1 = -beta*H0inv*d1;
69          H1inv = vertcat([B11,e1],[e1',beta]);
70          return
71       end
72    end
73
74  return
```

Figure 7.3.   Procedure Ψ_1, Psi1.m.

Discussion of Psi1.m

```
1  %      Psi2.m
2  function [ H1inv,singular,w ] = Psi2 ( H0inv,p,q )
3
4  %Psi2 implements Procedure Psi 2 of Chapter 7, text.
5  %A matrix with a known inverse (H0inv) has a row and
6  % column deleted from it. Procedure Psi2 finds
7  %inverse (H1inv) of the new matrix.  If the new matrix is
8  %singular, Psi2 returns a vector w in its null space.
9  %
10 tol = 1.e-6;
11 if q > 1 & q < p
12     beta = H0inv(q,q);
13     if abs(beta) <= tol              % H1 is singular
14         singular = 'true';
15         e1 = H0inv(1:q-1,q);
16         e2 = H0inv(q+1:p,q);
17         w = vertcat(e1,e2);          % w is in null(H1)
18         H1inv = [];
19         return
20
21     else                            % H1 is nonsingular
22         singular = 'false';
23         w = [];
24         e1 = H0inv(1:q-1,q);
25         e2 = H0inv(q+1:p,q);
26         e = vertcat(e1,e2);
27         B11 = H0inv(1:q-1,1:q-1);  B12 = H0inv(1:q-1,q+1:p);
28         B21 = H0inv(q+1:p,1:q-1);  B22 = H0inv(q+1:p,q+1:p);
29         H1inv = vertcat( [B11,B12],[B21,B22] ) ...
30                              - 1./beta*e*e' ;
31         return
32     end
33 end
34
35 if q == 1
36     beta = H0inv(1,1);
37     if abs(beta) <= tol                % H1 is singular
```

```
38        singular = 'true';
39        e2 = H0inv(2:p,1);
40        w = e2;                        % w is in null(H1)
41        H1inv = [ ];
42    else                              % H1 is nonsingular
43        singular = 'false';
44        w = [ ];
45        B22 = H0inv(2:p,2:p);   e2 = H0inv(2:p,1);
46        H1inv = B22 - (1./beta)*e2*e2';
47        return
48    end
49 end
50
51 if q == p
52    beta = H0inv(p,p)
53    if abs(beta) ≤ tol               % H1 is singular
54        singular = 'true';
55        e1 = H0inv(1:p-1,p);
56        w = e1;                        % w is in null(H1)
57        H1inv = [ ];
58    else                              % H1 is nonsingular
59        singular = 'false';
60        w = [ ];
61        B11 = H0inv(1:p-1,1:p-1)
62        e1 = H0inv(1:p-1,p)
63        H1inv = B11 - (1./beta)*e1*e1'
64        return
65    end
66 end
```

Figure 7.4. Procedure Ψ_2, Psi2.m.

Discussion of Psi2.m

```
1      %   Psi3.m
2      function [ H1inv ] = Psi3 ( H0inv,d,p,q )
3
4      %   Sherman–Morrison update for (A + uv')^{-1} where
5      %   u = [e_q, d], v = [d, e_q], e_q = q-th unit vector
6      %   H0inv, H1inv are (p,p)
7      tol = 1.e-6;
8      eq = zeros(p,1);
9      eq(q) = 1.;
10     u = [ eq d ];
11     v = [ d eq ];
12     sm = eye(2) + v'*H0inv*u;
13     detsm = det(sm);
14     H1inv = H0inv - H0inv*u*inv(eye(2) + v'*H0inv*u)*v'*H0inv;
15     return
```

```
16        end
```

Figure 7.5.   Procedure Ψ_3, Psi3.m.

Discussion of Psi3.m

```
1  %     Alg6a.m
2  function [x,K,fx,Hinv,msg,dual,iter,alg] = ...
3                       Alg6a(c,C,n,m,r,tol,x,K,actineq,A,b)
4  %     Alg6a is an implementation of the general Alg6 QP
5  %     algorithm where the matrix inverse updating is done
6  %     using Procedures Psi1, Psi2 and Psi3.
7  %     Initialization
8  alg = 'Algorithm 6a';
9  fx = c'*x + 0.5*( x'*C*x);
10 g = c + C*x;
11 Aj = vertcat(A(K,:));
12 H = vertcat([C,Aj'],[Aj,zeros(actineq+r,actineq+r)]);
13 Hinv = inv(H);
14 singular = 'false';
15 next = 'Step1';
16 for iter=0:10000
17     if strcmp(next,'Step1')
18                              % Step 1
19         if strcmp(singular,'false')
20             next = 'Step1p1';
21         else
22             next = 'Step1p2';
23         end
24     end
25     if strcmp(next,'Step1p1')
26                              % Step 1.1
27         rhs = vertcat(g,zeros(actineq+r,1));
28         sv = Hinv*rhs;
29         s = sv(1:n);
30         v = sv(n+1:n+actineq);
31         gamma = 0;
32         next = 'Step2';
33     end
34     if strcmp(next,'Step1p2')
35                              % Step 1.2
36         s = w(1:n);
37         s = 1./(A(kay,:)*s)*s;
38         gamma = 1;
39         next = 'Step2';
40     end
```

```
41      if strcmp(next,'Step2')
42                                  % Step 2
43          if gamma == 0;
44              sigtilde = 1.;
45          end
46          if gamma == 1
47              sigtilde = Inf;
48          end
49          sighat = Inf;
50          ell = 0;
51          for i=1:m
52              if ¬ ismember(i,K)
53                  temp = A(i,:)*s;
54                  if temp <0
55                      test = (A(i,:)*x − b(i))/temp;
56                      if test ≤ sighat
57                          sighat = test;
58                          ell = i;
59                      end
60                  end
61              end
62          end       %end for loop
63          if sigtilde == Inf & sighat == Inf
64              msg = 'objective function is inbounded from below';
65              return
66          else
67              sigma = min(sigtilde,sighat);
68              next = 'Step3';
69          end
70      end
71      if strcmp(next,'Step3')
72                                  % Step 3
73          x = x − sigma*s;
74          fx = c'*x + 0.5*( x'*C*x);
75          g = c + C*x;
76          if sigma == sighat
77              next = 'Step3p1';
78          else
79              next = 'Step3p2';
80          end
81          if strcmp(next,'Step3p1')
82                                  % Step 3.1
83              if strcmp(singular,'false')
84                  p = n + actineq + r;
85                  q = n + actineq;
86                  d1 = vertcat(A(ell,:)',zeros(actineq−1));
87                  gamma = 0.;
88                  d2 = zeros(actineq+r,1);
89                  [ Hinv,singular,w ] = ...
90                      Psi1 ( Hinv,d1,gamma,d2,p,q );
91                  K1 = K(1:actineq);
```

```
92                       K2 = K(actineq+1:actineq+r);
93                       K = horzcat(K1,ell,K2);
94                       actineq = actineq +1;
95                       next = 'Step1p1';
96                   else              % singular = 'true'
97                       posn = find(kay==K);
98                       d = vertcat( (A(ell,:)-A(kay,:))', ...
99                                           zeros(actineq+r,1));
100                      p = n + actineq + r;
101                      q = n + posn;
102                      [Hinv] = Psi3(Hinv,d,p,q);
103                      K(posn) = ell;
104                      singular = 'false';
105                      next = 'Step1p1';
106                  end
107              end
108              if strcmp(next,'Step3p2')
109                  u = zeros(m,1);
110                  for i=1:actineq
111                      u(K(i)) = - v(i);
112                  end
113                  minmult = min(u);
114                  kay = find(u == minmult,1);
115                  if minmult >= - tol
116                      dual = zeros(m+r,1);
117                      vv = sv(n+1:n+actineq+r);
118                      for i=1:actineq+r
119                          dual(K(i)) = - vv(i);
120                      end
121                      msg = 'Optimal solution obtained';
122                      return
123                  end
124                  posn = find(kay==K);
125                  [ Hinv,singular,w ] = ...
126                              Psi2 ( Hinv,n+actineq+r,n+posn );
127                  if strcmp(singular,'false')
128                      K(posn) = [];
129                      actineq = actineq - 1;
130                  end
131                  next = 'Step1';
132              end
133          end
134  end                       % end of for iter loop
```

Figure 7.6.   Algorithm 6a, Alg6a.m.

7.5 Exercises

7.1 Consider the problem

$$\text{minimize:} \quad -3x_1 - x_2 + \tfrac{1}{2}(x_1^2 + x_2^2)$$

$$\text{subject to:} \quad x_1 \qquad + x_3 \qquad = 2, \quad (1)$$

$$x_2 \qquad + x_4 = 2, \quad (2)$$

$$x_i \geq 0, \quad i = 1, \ldots, 4.$$

Solve this problem using Algorithm 6a (Alg6a.m, Figure 7.6) with starting point $x_0 = (\ 0,\ 2,\ 2, 0\ )'$.

Note that this problem must first be reformulated to match the format of (7.1).

7.2 Consider the problem

$$\text{minimize:} \quad -5x_1 + 2x_2$$

$$\text{subject to:} \quad -2x_1 + x_2 + x_3 \qquad = 2, \quad (1)$$

$$x_1 + 2x_2 \qquad + x_4 \qquad = 14, \quad (2)$$

$$4x_1 + 3x_2 \qquad + x_5 = 36, \quad (3)$$

$$x_i \geq 0, \quad i = 1, \ldots, 5.$$

Solve this problem using Algorithm 6a (Alg6a.m, Figure 7.4) with starting point $x_0 = (\ 2,\ 6,\ 0,\ 0,\ 10\ )'$.

Note that this problem must first be reformulated to match the format of (7.1).

7.3 Suppose Algorithm 6a (Alg6a, Figure 7.6), is applied to an LP; i.e., a QP with $C = 0$. Assume x_0 is a nondegenerate extreme point so that Assumption 7.1 is satisfied. Show that Algorithm 6a can be simplified to a typical LP solution which moves from one extreme point to an adjacent extreme point in such a way that the objective function is reduced at each iteration. Show that all matrix solves can be done with a smaller matrix which can be performed using Procedure Φ. In order to see the pattern, it may be helpful to modify the code for Alg6a so that each time Hinv is modified, inv(Hinv) is computed and printed. A good example to work with is solving the problem of Exercise 7.2 with Alg6a.

7.4 Algorithm 7 is general in the sense that it does not specify a method to solve the requisite linear equations.

(a) Using Procedures Ψ_1, Ψ_2 and Ψ_3, formulate an implementable version of Algorithm 7. Call this Algorithm 7a (analogous to Algorithm 6a).

(b) Implement this in a Matlab program. Use it to solve the problems of Examples 7.2 and 7.3.

7.5 Suppose that A is an (n , n) nonsingular matrix and that C is (n , n) symmetric and positive semidefinite. Show that

$$H = \begin{bmatrix} C & A' \\ A & 0 \end{bmatrix}$$

is nonsingular and that

$$H^{-1} = \begin{bmatrix} 0 & A^{-1} \\ (A^{-1})' & -(A^{-1})'CA^{-1} \end{bmatrix}.$$

7.6 Show that each of Procedures Ψ_1, Ψ_2 and Ψ_3 require approximately p^2 arithmetic operations.

7.7 Consider the optimization problem of Exercise 4.5. Let x_0 be any feasible point for this problem. Solve this problem using Algorithm 6 with x_0 as the starting point. Also solve this same problem using Algorithm 6a (Alg6a, Figure 7.6) and with the same starting point.

7.8 Complete the proof of Lemma 7.4.

7.9 Let A be an (n , n) nonsingular matrix. Let u and v be (n , k) matrices with $1 \leq k \leq n$. Prove that $(A + uv')$ is nonsingular if and only if $(I + v'A^{-1}u)$ is nonsingular.

7.10 Derive the results of Procedure Φ using the Generalized Sherman-Morrison Formula (Lemma 7.4) for a rank 1 change with $u = e_k$, and $v = d - d_k$, where e_k denotes the kth n-dimensional unit vector.

7.11 The algorithms in this book are iterative in nature and require a sequence of matrices $D_0, D_1, \ldots, D_j$ and their inverses $D_0^{-1}, D_1^{-1}, \ldots, D_j^{-1}$. In many cases, D'_{j+1} is obtained from D'_j by replacing column k with some new vector d. Procedure Φ can be used to obtain D_{j+1}^{-1} as $D_{j+1}^{-1} = \Phi(D_j^{-1} , d , k)$. Since computations performed on a digital computer are subject to roundoff errors, one might expect the computed value of each successive D_j^{-1} to contain greater and greater errors. Eventually perhaps, the computed value of D_j^{-1} might be so bad that using it would give quite erroneous results. The purpose of this exercise is to show that this possibility is most unlikely. Let H_j denote the computed value of D_j^{-1}. With no assumption on any possible relation between D_j^{-1} and H_j, let $H_{j+1} = \Phi(H_j, d, k)$. Assuming for simplicity that the computations for Procedure Φ in producing H_{j+1} are made with exact arithmetic, show that

$$(D_{j+1} - H_{j+1}^{-1}) = Q_k(D_j - H_j^{-1}),$$

where Q_k is obtained from the identity matrix by replacing the kth diagonal element with zero. This shows that no matter how badly H_j estimates D_j^{-1}, the kth row of H_{j+1} is exactly d and the remaining rows are identical to those of H_j.

7.12 The purpose of this exercise is to show how a QP may be solved without having an initial point, without having a QP algorithm but only having a parametric QP solver. Suppose the model problem is

$$\min\{\, c'x + \tfrac{1}{2}x'Cx \mid Ax \geq b + b,\ x \geq 0 \,\}.$$

Let q be a strictly positive vector and p be a strictly negative vector. Consider the related problem

$$\min\{\, (c + tq)'x + \tfrac{1}{2}x'Cx \mid Ax \geq b + tp,\ x \geq 0 \,\},$$

where the vectors p and q are any vectors satisfying $q > 0$ and $p < 0$. Show that $x = 0$ is optimal for all t satisfying $t \geq t_1$ and $t \geq t_2$ where $t_1 = \max\{-b_i,\ i = 1, 2, \ldots, m\,\}$ and $t_2 = \max\{-c_i,\ i = 1, 2, \ldots, n\,\}$. A parametric QP algorithm may now be used to reduce t to zero and we have an optimal solution to the original problem. Note that our parametric QP algorithms are formulated for the parameter to increase but it is a simple matter to reformulate them in terms of decreasing the parameter.

7.13 Repeat Exercise 5.8 with Algorithm 3 replaced with Algorithm 6.

Chapter 8

Simplex Method for QP and PQP

The simplex method for quadratic programming may be regarded as a generalization of the simplex method for linear programming. The model problem in both cases has linear equality constraints and nonnegativity restrictions on all variables. For the quadratic programming case, the method uses the structure of the model problem to obtain the Newton direction using an inverse matrix which may be considerably smaller than that required by Algorithm 6. The method was originally developed by Dantzig [8] and van de Panne and Whinston [16].

Section 8.1 will develop the method itself by applying Algorithm 6 to the required model problem and simplifying the intermediate calculations. Because the resulting algorithm is a special case of Algorithm 6, it will be equivalent to it in the sense of generating the same sequence of points. However, if the Hessian matrix for the model problem has low rank (i.e., the quadratic programming problem is closer to a linear programming problem) or if the optimal solution has many components with value zero, then the intermediate calculations will be correspondingly easier than for Algorithm 6.

Section 8.2 will develop the parametric extension of the method.

Throughout this chapter, we will assume that the objective function of the QP is convex.

8.1   Simplex Method for QP: Algorithm 8

Here we develop the simplex method for quadratic programming by applying Algorithm 6 to a model problem having linear equality constraints and nonnegativity constraints on all variables. Because of the form of the model problem, the intermediate calculations can be performed more efficiently in certain

cases. The derivation is analogous to that of the revised simplex method from
Algorithm 3 in Best and Ritter [1985].

In this section, we use the model problem

$$\min \{ c'x + \tfrac{1}{2}x'Cx \mid Ax = b, \ x \geq 0 \}, \tag{8.1}$$

where c and x are n-vectors, A is an (m, n) matrix,[1] C is an (n, n) symmetric,
positive semidefinite matrix, and b is an m-vector. From Theorem 4.6, the
optimality conditions for (8.1) are

$$\left.\begin{array}{rclcl}
Ax &=& b &,& x \geq 0, \\
-c - Cx &=& A'u - v &,& v \geq 0, \\
v'x &=& 0 &.&
\end{array}\right\} \tag{8.2}$$

Suppose we apply Algorithm 6 to (8.1). At any iteration j, we can divide
the components of x_j into two sets; basic and nonbasic. Nonbasic components
have value zero and correspond to active nonnegativity constraints. The remaining components are positive and are called basic. Suppose for simplicity
that the first r nonnegativity[2] constraints are inactive at x_j. Then the basic
variables are the first r and the nonbasic are the last $n - r$. Note that for
the revised simplex method of linear programming, there are always $r = m$
basic variables. However, for quadratic programming r will vary between m
and $n - m$.

We can partition x_j as

$$x_j = \begin{bmatrix} x_B \\ x_N \end{bmatrix},$$

where the "B" and "N" subscripts mean basic and nonbasic, respectively. The
data of (8.1) may be partitioned according to basic and nonbasic as follows.

$$A = [\, B, \, N \,], \quad C = \begin{bmatrix} C_{BB} & C_{BN} \\ C_{NB} & C_{NN} \end{bmatrix}, \quad c = \begin{bmatrix} c_B \\ c_N \end{bmatrix},$$

where B is the matrix consisting of the first r columns of A, N consists of the
remaining $n - r$ columns, C_{BB} is the top lefthand corner (r, r) submatrix
of C and C_{BN}, C_{NB}, and C_{NN} are the corresponding remaining partitions of
C. Let A_i denote the ith column of A. Then

$$B = [\, A_1, \ldots, A_r \,] \quad \text{and} \quad N = [\, A_{r+1}, \ldots, A_n \,].$$

---

[1]Previously, we used "m" to denote the number of **inequality** constraints. Throughout
this chapter, "m" denotes the number of **equality** constraints.

[2]Previously, we have used "r" to denote the number of equality constraints. In Section
7.1, we use "r" to denote the number of basic variables. Note that r will change from
iteration to iteration.

Since $x_N = 0$, we have $f(x_j) = c'_B x_B + \frac{1}{2} x'_B C_{BB} x_B$ and

$$g(x_j) = \begin{bmatrix} c_B \\ c_N \end{bmatrix} + \begin{bmatrix} C_{BB} x_B \\ C_{NB} x_B \end{bmatrix} \equiv \begin{bmatrix} g_B \\ g_N \end{bmatrix}.$$

Since the last $n - r$ nonnegativity constraints are the only active inequality constraints, the "A_j" for Algorithm 6 is

$$\begin{bmatrix} B & N \\ 0 & -I \end{bmatrix},$$

where I is the $(n - r, n - r)$ identity matrix. Writing

$$\begin{bmatrix} s_B \\ s_N \end{bmatrix}$$

for the "s_j" of Algorithm 6, the equations defining the Newton direction are

$$\begin{bmatrix} C_{BB} & C_{BN} & B' & 0 \\ C_{NB} & C_{NN} & N' & -I \\ B & N & 0 & 0 \\ 0 & -I & 0 & 0 \end{bmatrix} \begin{bmatrix} s_B \\ s_N \\ u \\ v_N \end{bmatrix} = \begin{bmatrix} g_B \\ g_N \\ 0 \\ 0 \end{bmatrix}. \tag{8.3}$$

These equations can be simplified. The fourth partition requires $s_N = 0$ and having obtained s_B and u from the first and third partitions, v_N is obtained from the second partition as

$$v_N = -g_N + C_{NB} s_B + N'u. \tag{8.4}$$

It remains to determine s_B and u from

$$\begin{bmatrix} C_{BB} & B' \\ B & 0 \end{bmatrix} \begin{bmatrix} s_B \\ u \end{bmatrix} = \begin{bmatrix} g_B \\ 0 \end{bmatrix}. \tag{8.5}$$

Let

$$H_B = \begin{bmatrix} C_{BB} & B' \\ B & 0 \end{bmatrix}$$

denote the coefficient matrix for (8.5). The linear equations (8.5) may be solved in a variety of ways such as maintaining a factorization of H_B. Assuming that H_B is nonsingular, another way is to maintain H_B^{-1} explicitly as we shall do here. Thus

$$\begin{bmatrix} s_B \\ u \end{bmatrix} = H_B^{-1} \begin{bmatrix} g_B \\ 0 \end{bmatrix}. \tag{8.6}$$

The utility of the simplex method for quadratic programming is a consequence of the fact that only the smaller coefficient matrix H_B need be inverted (or factorized) rather than the larger coefficient matrix (8.3). Because H_B is non-singular, the optimal step size $\tilde{\sigma}_B$ is unity. The maximum feasible step size $\hat{\sigma}_B$ is the largest value of σ such that

$$x_B - \sigma s_B \geq 0 .$$

If $s_B \leq 0$, then $\hat{\sigma}_B = +\infty$. Otherwise,

$$\hat{\sigma}_B = \min \left\{ \frac{(x_B)_i}{(s_B)_i} \mid \text{all } i = 1,\ldots,r \text{ with } (s_B)_i > 0 \right\} \qquad (8.7)$$

$$= \frac{(x_B)_l}{(s_B)_l} .$$

Setting $\sigma_B = \min \{ \tilde{\sigma}_B , \hat{\sigma}_B \}$, the new values of the basic variables are $x_B - \sigma_B s_B$. Suppose first that $\sigma_B = \hat{\sigma}_B$. Assume for simplicity that $l = r$; i.e., the last basic variable is reduced to zero and becomes nonbasic. Let $C = [\gamma_{ij}]$. Then

$$H_B = \begin{bmatrix} \gamma_{11} & \cdots & \gamma_{1r} & A_1' \\ \vdots & & \vdots & \vdots \\ \gamma_{r1} & \cdots & \gamma_{rr} & A_r' \\ A_1 & \cdots & A_r & 0 \end{bmatrix} . \qquad (8.8)$$

Since variable r has become nonbasic, H_B must be modified by deleting row and column r. We will show that the modified matrix is nonsingular [Lemma 8.1(b)]. Its inverse can be obtained by using Procedure Ψ_2; i.e.,

$$[H_B^{-1}, \text{ singular}, w] \leftarrow \Psi_2(H_B^{-1} , m + r , r) . \qquad (8.9)$$

In addition to replacing x_B with $x_B - \sigma_B s_B$, the last component of the modified x_B must be removed reflecting the change of variable r from basic to nonbasic. Thus

$$x_B \leftarrow ((x_B)_1,\ldots,(x_B)_{r-1})' .$$

Next suppose that $\sigma_B = 1$. Since s_B and u are already available from (8.6), the multipliers v_N for the active nonnegativity constraints can be calculated from (8.4). Alternatively, v_N may be calculated as follows. Replacing x_B with $x_B - \sigma_B s_B$ and computing the new gradient

$$\begin{bmatrix} g_B \\ g_N \end{bmatrix} ,$$

it follows from Taylor's series that

$$g_B \leftarrow g_B - C_{BB} s_B .$$

Comparing this with (8.5) and (8.2) (setting $v_B = 0$ to satisfy complementary slackness) we see that the u computed by (8.6) is the negative of that required for (8.2). Having replaced this u with its negative, v_N can be computed from (8.2) as

$$v_N = g_N + N'u . \qquad (8.10)$$

Having computed v_N from either (8.4) or (8.10), the complete vector of multipliers is

$$v = \begin{bmatrix} 0 \\ v_N \end{bmatrix} .$$

We next compute k such that

$$v_k = \min \{ v_i \,|\, i = 1, \ldots, n \}$$

and if $v_k \geq 0$, the current point is optimal. If $v_k < 0$, we proceed by deleting the kth nonnegativity constraint from the active set; i.e., nonbasic variable k becomes basic. For simplicity, assume that $k = r + 1$. Then the modified H_B becomes

$$H_B = \begin{bmatrix} \gamma_{11} & \cdots & \gamma_{1r} & \gamma_{1,r+1} & A'_1 \\ \vdots & & \vdots & \vdots & \vdots \\ \gamma_{r1} & \cdots & \gamma_{rr} & \gamma_{r,r+1} & A'_r \\ \gamma_{r+1,1} & \cdots & \gamma_{r+1,r} & \gamma_{r+1,r+1} & A'_{r+1} \\ A_1 & \cdots & A_r & A_{r+1} & 0 \end{bmatrix} . \qquad (8.11)$$

Let

$$d_1 = \begin{bmatrix} \gamma_{1,r+1} \\ \vdots \\ \gamma_{r,r+1} \end{bmatrix} .$$

Then the new H_B is obtained from the old H_B by inserting

$$\begin{bmatrix} d_1 \\ \gamma_{r+1,r+1} \\ A_{r+1} \end{bmatrix}$$

after row and column r of the old H_B. The new H_B^{-1} may be obtained, provided H_B is nonsingular, using Procedure Ψ_1:

$$[H_B^{-1}, \text{ singular }, w] \leftarrow \Psi_1(H_B^{-1} , d_1 , \gamma_{r+1,r+1} , A_{r+1} , m + r , r) .$$

In addition to replacing x_B with $x_B - \sigma_B s_B$, the modified x_B must be appended with a zero, the current value of the new basic variable $r + 1$. Thus

$$x_B \leftarrow (x'_B , 0)'$$

and we continue by finding the next Newton direction using the modified H_B^{-1}.

Invoking Procedure Ψ_1 will also tell us if the new H_B is singular. If it returns with singular $=$ "true," then the returned value of w is such that $H_B w = 0$. It is easy to show (Exercise 8.5) that the last m components of w are zero. Setting s_B to be the first $r + 1$ components it follows that

$$
\begin{bmatrix}
\gamma_{11} & \cdots & \gamma_{1r} & \gamma_{1,r+1} \\
\vdots & & \vdots & \vdots \\
\gamma_{r+1,1} & \cdots & \gamma_{r+1,r} & \gamma_{r+1,r+1}
\end{bmatrix} s_B = 0 \, ,
$$

$$
[\, A_1, \ldots, A_r \, , \, A_{r+1} \,] s_B = 0j
$$

and by construction of w in Procedure Ψ_1, $(s_B)_{r+1} = -1$. Therefore

$$
(-e_{r+1})' \begin{bmatrix} s_B \\ 0 \end{bmatrix} = 1
$$

and thus $(s_B' \, , \, 0)'$ is the uniquely determined search direction of Step 1.2 of Algorithm 6. Using this search direction results in either the conclusion that the problem is unbounded from below or a previously inactive constraint becomes active. In the latter case, this means that some basic variable becomes nonbasic. For simplicity, suppose that the index of this variable is r. Then the next H_B is

$$
H_B = \begin{bmatrix}
\gamma_{11} & \cdots & \gamma_{1,r-1} & \gamma_{1,r+1} & A_1' \\
\vdots & & \vdots & \vdots & \vdots \\
\gamma_{r-1,1} & \cdots & \gamma_{r-1,r-1} & \gamma_{r-1,r+1} & A_{r-1}' \\
\gamma_{r+1,1} & \cdots & \gamma_{r+1,r-1} & \gamma_{r+1,r+1} & A_{r+1}' \\
A_1 & \cdots & A_{r-1} & A_{r+1} & 0
\end{bmatrix} . \tag{8.12}
$$

Before proceeding, it is useful to summarize. We began with a nonsingular H_B as in (8.8). Variable $r + 1$ became basic giving the next H_B as in (8.11). Procedure Ψ_1 told us that this H_B was singular and gave us the corresponding search direction. Using this search direction led to a new active constraint giving the next H_B, (8.12). We will show that this H_B is nonsingular [Lemma 8.1(c)]. Thus we have a sequence of three consecutive H_B's which are nonsingular, singular, and nonsingular. Now the critical observation is that H_B (8.12) is obtained from the H_B (8.8) by replacing row and column r with

$$
\begin{bmatrix}
\gamma_{1,r+1} \\
\vdots \\
\gamma_{r-1,r+1} \\
\gamma_{r+1,r+1} \\
A_{r+1}
\end{bmatrix} .
$$

Defining

$$
d = \begin{bmatrix} \gamma_{1,r+1} - \gamma_{1,r} \\ \vdots \\ \gamma_{r-1,r+1} - \gamma_{r-1,r} \\ (\gamma_{r+1,r+1} - \gamma_{rr})/2 \\ A_{r+1} - A_r \end{bmatrix}.
$$

The inverse of H_B (8.12) may be obtained from the inverse of H_B (8.8) using Procedure Ψ_3:

$$
[\, H_B^{-1} \,] \leftarrow \Psi_3(\, H_B^{-1} \,,\, d \,,\, r + m \,,\, r \,).
$$

In addition to replacing x_B with $x_B - \sigma_B s_B$, the updated x_B must be further modified. Since $(s_B)_{r+1} = -1$, the new basic variable $r + 1$ has been increased from 0 to σ_B. Also, basic variable r has been reduced to zero so $(x_B)_r$ must be replaced with σ_B.

We have described the method under the simplifying assumption that the basic variables are the first r. In general, we require an index set associated with the basic variables. Let

$$
I_B = \{\, \beta_1, \ldots, \beta_{n_B} \,\},
$$

where n_B denotes the number of basic variables and $\beta_1, \ldots, \beta_{n_B}$ are the indices of the basic variables. Then

$$
H_B = \begin{bmatrix} \gamma_{\beta_1 \beta_1} & \cdots & \gamma_{\beta_1 \beta_{n_B}} & A'_{\beta_1} \\ \vdots & & \vdots & \vdots \\ \gamma_{\beta_{n_B} \beta_1} & \cdots & \gamma_{\beta_{n_B} \beta_{n_B}} & A'_{\beta_{n_B}} \\ A_{\beta_1} & \cdots & A_{\beta_{n_B}} & 0 \end{bmatrix}
$$

and

$$
g_B = \begin{bmatrix} (g(x_j))_{\beta_1} \\ \vdots \\ (g(x_j))_{\beta_{n_B}} \end{bmatrix},
$$

where the current point x_j is obtained from x_B as

$$
(x_j)_i = \begin{cases} (x_B)_k, & \text{if } i \in I_B \text{ and } \beta_k = i, \\ 0, & \text{if } i \notin I_B. \end{cases}
$$

s_B and u are still obtained from (8.6). The multipliers corresponding to the active inequality constraints are from (8.4)

$$
v_i = -(g(x_j))_i + \sum_{k=1}^{n_B} \gamma_{i\beta_k}(s_B)_k + A'_i u,
$$

for all $i \notin I_B$. Alternatively, following the discussion leading to (8.10), when x_{j+1} is a quasistationary point and u is replaced with $-u$, v_N can be calculated from

$$v_N = g_N(x_{j+1}) + N'u .$$

Setting $v_i = 0$ for all $i \in I_B$ gives the remaining multipliers.

The computation of the maximum feasible step size, (8.7), remains unchanged. However, if $\sigma_B = \hat{\sigma}_B$ it is variable β_l that is reduced to zero and becomes nonbasic. Procedure Ψ_2 is invoked as

$$[\, H_B^{-1} \, , \text{ singular }, \, w \,] \leftarrow \Psi_2(H_B^{-1} \, , \, m + n_B \, , \, l) \, ,$$

x_B is replaced with $x_B - \sigma_B s_B$ and then further modified as

$$x_B \leftarrow (\, (x_B)_1, \ldots, (x_B)_{l-1} \, , \, (x_B)_{l+1}, \ldots, (x_B)_{n_B} \,)' \, ,$$
$$I_B \leftarrow \{ \, \beta_1, \ldots, \beta_{l-1} \, , \, \beta_{l+1}, \ldots, \beta_{n_B} \, \} \, ,$$

and n_B is reduced by 1.

If $\sigma_B = 1$, we compute the smallest of the multipliers associated with the active nonnegativity constraints:

$$v_k = \min \{ \, v_i \mid i = 1, \ldots, n \, \} \, .$$

If $v_k \geq 0$, the current solution is optimal. Otherwise, variable k is to become basic. Let

$$f_1 = \begin{bmatrix} \gamma_{\beta_1 k} \\ \vdots \\ \gamma_{\beta_{n_B} k} \end{bmatrix} .$$

Then H_B is modified by inserting

$$\begin{bmatrix} f_1 \\ \gamma_{kk} \\ A_k \end{bmatrix}$$

after row and column n_B. The new H_B^{-1} may be obtained as

$$[\, H_B^{-1}, \text{ singular }, \, w \,] \leftarrow \Psi_1(\, H_B^{-1} \, , \, f_1 \, , \, \gamma_{kk} \, , \, A_k \, , \, m + n_B \, , \, n_B \,) \, .$$

If singular = "false," x_B is replaced with $x_B - \sigma_B s_B$ and then further modified as

$$x_B \leftarrow (x_B' \, , \, 0)' \, ,$$
$$I_B \leftarrow \{ \, \beta_1, \ldots, \beta_{n_B} \, , \, k \, \}$$

and n_B is increased by 1. If Procedure Ψ_1 returns with singular = "true," we set s_B to be the negative of the first $n_B + 1$ components of w. Following this search direction results in either the conclusion that the problem is unbounded from below or a previously inactive constraint becomes active. In the latter case, let l be as in (8.7) so that variable β_l becomes nonbasic. We now wish to update H_B by replacing row and column l with

$$
\begin{bmatrix}
f_1 \\
\gamma_{kk} \\
A_k
\end{bmatrix} .
$$

We will show [Lemma 8.1(c)] that the resulting H_B is nonsingular. The new H_B^{-1} can be computed using Procedure Ψ_3 as follows. Let

$$
d =
\begin{bmatrix}
\gamma_{\beta_1 k} - \gamma_{\beta_1 \beta_l} \\
\vdots \\
\gamma_{\beta_{l-1},k} - \gamma_{\beta_{l-1},\beta_l} \\
(\gamma_{kk} - \gamma_{\beta_l \beta_l})/2 \\
\gamma_{\beta_{l+1},k} - \gamma_{\beta_{l+1},\beta_l} \\
\vdots \\
\gamma_{\beta_{n_B},k} - \gamma_{\beta_{n_B}\beta_l} \\
A_k - A_{\beta_l}
\end{bmatrix}
$$

and invoke Procedure Ψ_3 as

$$
[\, H_B^{-1} \,] \leftarrow \Psi_3(H_B^{-1}, \, d, \, n_B + m, \, l\,).
$$

In addition to replacing x_B with $x_B - \sigma_B s_B$, the updated x_B must be further modified. Since $(s_B)_{n_B+1} = -1$, the new basic variable k has been increased from 0 to σ_B. Also, basic variable β_l has been reduced to zero so $(x_B)_l$ must be replaced with σ_B. Accordingly,

$$
I_B \leftarrow \{\, \beta_1, \ldots, \beta_{l-1}, \, k, \, \beta_{l+1}, \ldots, \beta_{n_B} \,\}
$$

and n_B remains unchanged.

Based on this description, we now give a detailed statement of the algorithm.

Algorithm 8: Simplex Method for Quadratic Programming

Model Problem:

$$
\begin{aligned}
\text{minimize:} \quad & c'x + \tfrac{1}{2}x'Cx \\
\text{subject to:} \quad & Ax = b, \\
& x \geq 0.
\end{aligned}
$$

Initialization:
Start with index set $I_B = \{ \beta_1, \ldots, \beta_{n_B} \}$, feasible point $(x'_B , x'_N)'$, matrix B having full row rank, nonsingular matrix H_B, and H_B^{-1}, where

$$H_B = \begin{bmatrix} C_{BB} & B' \\ B & 0 \end{bmatrix} , \quad B = [A_{\beta_1}, \ldots, A_{\beta_{n_B}}] , \quad C_{BB} = [\gamma_{\beta_1 \beta_j}]$$

and $(x'_B , x'_N)'$ satisfies $x_N = 0$ and $Bx_B = b$. Compute $f_B = c'_B x_B + \frac{1}{2} x'_B C_{BB} x_B$ and $g_B = c_B + C_{BB} x_B$. Set singular = "false."

Step 1:   Computation of Search Direction s_B.
If singular = "false," go to Step 1.1. If singular = "true," go to Step 1.2.

Step 1.1:
Compute

$$\begin{bmatrix} s_B \\ u \end{bmatrix} = H_B^{-1} \begin{bmatrix} g_B \\ 0 \end{bmatrix} .$$

Set $\gamma_B = 0$ and go to Step 2.

Step 1.2:
Compute

$$(s_B)_i = -w_i , \quad i = 1, \ldots, n_B .$$

Set $\gamma_B = 1$ and go to Step 2.

Step 2:   Computation of Step Size σ_B.
Set $\tilde{\sigma}_B = 1$ if $\gamma_B = 0$ and $\tilde{\sigma}_B = +\infty$ if $\gamma_B = 1$. If $s_B \leq 0$, set $\hat{\sigma}_B = +\infty$. Otherwise, compute the smallest index l and $\hat{\sigma}_B$ such that

$$\hat{\sigma}_B = \frac{(x_B)_l}{(s_B)_l} = \min \left\{ \frac{(x_B)_i}{(s_B)_i} \mid \text{all } i = 1, \ldots, n_B \text{ with } (s_B)_i > 0 \right\} .$$

If $\tilde{\sigma}_B = \hat{\sigma}_B = +\infty$, print the message "objective function is unbounded from below" and stop. Otherwise, set $\sigma_B = \min \{ \tilde{\sigma}_B , \hat{\sigma}_B \}$ and go to Step 3.

Step 3:   Update
If $\sigma_B = \hat{\sigma}_B$, go to Step 3.1. Otherwise, go to Step 3.2.

Step 3.1:
If singular = "false," replace x_B with $x_B - \sigma_B s_B$ then further replace x_B with $((x_B)_1, \ldots, (x_B)_{l-1} , (x_B)_{l+1}, \ldots, (x_B)_{n_B})'$. Update H_B^{-1} by invoking Procedure Ψ_2 as

$$[H_B^{-1}, \text{singular} , w] \leftarrow \Psi_2(H_B^{-1} , m + n_B , l) .$$

Replace I_B with $\{\ \beta_1, \ldots, \beta_{l-1}\ ,\ \beta_{l+1}, \ldots, \beta_{n_B}\ \}$, n_B with $n_B - 1$, f_B with $c'_B x_B + \frac{1}{2} x'_B C_{BB} x_B$, g_B with $c_B + C_{BB} x_B$, g_N with $c_N + C_{NB} x_B$, set $g = (g'_B\ ,\ g'_N)'$, and go to Step 1.

If singular is "true," replace x_B with $x_B - \sigma_B s_B$ then further replace $(x_B)_l$ with σ_B, set

$$
d = \begin{bmatrix}
\gamma_{\beta_1 k} - \gamma_{\beta_1 \beta_l} \\
\vdots \\
\gamma_{\beta_{l-1},k} - \gamma_{\beta_{l-1},\beta_l} \\
(\gamma_{kk} - \gamma_{\beta_l \beta_l})/2 \\
\gamma_{\beta_{l+1},k} - \gamma_{\beta_{l+1},\beta_l} \\
\vdots \\
\gamma_{\beta_{n_B},k} - \gamma_{\beta_{n_B}\beta_l} \\
A_k - A_{\beta_l}
\end{bmatrix},
$$

update H_B^{-1} by invoking Procedure Ψ_3 as

$$[\ H_B^{-1}\ ] \leftarrow \Psi_3(H_B^{-1}\ ,\ d\ ,\ n_B + m\ ,\ l\ ),$$

I_B with $\{\ \beta_1, \ldots, \beta_{l-1}\ ,\ k\ ,\ \beta_{l+1}, \ldots, \beta_{n_B}\ \}$, leave n_B unchanged, replace f_B with $c'_B x_B + \frac{1}{2} x'_B C_{BB} x_B$, g_B with $c_B + C_{BB} x_B$, g_N with $c_N + C_{NB} x_B$, set $g = (g'_B\ ,\ g'_N)'$, and go to Step 1.

Step 3.2:
Replace x_B with $x_B - \sigma_B s_B$, f_B with $c'_B x_B + \frac{1}{2} x'_B C_{BB} x_B$, g_B with $c_B + C_{BB} x_B$, g_N with $c_N + C_{NB} x_B$, and set $g = (g'_B\ ,\ g'_N)'$. Replace u with $-u$, set $v_i = 0$ for all $i \in I_B$ and $v_i = A'_i u + g_i$ for all $i \notin I_B$. Compute the smallest index k such that

$$v_k = \min\ \{\ v_i\ |\ i = 1, \ldots, n\ \}.$$

If $v_k \geq 0$, stop with optimal solution $(x'_B\ ,\ x'_N)'$. Otherwise, set $d_1 = (\gamma_{\beta_1 k}, \ldots, \gamma_{\beta_{n_B}k})'$ and invoke Procedure Ψ_1 with arguments

$$[\ H_B^{-1}, \text{singular}, w\ ] \leftarrow \Psi_1(H_B^{-1}\ ,\ d_1\ ,\ \gamma_{kk}\ ,\ A_k\ ,\ m + n_B\ ,\ n_B\ ).$$

If singular $=$ "false," further replace x_B with $(x'_B\ ,\ 0)'$, I_B with $\{\ \beta_1, \ldots, \beta_{n_B}\ ,\ k\ \}$, n_B with $n_B + 1$, further replace g_B with $(g'_B\ ,\ g_k)'$, delete g_k from g_N, and go to Step 1.

If singular $=$ "true" go to Step 1.

A Matlab program implementing the simplex method for quadratic programming is given in Figure 8.2. This function is called Alg8.m and is shown in Section 8.3 along with its companion function simpobj.m in Figure 8.3.

We illustrate the simplex method for quadratic programming in

Example 8.1

$$\begin{aligned}
\text{minimize:} \quad & -3x_1 - x_2 + \tfrac{1}{2}(x_1^2 + x_2^2) \\
\text{subject to:} \quad & x_1 \qquad + x_3 \qquad = 2, \qquad (1) \\
& \qquad x_2 \qquad + x_4 = 2, \qquad (2) \\
& x_i \geq 0, \quad i = 1,\ldots,4 .
\end{aligned}$$

Here

$$c = \begin{bmatrix} -3 \\ -1 \\ 0 \\ 0 \end{bmatrix}, \quad C = \begin{bmatrix} 1 & 0 & 0 & 0 \\ 0 & 1 & 0 & 0 \\ 0 & 0 & 0 & 0 \\ 0 & 0 & 0 & 0 \end{bmatrix}, \quad A = \begin{bmatrix} 1 & 0 & 1 & 0 \\ 0 & 1 & 0 & 1 \end{bmatrix},$$

$$b = \begin{bmatrix} 2 \\ 2 \end{bmatrix}, \quad \text{and} \quad n = 4 .$$

Initialization:

$$I_B = \{\, 2\,,\, 3 \,\}\,, \quad x_B = \begin{bmatrix} 2 \\ 2 \end{bmatrix}, \quad B = \begin{bmatrix} 0 & 1 \\ 1 & 0 \end{bmatrix}, \quad C_{BB} = \begin{bmatrix} 1 & 0 \\ 0 & 0 \end{bmatrix},$$

$$H_B = \begin{bmatrix} 1 & 0 & 0 & 1 \\ 0 & 0 & 1 & 0 \\ 0 & 1 & 0 & 0 \\ 1 & 0 & 0 & 0 \end{bmatrix}, \quad H_B^{-1} = \begin{bmatrix} 0 & 0 & 0 & 1 \\ 0 & 0 & 1 & 0 \\ 0 & 1 & 0 & 0 \\ 1 & 0 & 0 & -1 \end{bmatrix}, \quad f_B = 0,$$

$$g_B = \begin{bmatrix} 1 \\ 0 \end{bmatrix}.$$

Iteration 0

Step 1: H_B is nonsingular. Transfer to Step 1.1.

Step 1.1: $s_B = \begin{bmatrix} 0 \\ 0 \end{bmatrix}$, $u = \begin{bmatrix} 0 \\ 1 \end{bmatrix}$, $\gamma_B = 0$.

Step 2: $\tilde{\sigma}_B = 1$,

$\quad\quad\quad \hat{\sigma}_B = +\infty,$

$\quad\quad\quad \sigma_B = \min\{\,1\,,\, +\infty\} = 1$.

Step 3: Transfer to Step 3.2.

Step 3.2: $x_B \leftarrow \begin{bmatrix} 2 \\ 2 \end{bmatrix}$, $f_B = 0$,

$$g_B = \begin{bmatrix} 1 \\ 0 \end{bmatrix}, \quad g_N = \begin{bmatrix} -3 \\ 0 \end{bmatrix},$$

$$u \leftarrow \begin{bmatrix} 0 \\ -1 \end{bmatrix},$$

$v_1 = \min\{-3, 0, 0, -1\} = -3$, $k = 1$,

$$d_1 = \begin{bmatrix} 0 \\ 0 \end{bmatrix}, \quad \gamma_{11} = 1, \quad A_1 = \begin{bmatrix} 1 \\ 0 \end{bmatrix},$$

invoking Procedure Ψ_1 as
$[H_B^{-1}, \text{ singular}, w] \leftarrow \Psi_1(H_B^{-1}, d_1, \gamma_{11}, A_1, 4, 2)$
gives singular = "false,"

$$x_B \leftarrow \begin{bmatrix} 2 \\ 2 \\ 0 \end{bmatrix}, \quad H_B^{-1} \leftarrow \begin{bmatrix} 0 & 0 & 0 & 0 & 1 \\ 0 & 1 & -1 & 1 & 0 \\ 0 & -1 & 1 & 0 & 0 \\ 0 & 1 & 0 & 0 & 0 \\ 1 & 0 & 0 & 0 & -1 \end{bmatrix},$$

$$I_B \leftarrow \{2, 3, 1\}, \quad g_B \leftarrow \begin{bmatrix} 1 \\ 0 \\ -3 \end{bmatrix}, \quad g_N = [0],$$

$$B \leftarrow \begin{bmatrix} 0 & 1 & 1 \\ 1 & 0 & 0 \end{bmatrix}, \quad H_B \leftarrow \begin{bmatrix} 1 & 0 & 0 & 0 & 1 \\ 0 & 0 & 0 & 1 & 0 \\ 0 & 0 & 1 & 1 & 0 \\ 0 & 1 & 1 & 0 & 0 \\ 1 & 0 & 0 & 0 & 0 \end{bmatrix},$$

$n_B \leftarrow 3$.

Iteration 1

Step 1: H_B is nonsingular. Transfer to Step 1.1.

Step 1.1: $s_B = \begin{bmatrix} 0 \\ 3 \\ -3 \end{bmatrix}$, $u = \begin{bmatrix} 0 \\ 1 \end{bmatrix}$, $\gamma_B = 0$.

Step 2: $\tilde{\sigma}_B = 1$,

$$\hat{\sigma}_B = \min\left\{-, \frac{2}{3}, -\right\} = \frac{2}{3}, \quad l = 2,$$

$$\sigma_B = \min \left\{ 1 , \frac{2}{3} \right\} = \frac{2}{3} .$$

Step 3:   Transfer to Step 3.1.

Step 3.1:   singular is "false,"

$$x_B \leftarrow \begin{bmatrix} 2 \\ 2 \\ 0 \end{bmatrix} - \frac{2}{3} \begin{bmatrix} 0 \\ 3 \\ -3 \end{bmatrix} = \begin{bmatrix} 2 \\ 0 \\ 2 \end{bmatrix} , \quad x_B \leftarrow \begin{bmatrix} 2 \\ 2 \end{bmatrix} ,$$

$$[\, H_B^{-1}, \text{singular} \, , \, w \,] \leftarrow \Psi_2(H_B^{-1} , \, 5 , \, 2 \,) ,$$

$$H_B^{-1} = \begin{bmatrix} 0 & 0 & 0 & 1 \\ 0 & 0 & 1 & 0 \\ 0 & 1 & -1 & 0 \\ 1 & 0 & 0 & -1 \end{bmatrix} ,$$

$$I_B \leftarrow \{ \, 2 \, , \, 1 \, \} , \quad B \leftarrow \begin{bmatrix} 0 & 1 \\ 1 & 0 \end{bmatrix} ,$$

$$H_B \leftarrow \begin{bmatrix} 1 & 0 & 0 & 1 \\ 0 & 1 & 1 & 0 \\ 0 & 1 & 0 & 0 \\ 1 & 0 & 0 & 0 \end{bmatrix} , \quad n_B \leftarrow 2 , \quad f_B = -4 ,$$

$$g_B = \begin{bmatrix} 1 \\ -1 \end{bmatrix} , \quad g_N = \begin{bmatrix} 0 \\ 0 \end{bmatrix} .$$

Iteration 2

Step 1: H_B is nonsingular. Transfer to Step 1.1.

Step 1.1: $s_B = \begin{bmatrix} 0 \\ 0 \end{bmatrix} , \quad u = \begin{bmatrix} -1 \\ 1 \end{bmatrix} , \quad \gamma_B = 0 .$

Step 2: $\tilde{\sigma}_B = 1 ,$

$\hat{\sigma}_B = +\infty ,$

$\sigma_B = \min \{ \, 1 \, , \, +\infty \} = 1 .$

Step 3:   Transfer to Step 3.2.

Step 3.2: $x_B \leftarrow \begin{bmatrix} 2 \\ 2 \end{bmatrix} , \quad f_B = -4 ,$

$$g_B = \begin{bmatrix} 1 \\ -1 \end{bmatrix} , \quad g_N = \begin{bmatrix} 0 \\ 0 \end{bmatrix} ,$$

$$u \leftarrow \begin{bmatrix} 1 \\ -1 \end{bmatrix} ,$$

$$v_4 = \min \{ 0 , 0 , 1 , -1 \} = -1 , \quad k = 4 ,$$

$$d_1 = \begin{bmatrix} 0 \\ 0 \end{bmatrix} , \quad \gamma_{44} = 0 , \quad A_4 = \begin{bmatrix} 0 \\ 1 \end{bmatrix} ,$$

invoking Procedure Ψ_1 as
$[H_B^{-1}, \text{singular} , w] \leftarrow \Psi_1(H_B^{-1} , d_1 , \gamma_{44} , A_4 , 4 , 2)$
gives singular $=$ "false,"

$$x_B \leftarrow \begin{bmatrix} 2 \\ 2 \\ 0 \end{bmatrix} , \quad H_B^{-1} \leftarrow \begin{bmatrix} 1 & 0 & -1 & 0 & 0 \\ 0 & 0 & 0 & 1 & 0 \\ -1 & 0 & 1 & 0 & 1 \\ 0 & 1 & 0 & -1 & 0 \\ 0 & 0 & 1 & 0 & 0 \end{bmatrix} ,$$

$$I_B \leftarrow \{ 2 , 1 , 4 \} , \quad g_B = \begin{bmatrix} 1 \\ -1 \\ 0 \end{bmatrix} , \quad g_N = [0] ,$$

$$B \leftarrow \begin{bmatrix} 0 & 1 & 0 \\ 1 & 0 & 1 \end{bmatrix} , \quad H_B \leftarrow \begin{bmatrix} 1 & 0 & 0 & 0 & 1 \\ 0 & 1 & 0 & 1 & 0 \\ 0 & 0 & 0 & 0 & 1 \\ 0 & 1 & 0 & 0 & 0 \\ 1 & 0 & 1 & 0 & 0 \end{bmatrix} ,$$

$$n_B \leftarrow 3 .$$

Iteration 3

Step 1: H_B is nonsingular. Transfer to Step 1.1.

Step 1.1: $s_B = \begin{bmatrix} 1 \\ 0 \\ -1 \end{bmatrix} , \quad u = \begin{bmatrix} -1 \\ 0 \end{bmatrix} , \quad \gamma_B = 0 .$

Step 2: $\tilde{\sigma}_B = 1 ,$

$$\hat{\sigma}_B = \min \left\{ \frac{2}{1} , - , - \right\} = 2 , \quad l = 1 ,$$

$$\sigma_B = \min \{ 1 , 2 \} = 1 .$$

Step 3:   Transfer to Step 3.2.

Step 3.2: $x_B \leftarrow \begin{bmatrix} 2 \\ 2 \\ 0 \end{bmatrix} - \begin{bmatrix} 1 \\ 0 \\ -1 \end{bmatrix} = \begin{bmatrix} 1 \\ 2 \\ 1 \end{bmatrix}$, $f_B = \dfrac{-9}{2}$,

$$g_B = \begin{bmatrix} 0 \\ -1 \\ 0 \end{bmatrix} , \quad g_N = [\, 0 \,] ,$$

$$u \leftarrow \begin{bmatrix} 1 \\ 0 \end{bmatrix} ,$$

$$v_1 = \min \{\, 0 \,,\, 0 \,,\, 1 \,,\, 0 \,\} = 0 , \quad k = 1 ,$$

$v_1 \geq 0$, stop with optimal solution $x = (2 \,,\, 1 \,,\, 0 \,,\, 1)'$.

$\diamond$

A computer program which implements the simplex method for quadratic programming (Algorithm 8) is given in Section 8.3, Figure 8.2. The calling program to solve Example 8.1 is given in Figure 8.4. The output from applying this program to the problem of Example 8.1 is shown in Figure 8.5.

The problem of Example 8.1 has 4 variables thus making impossible a direct geometric interpretation of the steps of the algorithm. However, since x_3 and x_4 do not appear in the objective function, these two variables may be eliminated giving the equivalent problem

$$
\begin{aligned}
\text{minimize:} \quad & -3x_1 - x_2 + \tfrac{1}{2}(x_1^2 + x_2^2) \\
\text{subject to:} \quad & x_1 \qquad\quad \leq 2, \quad (1) \\
& \qquad\quad x_2 \leq 2, \quad (2) \\
& -x_1 \qquad\quad \leq 0, \quad (3) \\
& \qquad -x_2 \leq 0. \quad (4)
\end{aligned}
$$

The geometry of the problem as well as the progress of the algorithm is shown in Figure 8.1.

In Example 8.1, Step 3.1 never required the use of Procedure Ψ_3. In general, Ψ_3 will tend to be used when C has low rank. An extreme case occurs when $C = 0$ for which the problem reduces to a linear programming problem. Although we do not recommend the simplex method for quadratic programming to solve an LP of this form (the revised simplex method for linear programming is obviously more appropriate), it is nonetheless illustrative to do so.

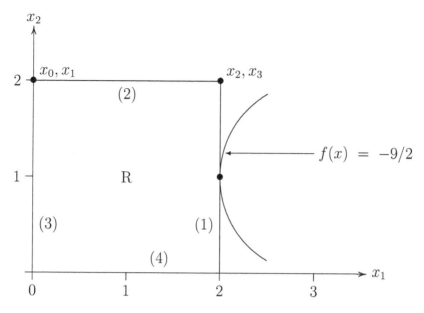

Figure 8.1. Geometry of Example 8.1.

Example 8.2

$$\begin{aligned}
\text{minimize:} \quad & -5x_1 + 2x_2 \\
\text{subject to:} \quad & -2x_1 + x_2 + x_3 && = 2, && (1) \\
& x_1 + 2x_2 \quad + x_4 && = 14, && (2) \\
& 4x_1 + 3x_2 \quad\quad + x_5 && = 36, && (3) \\
& x_i \geq 0, \quad i = 1, \ldots, 5.
\end{aligned}$$

Here

$$c = \begin{bmatrix} -5 \\ 2 \\ 0 \\ 0 \\ 0 \end{bmatrix}, \quad C = \begin{bmatrix} 0 & 0 & 0 & 0 & 0 \\ 0 & 0 & 0 & 0 & 0 \\ 0 & 0 & 0 & 0 & 0 \\ 0 & 0 & 0 & 0 & 0 \\ 0 & 0 & 0 & 0 & 0 \end{bmatrix}, \quad A = \begin{bmatrix} -2 & 1 & 1 & 0 & 0 \\ 1 & 2 & 0 & 1 & 0 \\ 4 & 3 & 0 & 0 & 1 \end{bmatrix},$$

$$b = \begin{bmatrix} 2 \\ 14 \\ 36 \end{bmatrix}, \quad \text{and } n = 5.$$

Initialization:

$$I_B = \{1, 2, 5\}, \quad x_B = \begin{bmatrix} 2 \\ 6 \\ 10 \end{bmatrix},$$

$$H_B^{-1} = \begin{bmatrix} 0 & 0 & 0 & -2/5 & 1/5 & 0 \\ 0 & 0 & 0 & 1/5 & 2/5 & 0 \\ 0 & 0 & 0 & 1 & -2 & 1 \\ -2/5 & 1/5 & 1 & 0 & 0 & 0 \\ 1/5 & 2/5 & -2 & 0 & 0 & 0 \\ 0 & 0 & 1 & 0 & 0 & 0 \end{bmatrix}, \quad f_B = 2, \quad g_B = \begin{bmatrix} -5 \\ 2 \\ 0 \end{bmatrix}.$$

Iteration 0

Step 1: H_B is nonsingular. Transfer to Step 1.1.

Step 1.1: $s_B = \begin{bmatrix} 0 \\ 0 \\ 0 \end{bmatrix}, \quad u = \begin{bmatrix} 12/5 \\ -1/5 \\ 0 \end{bmatrix}, \quad \gamma_B = 0.$

Step 2: $\tilde{\sigma}_B = 1$,

$\hat{\sigma}_B = +\infty$,

$\sigma_B = \min\{1, +\infty\} = 1$.

Step 3: Transfer to Step 3.2.

Step 3.2: $x_B \leftarrow \begin{bmatrix} 2 \\ 6 \\ 10 \end{bmatrix}, \quad f_B = 2$,

$g_B = \begin{bmatrix} -5 \\ 2 \\ 0 \end{bmatrix}, \quad g_N = \begin{bmatrix} 0 \\ 0 \end{bmatrix}$,

$u \leftarrow \begin{bmatrix} -12/5 \\ 1/5 \\ 0 \end{bmatrix}$,

$v_3 = \min\left\{0, 0, \dfrac{-12}{5}, \dfrac{1}{5}, 0\right\} = \dfrac{-12}{5}, \quad k = 3$,

$d_1 = \begin{bmatrix} 0 \\ 0 \\ 0 \end{bmatrix}, \quad \gamma_{33} = 0, \quad A_3 = \begin{bmatrix} 1 \\ 0 \\ 0 \end{bmatrix}$,

invoking Procedure Ψ_1 as
$[H_B^{-1}, \text{singular}, w] \leftarrow \Psi_1(H_B^{-1}, d_1, \gamma_{33}, A_3, 6, 3)$
gives singular = "true."

Iteration 1

Step 1: H_B is singular. Transfer to Step 1.2.

Step 1.2: $s_B = \begin{bmatrix} -2/5 \\ 1/5 \\ 1 \end{bmatrix}$, $\gamma_B = 1$.

Step 2: $\tilde{\sigma}_B = +\infty$,

$$\hat{\sigma}_B = \min \left\{ - , \frac{6}{1/5} , \frac{10}{1} \right\} = 10 , \quad l = 3 ,$$

$$\sigma_B = \min \{ +\infty, 10 \} = 10 .$$

Step 3: Transfer to Step 3.1.

Step 3.1: singular is "true,"

$$x_B \leftarrow \begin{bmatrix} 2 \\ 6 \\ 10 \end{bmatrix} - 10 \begin{bmatrix} -2/5 \\ 1/5 \\ 1 \end{bmatrix} = \begin{bmatrix} 6 \\ 4 \\ 0 \end{bmatrix} , \quad x_B \leftarrow \begin{bmatrix} 6 \\ 4 \\ 10 \end{bmatrix} ,$$

$$d = \begin{bmatrix} 0 \\ 0 \\ 0 \\ 1 \\ 0 \\ -1 \end{bmatrix} ,$$

invoking Procedure Ψ_3 as
$$[H_B^{-1}] \leftarrow \Psi_3(H_B^{-1} , d , 3 , 3) ,$$

$$H_B^{-1} = \begin{bmatrix} 0 & 0 & 0 & 0 & -3/5 & 2/5 \\ 0 & 0 & 0 & 0 & 4/5 & -1/5 \\ 0 & 0 & 0 & 1 & -2 & 1 \\ 0 & 0 & 1 & 0 & 0 & 0 \\ -3/5 & 4/5 & -2 & 0 & 0 & 0 \\ 2/5 & -1/5 & 1 & 0 & 0 & 0 \end{bmatrix} ,$$

$$I_B \leftarrow \{ 1 , 2 , 3 \} , \quad n_B = 3 , \quad f_B = -22 ,$$

$$g_B = \begin{bmatrix} -5 \\ 2 \\ 0 \end{bmatrix} , \quad g_N = \begin{bmatrix} 0 \\ 0 \end{bmatrix} .$$

Iteration 2

Step 1: H_B is nonsingular. Transfer to Step 1.1.

Step 1.1: $s_B = \begin{bmatrix} 0 \\ 0 \\ 0 \end{bmatrix}$, $u = \begin{bmatrix} 0 \\ 23/5 \\ -12/5 \end{bmatrix}$, $\gamma_B = 0$.

Step 2: $\tilde{\sigma}_B = 1$,

$\hat{\sigma}_B = +\infty$,

$\sigma_B = \min\{1, +\infty\} = 1$.

Step 3:   Transfer to Step 3.2.

Step 3.2: $x_B \leftarrow \begin{bmatrix} 6 \\ 4 \\ 10 \end{bmatrix}$, $f_B = -22$,

$g_B = \begin{bmatrix} -5 \\ 2 \\ 0 \end{bmatrix}$, $g_N = \begin{bmatrix} 0 \\ 0 \end{bmatrix}$,

$u \leftarrow \begin{bmatrix} 0 \\ -23/5 \\ 12/5 \end{bmatrix}$,

$v_4 = \min\left\{0, 0, 0, \dfrac{-23}{5}, \dfrac{12}{5}\right\} = \dfrac{-23}{5}$, $k = 4$,

$d_1 = \begin{bmatrix} 0 \\ 0 \\ 0 \end{bmatrix}$, $\gamma_{44} = 0$, $A_4 = \begin{bmatrix} 0 \\ 1 \\ 0 \end{bmatrix}$,

invoking Procedure Ψ_1 as
$[\, H_B^{-1},\ \text{singular},\ w\,] \leftarrow \Psi_1(H_B^{-1},\ d_1,\ \gamma_{44},\ A_4, 6,\ 3\,)$,
gives singular = "true."

Iteration 3

Step 1: H_B is singular. Transfer to Step 1.2.

Step 1.2: $s_B = \begin{bmatrix} -3/5 \\ 4/5 \\ -2 \end{bmatrix}$, $\gamma_B = 1$.

Step 2: $\tilde{\sigma}_B = +\infty$,

$$\hat{\sigma}_B = \min \left\{ - , \; \frac{4}{4/5} , \; - \right\} = 5 , \quad l = 2 ,$$

$$\sigma_B = \min \{ +\infty , \; 5 \} = 5 .$$

Step 3:   Transfer to Step 3.1.

Step 3.1:   singular is "true,"

$$x_B \leftarrow \begin{bmatrix} 6 \\ 4 \\ 10 \end{bmatrix} - 5 \begin{bmatrix} -3/5 \\ 4/5 \\ -2 \end{bmatrix} = \begin{bmatrix} 9 \\ 0 \\ 20 \end{bmatrix} , \quad x_B \leftarrow \begin{bmatrix} 9 \\ 5 \\ 20 \end{bmatrix} ,$$

$$d = \begin{bmatrix} 0 \\ 0 \\ 0 \\ -1 \\ -1 \\ -3 \end{bmatrix} ,$$

$$[\, H_B^{-1} \,] \leftarrow \Psi_3 (H_B^{-1} , \; d , \; 3 , \; 2 , \;) ,$$

$$H_B^{-1} = \begin{bmatrix} 0 & 0 & 0 & 0 & 0 & 1/4 \\ 0 & 0 & 0 & 0 & 1 & -1/4 \\ 0 & 0 & 0 & 1 & 0 & 1/2 \\ 0 & 0 & 1 & 0 & 0 & 0 \\ 0 & 1 & 0 & 0 & 0 & 0 \\ 1/4 & -1/4 & 1/2 & 0 & 0 & 0 \end{bmatrix} ,$$

$$I_B \leftarrow \{ 1 , \; 4 , \; 3 \} , \quad n_B = 3 , \quad f_B = -45 ,$$

$$g_B = \begin{bmatrix} -5 \\ 0 \\ 0 \end{bmatrix} , \quad g_N = \begin{bmatrix} 0 \\ 0 \end{bmatrix} .$$

Iteration 4

Step 1: H_B is nonsingular. Transfer to Step 1.1.

Step 1.1: $s_B = \begin{bmatrix} 0 \\ 0 \\ 0 \end{bmatrix} , \quad u = \begin{bmatrix} 0 \\ 0 \\ -5/4 \end{bmatrix} , \quad \gamma_B = 0 .$

Step 2: $\tilde{\sigma}_B = 1 ,$

$$\hat{\sigma}_B = +\infty ,$$

$$\sigma_B = \min \{ 1 , \; +\infty \} = 1 .$$

Step 3:   Transfer to Step 3.2.

Step 3.2: $x_B \leftarrow \begin{bmatrix} 9 \\ 5 \\ 20 \end{bmatrix}$, $f_B = -45$,

$$g_B = \begin{bmatrix} -5 \\ 0 \\ 0 \end{bmatrix} , \quad g_N = \begin{bmatrix} 0 \\ 0 \end{bmatrix} ,$$

$$u \leftarrow \begin{bmatrix} 0 \\ 0 \\ 5/4 \end{bmatrix} ,$$

$$v_1 = \min \left\{ 0 , \frac{23}{4} , 0 , 0 , \frac{5}{4} \right\} = 0 , \quad k = 1 .$$

$v_1 \geq 0$, stop with optimal solution $x = (9 , 0 , 20 , 5 , 0)'$.

$\Diamond$

The problem of Example 8.2 is solved by the Matlab program Alg8.m (Figure 8.2, Section 8.3). The program formulating the data for this example is shown in Figure 8.6 and the resulting output is given in Figure 8.7.

With analogous initial data, the problem of Example 8.2 was solved by the revised simplex method for linear programming in Example 4.2 of Best and Ritter [6]. The reader may find it useful to compare the computations.

The properties of the simplex method for quadratic programming are a consequence of the following lemma. The lemma is analogous to Lemma 7.1. In order to formulate the lemma, let

$$I_B = \{ \beta_1, \ldots, \beta_r \} , \quad B = [A_{\beta_1}, \ldots, A_{\beta_r}] , \quad C_{BB} = [\gamma_{\beta_i \beta_j}] ,$$

and

$$H_B = \begin{bmatrix} C_{BB} & B' \\ B & 0 \end{bmatrix} .$$

Lemma 8.1.

(a) Suppose that B has full row rank. Then H_B is nonsingular if and only if $s'_B C_{BB} s_B > 0$ for all $s_B \neq 0$ such that $B s_B = 0$.

(b) Suppose that B_0 has full row rank and that H_{B_0} is nonsingular. For some l with $1 \leq l \leq r$ suppose that there is an s_{B_0} such that $B_0 s_{B_0} = 0$ and $(s_{B_0})_l \neq 0$. Then B_1 has full row rank and H_{B_1} is nonsingular, where $I_B = \{ \beta_1, \ldots, \beta_{l-1} , \beta_{l+1}, \ldots, \beta_r \}$.

(c) Suppose that B_0 has full row rank. Let k be such that $k \notin I_{B_0}$. Let $I_{B_1} = \{ \beta_1, \ldots, \beta_r, k \}$. Then H_{B_1} is singular if and only if there exists a unique $(r + 1)$-vector s_{B_1} such that $B_1 s_{B_1} = 0$, $C_{B_1 B_1} s_{B_1} = 0$, and $(s_{B_1})_{r+1} = -1$. Furthermore, if H_{B_1} is singular then B_2 has full row rank and H_{B_2} is nonsingular, where $I_{B_2} = \{ \beta_1, \ldots, \beta_{l-1}, k, \beta_{l+1}, \ldots, \beta_r \}$, l is any index satisfying $1 \leq l \leq r$ and $(s_{B_1})_l \neq 0$, and s_{B_1} is the uniquely determined $(r + 1)$-vector of the previous statement.

The proof of Lemma 8.1 mirrors that of Lemma 7.1 and is left as an exercise (Exercise 8.4).

There are several consequences of Lemma 8.1 to the simplex method for quadratic programming. From the initialization, we start with an H_B for which H_B has full row rank and H_B is nonsingular. Generally, Lemma 8.1(b) asserts that whenever B has full row rank, H_B is nonsingular, and a basic variable becomes nonbasic, then the new B has full row rank and the new H_B is nonsingular. Thus invoking Procedure Ψ_2 in the first part of Step 3.1 will always result in singular being "false." Invoking Procedure Ψ_1 in Step 3.2 may result in singular being "true." If so, the corresponding s_B obtained from w in Step 1.2 at the next iteration must be the uniquely determined s_{B_1} of Lemma 8.1(c). By definition of σ_B in the subsequent step size calculation (Step 2), $(s_B)_l < 0$ so that in particular, $(s_B)_l \neq 0$. Lemma 8.1(c) then asserts that the next H_B defined by Step 3.1 must be nonsingular. Consequently Ψ_3 will always return with singular being "false."

We derived the simplex method for quadratic programming by applying Algorithm 6 to the model problem (8.1) and then observing that because of the structure of (8.1), the linear equations (8.3) could be solved by solving the much smaller subsystem (8.6). Depending on (among other things) the rank of C (see Exercise 8.1) the system of equations (8.6) may be considerably smaller than that of (8.3), and this is the potential utility of the method. However, even though the calculations for the present method are explicitly different than those for Algorithm 6, they are implicitly identical. Thus Theorem 7.1 also applies to the present method and we have

Theorem 8.1 *Let the simplex method for quadratic programming be applied to the model problem* $\min \{ c'x + \frac{1}{2}x'Cx \mid Ax = b, x \geq 0 \}$*. If each quasistationary point obtained by the algorithm is nondegenerate, then the method terminates with either an optimal solution* (x_B, x_N) *or the information that the problem is unbounded from below. In the former case, the algorithm terminates with multipliers* u *and* v *such that* (x_B, x_N) *together with* (u, v) *are an optimal solution for the dual problem.*

8.2   Simplex Method for PQP: Algorithm 9

In this section, we derive the parametric extension of the simplex method for quadratic programming. We do so by specializing Algorithm 7 to a model problem having linear equality constraints and nonnegativity restrictions on all variables. The development is analogous to the derivation of the simplex method for quadratic programming from Algorithm 7.

Our model problem is

$$\min \{ (c + tq)'x + \tfrac{1}{2}x'Cx \mid Ax = b + tp, \ x \geq 0 \}, \qquad (8.13)$$

where A, b, c, and C are as in (8.1), q is an n-vector, and p is an m-vector. Given $\bar{t} \geq \underline{t}$, (8.13) is to be solved for all t satisfying $\underline{t} \leq t \leq \bar{t}$. Let $t_0 = \underline{t}$ and suppose that (8.13) has been solved for $t = t_0{}^3$ by the simplex method for quadratic programming. Let $I_B = \{ \beta_1, \ldots, \beta_{n_B} \}$ and H_B^{-1} be the associated final data where

$$H_B = \begin{bmatrix} C_{BB} & B' \\ B & 0 \end{bmatrix}, \quad B = \begin{bmatrix} A_{\beta_1}, \ldots, A_{\beta_{n_B}} \end{bmatrix},$$

and

$$C_{BB} = \begin{bmatrix} \gamma_{\beta_i \beta_j} \end{bmatrix}.$$

The first equations to be solved in Step 1 of Algorithm 7 are

$$H_j \begin{bmatrix} h_{1j} \\ g_{1j} \end{bmatrix} = \begin{bmatrix} -c \\ b_j \end{bmatrix}.$$

Writing $(h'_{1B}, h'_{1N})'$ for h_{1j}, partitioning g_{1j} as $(g'_{1B}, v'_{1N})'$, c as $(c'_B, c'_N)'$, b_j as $(b', 0')'$, and replacing H_j with the coefficient matrix of (8.3), we obtain

$$\begin{bmatrix} C_{BB} & C_{BN} & B' & 0 \\ C_{NB} & C_{NN} & N' & -I \\ B & N & 0 & 0 \\ 0 & -I & 0 & 0 \end{bmatrix} \begin{bmatrix} h_{1B} \\ h_{1N} \\ g_{1B} \\ v_{1N} \end{bmatrix} = \begin{bmatrix} -c_B \\ -c_N \\ b \\ 0 \end{bmatrix}.$$

The fourth partition implies that $h_{1N} = 0$. Having obtained h_{1B} and g_{1B}, v_{1N} may be obtained from the third partition:

$$v_{1N} = c_N + C_{NB}h_{1B} + N'g_{1B}. \qquad (8.14)$$

From the following equations we obtain h_{1B} and g_{1B}

$$C_{BB}\, h_{1B} + B'g_{1B} = -c_B,$$
$$B\, h_{1B} \qquad\qquad = b,$$

---

[3]We are implicitly assuming that (8.13) possesses an optimal solution for $t = t_0$. It is possible that there is no feasible solution or that the problem is unbounded from below for $t = t_0$. We assume that neither of these possibilities apply.

or,

$$\begin{bmatrix} h_{1B} \\ g_{1B} \end{bmatrix} = H_B^{-1} \begin{bmatrix} -c_B \\ b \end{bmatrix} .$$

Similarly, the second set of equations of Step 1 of Algorithm 7 may be solved as

$$\begin{bmatrix} h_{2B} \\ g_{2B} \end{bmatrix} = H_B^{-1} \begin{bmatrix} -q_B \\ p \end{bmatrix} ,$$

with $h_{2N} = 0$ and

$$v_{2N} = q_N + C_{NB} h_{2B} + N' g_{2B} . \tag{8.15}$$

For t sufficiently close to t_0, the optimal solution for (8.13) is

$$x_B(t) = h_{1B} + t h_{2B}$$

together with $x_N = 0$. The multipliers for the equality constraints are

$$u(t) = g_{1B} + t g_{2B}$$

and those for the nonnegativity constraints are

$$v(t) = w_1 + t w_2 , \tag{8.16}$$

where

$$(w_1)_i = (w_2)_i = 0 , \quad \text{for all } i \in I_B$$

and from (8.14) and (8.15), for all $i \notin I_B$,

$$(w_1)_i = c_i + \sum_{k=1}^{n_B} \gamma_{i\beta_k} (h_{1B})_k + A'_i g_{1B} ,$$

$$(w_2)_i = q_i + \sum_{k=1}^{n_B} \gamma_{i\beta_k} (h_{2B})_k + A'_i g_{2B} .$$

Now $x_B(t)$ is feasible for (8.13) provided that

$$h_{1B} + t h_{2B} \geq 0 .$$

Let $\hat{t}_1$ denote the largest value of t for which $x_B(t)$ is feasible for (8.13). If $h_{2B} \geq 0$, then $\hat{t}_1 = +\infty$. Otherwise,

$$\hat{t}_1 = \min \left\{ \frac{-(h_{1B})_i}{(h_{2B})_i} \,\middle|\, \text{all } i = 1, \ldots, n_B \text{ with } (h_{2B})_i < 0 \right\}$$

$$= \frac{-(h_{1B})_l}{(h_{2B})_l} .$$

$x_B(t)$ remains optimal provided that $v(t)$ in (8.16) remains nonnegative. Let $\tilde{t}_1$ denote the largest value of t for which $v(t) \geq 0$. If $w_2 \geq 0$, then $\tilde{t}_1 = +\infty$. Otherwise,

$$\tilde{t}_1 = \min \left\{ \frac{-(w_1)_i}{(w_2)_i} \mid \text{all } i = 1, \ldots, n \text{ with } (w_2)_i < 0 \right\}$$

$$= \frac{-(w_1)_k}{(w_2)_k} .$$

Setting $t_1 = \min \{ \hat{t}_1 , \tilde{t}_1 , \bar{t} \}$, it follows that $x_B(t)$ together with $x_N = 0$ is optimal for (8.13) with multiplier vector $u(t)$ for the equality constraints and multiplier vector $v(t)$ for the nonnegativity constraints for all t with $t_0 \leq t \leq t_1$.

If $t_1 = \bar{t}$, we are done. Otherwise, we proceed by updating I_B and H_B according to Step 2 of Algorithm 7. Suppose first that $t_1 = \hat{t}_1$. Then variable β_l has become nonbasic and we would like to modify H_B by deleting row and column l provided that this does not result in singularity. Suppose that we invoke Procedure Ψ_2 as

$$[H_B^{-1}, \text{singular}, w] \leftarrow \Psi_2(H_B^{-1}, m + n_B, l) .$$

If singular $=$ "false," then we proceed as in Step 2.1.1 of Algorithm 7 using H_B^{-1} as modified by Procedure Ψ_2 above , I_B with $\{ \beta_1, \ldots, \beta_{l-1} , \beta_{l+1}, \ldots, \beta_{n_B} \}$, n_B with $n_B - 1$, and return to Step 1. If singular $=$ "true," then from Lemma 8.1 the rows of

$$[A_{\beta_1}, \ldots, A_{\beta_{l-1}} , A_{\beta_{l+1}}, \ldots, A_{\beta_{n_B}}]$$

are linearly dependent. Thus there is a m-vector ξ with

$$A'_{\beta_i} \xi = 0 , \quad i = 1, \ldots, n_B , \quad i \neq l .$$

From Lemma 8.1 and the definition of Procedure Ψ_2

$$\xi_i = -w_{n_B - 1 + i} , \quad i = 1, \ldots, m , \tag{8.17}$$

where w has been obtained by invoking Procedure Ψ_2. That is, ξ is just the negative of the last m components of w.[4] In order to perform the calculations of Step 2.1.2 of Algorithm 7, we must solve

$$H_j \begin{bmatrix} s_j \\ \xi_j \end{bmatrix} = \begin{bmatrix} a_l \\ 0 \end{bmatrix} .$$

---

[4]Here it is irrelevant whether we use w or $-w$. Shortly it will be convenient to have worked with $-w$.

In the present context, $a_l = -e_{\beta_l}$, H_j is the coefficient matrix of (8.3), and letting $s_j = (s'_B, s'_N)'$ and $\xi_j = (\xi'_1, \xi'_N)'$ the equations to be solved are

$$\begin{bmatrix} C_{BB} & C_{BN} & B' & 0 \\ C_{NB} & C_{NN} & N' & -I \\ B & N & 0 & 0 \\ 0 & -I & 0 & 0 \end{bmatrix} \begin{bmatrix} s_B \\ s_N \\ \xi_1 \\ \xi_N \end{bmatrix} = \begin{bmatrix} -e_l \\ 0 \\ 0 \\ 0 \end{bmatrix} , \qquad (8.18)$$

where e_l is the lth unit vector of dimension n_B. The fourth partition implies that $s_N = 0$ so that the first and third partitions become

$$\begin{bmatrix} C_{BB} & B' \\ B & 0 \end{bmatrix} \begin{bmatrix} s_B \\ \xi_1 \end{bmatrix} = \begin{bmatrix} -e_l \\ 0 \end{bmatrix} .$$

Thus

$$\begin{bmatrix} s_B \\ \xi_1 \end{bmatrix} = H_B^{-1} \begin{bmatrix} -e_l \\ 0 \end{bmatrix}$$

and $(s'_B, \xi'_1)'$ is just the negative of the lth column of H_B^{-1}. Since $s_B = 0$, it follows that ξ_1 is just the negative of the last m components of the lth column of H_B^{-1}. But from the definition of Procedure Ψ_2, ξ from (8.17) is identical to ξ_1 from (8.18). Thus when Procedure Ψ_2 returns with singular = "true," ξ_1 is immediately available as the last m components of w.

Having thus obtained ξ_1, ξ_N may be obtained from the second partition of (8.18):

$$\xi_N = N'\xi_1 .$$

The tests in Step 2.1.2 of Algorithm 7 apply only to the active inequality constraints. If $\xi_N \leq 0$, or equivalently

$$A'_i \xi_1 \leq 0 , \quad \text{for all } i \notin I_B ,$$

then the problem has no feasible solution for $t > t_{j+1}$. Otherwise, we compute the smallest index k such that

$$\frac{(w_1)_k + t_{j+1}(w_2)_k}{A'_k \xi_1} = \min\left\{ \frac{(w_1)_i + t_{j+1}(w_2)_i}{A'_i \xi_1} \mid \text{all } i \notin I_B \text{ with} A'_i \xi_1 > 0 \right\}.$$

Then variable k becomes basic and variable β_l becomes nonbasic. Letting

$$d = \begin{bmatrix} \gamma_{\beta_1 k} - \gamma_{\beta_1 \beta_l} \\ \vdots \\ \gamma_{\beta_{l-1},k} - \gamma_{\beta_{l-1},\beta_l} \\ (\gamma_{kk} - \gamma_{\beta_l \beta_l})/2 \\ \gamma_{\beta_{l+1},k} - \gamma_{\beta_{l+1},\beta_l} \\ \vdots \\ \gamma_{\beta_{n_B} k} - \gamma_{\beta_{n_B} \beta_l} \\ A_k - A_{\beta_l} \end{bmatrix} ,$$

and we invoke Ψ_3 as

$$[\,H_B^{-1}\,] \;\leftarrow\; \Psi_3(\,H_B^{-1},\, d,\, m+n_B,\, l\,)\,,$$

I_B with $\{\,\beta_1,\ldots,\beta_{l-1}\,,\ k\,,\ \beta_{l+1},\ldots,\beta_{n_B}\,\}$, n_B remains unchanged, and we return to Step 1. This completes the discussion for the case of $t_1 \;=\; \hat{t}_1$.

Next consider the case of $t_1 \;=\; \tilde{t}_1$. In this case we would like to make variable k a basic variable. To do so we would augment H_B with a new row and column containing the relevant data associated with variable k. This will be fine provided that it does not result in singularity. Suppose that we invoke Procedure Ψ_1 as

$$[\,H_B^{-1},\, \text{singular},\, w\,] \;\leftarrow\; \Psi_1(\,H_B^{-1},\, d_1,\, \gamma_{kk}, A_k,\, m+n_B,\, n_B\,),$$

where

$$d_1 \;=\; (\gamma_{\beta_1 k},\ldots,\gamma_{\beta_{n_B} k})'\,.$$

If singular $=$ "false," we proceed as in Step 2.2.1 of Algorithm 7 with H_B^{-1} having just been updated by Procedure Ψ_1, I_B with $\{\,\beta_1,\ldots,\beta_{n_B}\,,\ k\,\}$, n_B with $n_B + 1$, and return to Step 1. If singular $=$ "true," it follows from Lemma 8.1 that there is an $s_{B_1} \neq 0$ satisfying

$$C_{B_1 B_1} s_{B_1} \;=\; 0 \quad \text{and}$$
$$B_1 s_{B_1} \;=\; 0\,,$$

where

$$B_1 \;=\; [\,B\,,\ A_k\,] \quad \text{and} \quad C_{B_1 B_1} \;=\; \begin{bmatrix} C_{BB} & d_1 & B' \\ d_1' & \gamma_{kk} & A_k' \\ B & A_k & 0 \end{bmatrix}.$$

From Procedure Ψ_1, s_{B_1} is the first $n_B + 1$ components of w. Indeed, s_{B_1} may be written

$$s_{B_1} \;=\; (s_B'\,,\ -1)'\,, \quad \text{where}$$
$$(s_B)_i \;=\; w_i\,,\quad i \;=\; 1,\ldots,n_B\,.$$

Proceeding as in Step 2.2.2 of Algorithm 7, we wish to determine how large σ may be in order that

$$\begin{bmatrix} x_B(t_{j+1}) \\ 0 \end{bmatrix} - \sigma \begin{bmatrix} s_B \\ -1 \end{bmatrix}$$

remains feasible. If $s_B \;\leq\; 0$ the problem is unbounded from below for all $t > t_{j+1}$. Otherwise, let σ_B and l be such that

$$\sigma_B \;=\; \min\left\{ \frac{(h_{1B})_i + t_{j+1}(h_{2B})_i}{(s_B)_i}\ \middle|\ \text{all } i \;=\; 1,\ldots,n_B \text{ with } (s_B)_i > 0 \right\}$$
$$=\; \frac{(h_{1B})_l + t_{j+1}(h_{2B})_l}{(s_B)_l}\,.$$

Thus basic variable β_l is reduced to zero and so becomes nonbasic, nonbasic variable k has increased to σ_B and becomes basic. H_B is now modified by replacing row and column l containing variable β_l's data with that for variable k. Let

$$
d = \begin{bmatrix}
\gamma_{\beta_1 k} - \gamma_{\beta_1 \beta_l} \\
\vdots \\
\gamma_{\beta_{l-1},k} - \gamma_{\beta_{l-1},\beta_l} \\
(\gamma_{kk} - \gamma_{\beta_l \beta_l})/2 \\
\gamma_{\beta_{l+1},k} - \gamma_{\beta_{l+1},\beta_l} \\
\vdots \\
\gamma_{\beta_{n_B} k} - \gamma_{\beta_{n_B} \beta_l} \\
A_k - A_{\beta_l}
\end{bmatrix} .
$$

Now invoke Procedure Ψ_3 as

$$
[\, H_B^{-1} \,] \leftarrow (\, H_B^{-1},\ d,\ m + n_B,\ l\,) ,
$$

update I_B with $\{\, \beta_1, \ldots, \beta_{l-1}\ ,\ k\ ,\ \beta_{l+1}, \ldots, \beta_{n_B}\, \}$, leave n_B unchanged, and go to Step 1.

Assembling the various steps, we now have

Algorithm 9: Simplex Method for Parametric QP

Model Problem:

$$
\begin{aligned}
\text{minimize:} \quad & (c + tq)'x + \tfrac{1}{2}x'Cx \\
\text{subject to:} \quad & Ax = b + tp, \\
& x \geq 0,
\end{aligned}
$$

for $\underline{t} \leq t \leq \bar{t}$.

Initialization:
Set $t_0 = \underline{t}$. Use the simplex method for quadratic programming to solve the problem for $t = t_0$. Let $I_B = \{\, \beta_1, \ldots, \beta_{n_B}\, \}$ and H_B^{-1} be the associated final data where

$$
H_B = \begin{bmatrix} C_{BB} & B' \\ B & 0 \end{bmatrix} , \quad B = [\, A_{\beta_1}, \ldots, A_{\beta_{n_B}}\,] , \quad \text{and}
$$

$$
C_{BB} = [\, \gamma_{\beta_i \beta_j}\,] .
$$

Set $j = 0$ and go to Step 1.

Step 1: **Computation of Optimal Solution, Multipliers, and End of Interval.**

Compute h_{1B}, h_{2B}, g_{1B}, and g_{2B} according to

$$\begin{bmatrix} h_{1B} \\ g_{1B} \end{bmatrix} = H_B^{-1} \begin{bmatrix} -c_B \\ b \end{bmatrix}, \quad \begin{bmatrix} h_{2B} \\ g_{2B} \end{bmatrix} = H_B^{-1} \begin{bmatrix} -q_B \\ p \end{bmatrix}.$$

For all $i \notin I_B$, compute

$$(w_1)_i = c_i + \sum_{k=1}^{n_B} \gamma_{i\beta_k}(h_{1B})_k + A_i' g_{1B},$$

$$(w_2)_i = q_i + \sum_{k=1}^{n_B} \gamma_{i\beta_k}(h_{2B})_k + A_i' g_{2B},$$

and for all $i \in I_B$ set

$$(w_1)_l - (w_2)_i = 0.$$

If $h_{2B} \geq 0$ set $\hat{t}_{j+1} = +\infty$. Otherwise, compute the smallest index l and $\hat{t}_{j+1}$ such that

$$\hat{t}_{j+1} = \frac{-(h_{1B})_l}{(h_{2B})_l}$$

$$= \min \left\{ \frac{-(h_{1B})_i}{(h_{2B})_i} \,\Big|\, \text{all } i = 1,\ldots,n_B \text{ with } (h_{2B})_i < 0 \right\}.$$

If $w_2 \geq 0$ set $\tilde{t}_{j+1} = +\infty$. Otherwise, compute the smallest index k and $\tilde{t}_{j+1}$ such that

$$\tilde{t}_{j+1} = \frac{-(w_1)_k}{(w_2)_k}$$

$$= \min \left\{ \frac{-(w_1)_i}{(w_2)_i} \,\Big|\, \text{all } i = 1,\ldots,n \text{ with } (w_2)_i < 0 \right\}.$$

Set $t_{j+1} = \min \{ \hat{t}_{j+1}, \tilde{t}_{j+1}, \bar{t} \}$. Print "the optimal solution is $x_B(t) = h_{1B} + t h_{2B}$, $x_N = 0$ with multipliers $u(t) = -g_{1B} - t g_{2B}$ and $v(t) = w_1 + t w_2$ for all t with $t_j \leq t \leq t_{j+1}$." If $t_{j+1} = \bar{t}$, stop. Otherwise, go to Step 2.

Step 2:   **Computation of New I_B and H_B^{-1}.**
If $t_{j+1} = \hat{t}_{j+1}$ then go to Step 2.1, and otherwise, go to Step 2.2.

Step 2.1:   **New Nonbasic Variable (New Active Constraint).**
Invoke Procedure Ψ_2 as

$$[\, H_B^{-1}, \text{ singular}, w \,] \leftarrow \Psi_2(\, H_B^{-1}, m+n_B, l \,).$$

If singular is "false," go to Step 2.1.1. If singular is "true," go to Step 2.1.2.

Step 2.1.1:
Use the updated H_B^{-1} just obtained in Step 2.1, replace I_B with $\{ \beta_1, \ldots, \beta_{l-1},$ $\beta_{l+1}, \ldots, \beta_{n_B} \}$, n_B with $n_B - 1$, and j with $j + 1$. Go to Step 1.

Step 2.1.2:
Set

$$\xi_i = -w_{n_B-1+i}, \quad i = 1, \ldots, m,$$

and let $\xi = (\xi_1, \ldots, \xi_m)'$. If $A_i'\xi \leq 0$, for all $i \notin I_B$, print "there is no feasible solution for $t > t_{j+1}$" and stop. Otherwise, compute the smallest index k such that

$$\frac{(w_1)_k + t_{j+1}(w_2)_k}{A_k'\xi} = \min\left\{ \frac{(w_1)_i + t_{j+1}(w_2)_i}{A_i'\xi} \mid \text{all } i \notin I_B \text{ with } A_i'\xi > 0 \right\}.$$

Let

$$d = \begin{bmatrix} \gamma_{\beta_1 k} - \gamma_{\beta_1 \beta_l} \\ \vdots \\ \gamma_{\beta_{l-1},k} - \gamma_{\beta_{l-1},\beta_l} \\ (\gamma_{kk} - \gamma_{\beta_l \beta_l})/2 \\ \gamma_{\beta_{l+1},k} - \gamma_{\beta_{l+1},\beta_l} \\ \vdots \\ \gamma_{\beta_{n_B} k} - \gamma_{\beta_{n_B} \beta_l} \\ A_k - A_{\beta_l} \end{bmatrix}.$$

Invoke Procedure Ψ_3 as

$$[H_B^{-1}] \leftarrow \Psi_3(H_B^{-1}, d, m+n_B, l),$$

update I_B with $\{ \beta_1, \ldots, \beta_{l-1}, k, \beta_{l+1}, \ldots, \beta_{n_B} \}$, leave n_B unchanged, replace j with $j + 1$, and go to Step 1.

Step 2.2: **New Basic Variable (Drop Active Constraint).**
Let $d_1 = (\gamma_{\beta_1 k}, \ldots, \gamma_{\beta_{n_B} k})'$. Invoke Procedure Ψ_1 as

$$[H_B^{-1}, \text{singular}, w] \leftarrow \Psi_1(H_b^{-1}, d_1, \gamma_{kk}, A_k, m+n_B, n_B).$$

If singular is "false," go to Step 2.2.1. If singular is "true," go to Step 2.2.2.

Step 2.2.1:
Using the updated H_B^{-1} from Step 2.2, replace I_B with $\{ \beta_1, \ldots, \beta_{n_B}, k \}$, n_B with $n_B + 1$, j with $j + 1$, and go to Step 1.

Step 2.2.2:

Set $(s_B)_i = w_i$, $i = 1, \ldots, n_B$. If $s_B \leq 0$ print "the objective function is unbounded from below for $t > t_{j+1}$" and stop. Otherwise, compute σ_B and l such that

$$\sigma_B = \frac{(h_{1B})_l + t_{j+1}(h_{2B})_l}{(s_B)_l}$$

$$= \min \left\{ \frac{(h_{1B})_i + t_{j+1}(h_{2B})_i}{(s_B)_i} \,\middle|\, \text{all } i = 1, \ldots, n_B \text{ with } (s_B)_i > 0 \right\}.$$

Set

$$d = \begin{bmatrix} \gamma_{\beta_1 k} - \gamma_{\beta_1 \beta_l} \\ \vdots \\ \gamma_{\beta_{l-1},k} - \gamma_{\beta_{l-1},\beta_l} \\ (\gamma_{kk} - \gamma_{\beta_l \beta_l})/2 \\ \gamma_{\beta_{l+1},k} - \gamma_{\beta_{l+1},\beta_l} \\ \vdots \\ \gamma_{\beta_{n_B} k} - \gamma_{\beta_{n_B} \beta_l} \\ A_k - A_{\beta_l} \end{bmatrix},$$

invoke Procedure Ψ_3 as

$$[\, H_B^{-1} \,] \leftarrow \Psi_3(\, H_B^{-1}, \, d, \, m + n_B, \, l \,),$$

replace I_B with $\{ \beta_1, \ldots, \beta_{l-1}, k, \beta_{l+1}, \ldots, \beta_{n_B} \}$, leave n_B unchanged, replace j with $j + 1$, and go to Step 1.

A computer program which implements the simplex method for parametric quadratic programming (Algorithm 9, Alg9.m) is given in Section 8.3, Figure 8.8.

The steps of the algorithm are illustrated in

Example 8.3

$$\begin{aligned} \text{minimize:} \quad & tx_1 - (1 - t)x_2 + \tfrac{1}{2}(4x_1^2 + x_2^2 + 4x_1 x_2) \\ \text{subject to:} \quad & -\, 2x_1 - x_2 + x_3 = -2, \quad (1) \\ & x_2 + x_4 = 3, \quad (2) \\ & x_i \geq 0, \quad i = 1, \ldots, 4, \end{aligned}$$

for all t with $t \geq 0$.
Here

$$c = \begin{bmatrix} 0 \\ -1 \\ 0 \\ 0 \end{bmatrix}, \quad q = \begin{bmatrix} 1 \\ 1 \\ 0 \\ 0 \end{bmatrix}, \quad C = \begin{bmatrix} 4 & 2 & 0 & 0 \\ 2 & 1 & 0 & 0 \\ 0 & 0 & 0 & 0 \\ 0 & 0 & 0 & 0 \end{bmatrix},$$

$$A = \begin{bmatrix} -2 & -1 & 1 & 0 \\ 0 & 1 & 0 & 1 \end{bmatrix}, \quad b = \begin{bmatrix} -2 \\ 3 \end{bmatrix}, \quad p = \begin{bmatrix} 0 \\ 0 \end{bmatrix}, \quad \underline{t} = 0,$$

$\bar{t} = +\infty,$ and $n = 4$.

Initialization:

$$t_0 = 0, \quad I_B = \{2, 4\}, \quad H_B^{-1} = \begin{bmatrix} 0 & 0 & -1 & 0 \\ 0 & 0 & 1 & 1 \\ -1 & 1 & -1 & 0 \\ 0 & 1 & 0 & 0 \end{bmatrix},$$

$$H_B = \begin{bmatrix} 1 & 0 & -1 & 1 \\ 0 & 0 & 0 & 1 \\ -1 & 0 & 0 & 0 \\ 1 & 1 & 0 & 0 \end{bmatrix}, \quad B = \begin{bmatrix} -1 & 0 \\ 1 & 1 \end{bmatrix}, \quad C_{BB} = \begin{bmatrix} 1 & 0 \\ 0 & 0 \end{bmatrix},$$

$j = 0$.

Iteration 0

Step 1: $c_B = (-1, 0)', \quad q_B = (1, 0)',$

$$h_{1B} = \begin{bmatrix} 2 \\ 1 \end{bmatrix}, \quad h_{2B} = \begin{bmatrix} 0 \\ 0 \end{bmatrix},$$

$$g_{1B} = \begin{bmatrix} 1 \\ 0 \end{bmatrix}, \quad g_{2B} = \begin{bmatrix} 1 \\ 0 \end{bmatrix},$$

$$w_1 = \begin{bmatrix} 2 \\ 0 \\ 1 \\ 0 \end{bmatrix}, \quad w_2 = \begin{bmatrix} -1 \\ 0 \\ 1 \\ 0 \end{bmatrix},$$

$\hat{t}_1 = +\infty,$

$$\tilde{t}_1 = \min\left\{ \frac{-2}{-1}, -, -, - \right\} = 2, \quad k = 1,$$

$$t_1 = \min\{ +\infty, 2, +\infty \} = 2.$$

The optimal solution is $x_B(t) = \begin{bmatrix} 2 \\ 1 \end{bmatrix} + t \begin{bmatrix} 0 \\ 0 \end{bmatrix},$

$x_N = 0$ with multipliers $u(t) = -\begin{bmatrix} 1 \\ 0 \end{bmatrix} - t \begin{bmatrix} 1 \\ 0 \end{bmatrix}$ and

$$v(t) = \begin{bmatrix} 2 \\ 0 \\ 1 \\ 0 \end{bmatrix} + t \begin{bmatrix} -1 \\ 0 \\ 1 \\ 0 \end{bmatrix} \quad \text{for all } t \text{ with } 0 \leq t \leq 2 .$$

Step 2:   Transfer to Step 2.2.

Step 2.2: $d_1 = \begin{bmatrix} 2 \\ 0 \end{bmatrix}$, $\gamma_{11} = 4$, $A_1 = \begin{bmatrix} -2 \\ 0 \end{bmatrix}$,

invoking Procedure Ψ_1 as

$[\, H_B^{-1}, \text{ singular}, \ w \,] \leftarrow \Psi_1(H_B^{-1} , d_1 , \gamma_{11} , A_1 , 4 , 2)$,

gives singular $=$ "true." Transfer to Step 2.2.2.

Step 2.2.2: $s_B = \begin{bmatrix} 2 \\ -2 \end{bmatrix}$,

$$\sigma_B = \min \left\{ \frac{2}{2} , - \right\} = 1 , \ l = 1 ,$$

$$d = \begin{bmatrix} 3/2 \\ 0 \\ -1 \\ -1 \end{bmatrix} ,$$

$[\, H_B^{-1} \,] \leftarrow \Psi_3(H_B^{-1} , d , 4 , 1)$,

$$H_B^{-1} = \begin{bmatrix} 0 & 0 & -1/2 & 0 \\ 0 & 0 & 0 & 1 \\ -1/2 & 0 & -1 & 0 \\ 0 & 1 & 0 & 0 \end{bmatrix} , \ I_B \leftarrow \{\, 1 , 4 \,\} ,$$

$n_B = 2 , \ j = 1 .$

Iteration 1

Step 1: $c_B = (0 , 0)'$, $q_B = (1 , 0)'$,

$$h_{1B} = \begin{bmatrix} 1 \\ 3 \end{bmatrix} , \ h_{2B} = \begin{bmatrix} 0 \\ 0 \end{bmatrix} ,$$

$$g_{1B} = \begin{bmatrix} 2 \\ 0 \end{bmatrix} , \ g_{2B} = \begin{bmatrix} 1/2 \\ 0 \end{bmatrix} ,$$

$$w_1 = \begin{bmatrix} 0 \\ -1 \\ 2 \\ 0 \end{bmatrix} , \quad w_2 = \begin{bmatrix} 0 \\ 1/2 \\ 1/2 \\ 0 \end{bmatrix} ,$$

$$\hat{t}_2 = +\infty ,$$

$$\tilde{t}_2 = +\infty ,$$

$$t_2 = \min \{ +\infty, +\infty, +\infty \} = +\infty .$$

The optimal solution is $x_B(t) = \begin{bmatrix} 1 \\ 3 \end{bmatrix} + t \begin{bmatrix} 0 \\ 0 \end{bmatrix} , \quad x_N = 0$

with multipliers $u(t) = - \begin{bmatrix} -2 \\ 0 \end{bmatrix} - t \begin{bmatrix} -1/2 \\ 0 \end{bmatrix}$ and

$$v(t) = \begin{bmatrix} 0 \\ -1 \\ 2 \\ 0 \end{bmatrix} + t \begin{bmatrix} 0 \\ 1/2 \\ 1/2 \\ 0 \end{bmatrix} \quad \text{for all } t \text{ with } 2 \leq t \leq +\infty.$$

$t_2 = +\infty$; stop.

A computer program which implements the simplex method for parametric quadratic programming (Algorithm 9, Alg9.m) is given in Figure 8.8, Section 8.3. The calling program to solve Example 8.3 is given in Figure 8.9. The output from applying this program to the problem of Example 8.3 is shown in Figure 8.10.

We further illustrate the simplex method for parametric quadratic programming in

Example 8.4

$$\begin{aligned}
\text{minimize:} \quad & tx_1 - (1 - t)x_2 + \tfrac{1}{2}(4x_1^2 + x_2^2 + 4x_1x_2) \\
\text{subject to:} \quad & -2x_1 - x_2 + x_3 = -2, \quad (1) \\
& 4x_1 + x_2 + x_4 = 6 - \tfrac{1}{2}t, \quad (2) \\
& x_i \geq 0, \quad i = 1, \ldots, 4 ,
\end{aligned}$$

for all t with $t \geq 0$.
Here

$$c = \begin{bmatrix} 0 \\ -1 \\ 0 \\ 0 \end{bmatrix} , \quad q = \begin{bmatrix} 1 \\ 1 \\ 0 \\ 0 \end{bmatrix} , \quad C = \begin{bmatrix} 4 & 2 & 0 & 0 \\ 2 & 1 & 0 & 0 \\ 0 & 0 & 0 & 0 \\ 0 & 0 & 0 & 0 \end{bmatrix} ,$$

$$A = \begin{bmatrix} -2 & -1 & 1 & 0 \\ 4 & 1 & 0 & 1 \end{bmatrix}, \quad b = \begin{bmatrix} -2 \\ 6 \end{bmatrix}, \quad p = \begin{bmatrix} 0 \\ -1/2 \end{bmatrix},$$

$\underline{t} = 0$, $\bar{t} = +\infty$, and $n = 4$.

Initialization:

$$t_0 = 0, \quad I_B = \{2, 4\}, \quad H_B^{-1} = \begin{bmatrix} 0 & 0 & -1 & 0 \\ 0 & 0 & 1 & 1 \\ -1 & 1 & -1 & 0 \\ 0 & 1 & 0 & 0 \end{bmatrix},$$

$$H_B = \begin{bmatrix} 1 & 0 & -1 & 1 \\ 0 & 0 & 0 & 1 \\ -1 & 0 & 0 & 0 \\ 1 & 1 & 0 & 0 \end{bmatrix}, \quad B = \begin{bmatrix} -1 & 0 \\ 1 & 1 \end{bmatrix}, \quad C_{BB} = \begin{bmatrix} 1 & 0 \\ 0 & 0 \end{bmatrix},$$

$j = 0$.

Iteration 0

Step 1: $c_B = (-1, 0)'$, $q_B = (1, 0)'$,

$$h_{1B} = \begin{bmatrix} 2 \\ 4 \end{bmatrix}, \quad h_{2B} = \begin{bmatrix} 0 \\ -1/2 \end{bmatrix},$$

$$g_{1B} = \begin{bmatrix} 1 \\ 0 \end{bmatrix}, \quad g_{2B} = \begin{bmatrix} 1 \\ 0 \end{bmatrix},$$

$$w_1 = \begin{bmatrix} 2 \\ 0 \\ 1 \\ 0 \end{bmatrix}, \quad w_2 = \begin{bmatrix} -1 \\ 0 \\ 1 \\ 0 \end{bmatrix},$$

$$\hat{t}_1 = \min\left\{-, \frac{-4}{-1/2}\right\} = 8, \quad l = 2,$$

$$\tilde{t}_1 = \min\left\{\frac{-2}{-1}, -, -, -\right\} = 2, \quad k = 1,$$

$$t_1 = \min\{8, 2, +\infty\} = 2.$$

The optimal solution is $x_B(t) = \begin{bmatrix} 2 \\ 4 \end{bmatrix} + t \begin{bmatrix} 0 \\ -1/2 \end{bmatrix}$,

$x_N = 0$ with multipliers $u(t) = -\begin{bmatrix} 1 \\ 0 \end{bmatrix} - t \begin{bmatrix} 1 \\ 0 \end{bmatrix}$ and

$$v(t) = \begin{bmatrix} 2 \\ 0 \\ 1 \\ 0 \end{bmatrix} + t \begin{bmatrix} -1 \\ 0 \\ 1 \\ 0 \end{bmatrix} \quad \text{for all } t \text{ with } 0 \le t \le 2 .$$

Step 2: Transfer to Step 2.2.

Step 2.2: $d_1 = \begin{bmatrix} 2 \\ 0 \end{bmatrix}$, $\gamma_{11} = 4$, $A_1 = \begin{bmatrix} -2 \\ 4 \end{bmatrix}$,

invoking Procedure Ψ_1 as

$[\, H_B^{-1}, \text{ singular}, \ w \,] \leftarrow \Psi_1(H_B^{-1} , \ d_1 , \ \gamma_{11} , \ A_1 , 4 , \ 2 \,) ,$

gives singular $=$ "true." Transfer to Step 2.2.2.

Step 2.2.2: $s_B = \begin{bmatrix} 2 \\ 2 \end{bmatrix}$,

$$\sigma_B = \min \left\{ \frac{2}{2} , \frac{3}{2} \right\} = 1 , \ l = 1 ,$$

$$d = \begin{bmatrix} 3/2 \\ 0 \\ -1 \\ 3 \end{bmatrix} ,$$

$[\, H_B^{-1} \,] \leftarrow \Psi_3(H_B^{-1} , \ d , \ 4 , \ 1 \,),$

$$H_B^{-1} = \begin{bmatrix} 0 & 0 & -1/2 & 0 \\ 0 & 0 & 2 & 1 \\ -1/2 & 2 & -1 & 0 \\ 0 & 1 & 0 & 0 \end{bmatrix} , \ I_B \leftarrow \{\, 1 , 4 \,\} ,$$

$n_B = 2 , \ j = 1 .$

Iteration 1

Step 1: $c_B = (0 , 0)'$, $q_B = (1 , 0)'$,

$$h_{1B} = \begin{bmatrix} 1 \\ 2 \end{bmatrix} , \ h_{2B} = \begin{bmatrix} 0 \\ -1/2 \end{bmatrix} ,$$

$$g_{1B} = \begin{bmatrix} 2 \\ 0 \end{bmatrix} , \ g_{2B} = \begin{bmatrix} 1/2 \\ 0 \end{bmatrix} ,$$

$$w_1 = \begin{bmatrix} 0 \\ -1 \\ 2 \\ 0 \end{bmatrix} , \quad w_2 = \begin{bmatrix} 0 \\ 1/2 \\ 1/2 \\ 0 \end{bmatrix} ,$$

$$\hat{t}_2 = \min \left\{ - , \frac{-2}{-1/2} \right\} = 4 , \quad l = 2 ,$$

$$\tilde{t}_2 = +\infty ,$$

$$t_2 = \min \{ 4 , +\infty, +\infty \} = 4 .$$

The optimal solution is $x_B(t) = \begin{bmatrix} 1 \\ 2 \end{bmatrix} + t \begin{bmatrix} 0 \\ -1/2 \end{bmatrix} ,$

$x_N = 0$ with multipliers $u(t) = -\begin{bmatrix} 2 \\ 0 \end{bmatrix} - t \begin{bmatrix} 1/2 \\ 0 \end{bmatrix} ,$

and $v(t) = \begin{bmatrix} 0 \\ -1 \\ 2 \\ 0 \end{bmatrix} + t \begin{bmatrix} 0 \\ 1/2 \\ 1/2 \\ 0 \end{bmatrix}$ for all t with $2 \le t \le 4 .$

Step 2:   Transfer to Step 2.1.

Step 2.1:   Invoking Procedure Ψ_2 as
$[H_B^{-1} , \text{ singular, } w] \leftarrow \Psi_2(H_B^{-1}, 4, 2) ,$
gives singular $=$ "true." Transfer to Step 2.1.2.

Step 2.1.2: $\xi = \begin{bmatrix} -2 \\ -1 \end{bmatrix} ,$

$$\frac{(w_1)_1 + t_2(w_2)_1}{A_1' \xi} = \min \left\{ \frac{1}{1} , - \right\} = 1 , \quad k = 2 ,$$

$$d = \begin{bmatrix} 2 \\ 1/2 \\ -1 \\ 0 \end{bmatrix} ,$$

$[H_B^{-1}] \leftarrow \Psi_3(H_B^{-1} , d , 4 , 2) ,$

$$H_B^{-1} = \begin{bmatrix} 0 & 0 & 1/2 & 1/2 \\ 0 & 0 & -2 & -1 \\ 1/2 & -2 & -1 & 0 \\ 1/2 & -1 & 0 & 0 \end{bmatrix} , \quad I_B \leftarrow \{ 1 , 2 \} ,$$

$n_B = 2 , \quad j = 2 .$

Iteration 2

Step 1: $c_B = (0 \,, \, -1)' \,, \quad q_B = (1 \,, \, 1)' \,,$

$$h_{1B} = \begin{bmatrix} 2 \\ -2 \end{bmatrix} \,, \quad h_{2B} = \begin{bmatrix} -1/4 \\ 1/2 \end{bmatrix} \,,$$

$$g_{1B} = \begin{bmatrix} 0 \\ -1 \end{bmatrix} \,, \quad g_{2B} = \begin{bmatrix} 3/2 \\ 1/2 \end{bmatrix} \,,$$

$$w_1 = \begin{bmatrix} 0 \\ 0 \\ 0 \\ -1 \end{bmatrix} \,, \quad w_2 = \begin{bmatrix} 0 \\ 0 \\ 3/2 \\ 1/2 \end{bmatrix} \,,$$

$$\hat{t}_3 = \min \left\{ \frac{-2}{-1/4} \,, \, - \right\} = 8 \,, \quad l = 1 \,,$$

$$\tilde{t}_3 = +\infty \,,$$

$$t_3 = \min \{ 8 \,, \, +\infty, \, +\infty \} = 8 \,.$$

The optimal solution is $x_B(t) = \begin{bmatrix} 2 \\ -2 \end{bmatrix} + t \begin{bmatrix} -1/4 \\ 1/2 \end{bmatrix} \,,$

$x_N = 0$ with multipliers $u(t) = - \begin{bmatrix} 0 \\ -1 \end{bmatrix} - t \begin{bmatrix} 3/2 \\ 1/2 \end{bmatrix}$ and

$$v(t) = \begin{bmatrix} 0 \\ 0 \\ 0 \\ -1 \end{bmatrix} + t \begin{bmatrix} 0 \\ 0 \\ 3/2 \\ 1/2 \end{bmatrix} \quad \text{for all } t \text{ with } 4 \leq t \leq 8 \,.$$

Step 2:   Transfer to Step 2.1.

Step 2.1:   Invoking Procedure Ψ_2 as
$[\, H_B^{-1}, \text{ singular}, \, w \,] \leftarrow \Psi_2(\, H_B^{-1}, \, 4, \, 1 \,) \,,$
gives singular $=$ "true." Transfer to Step 2.1.2.

Step 2.1.2: $\xi = \begin{bmatrix} -1/2 \\ -1/2 \end{bmatrix} \,.$

$A_i' \xi \leq 0, \quad i = 3 \,, \, 4 \,;$ stop, the problem has no feasible solution for $t > t_3 = 8 \,.$

$\Diamond$

Example 8.4 has been solved using the computer program Alg9. The routine which shows the data being set up is shown in Figure 8.11 and the output is shown in Figure 8.12.

Example 8.5 Consider the portfolio optimization problem:

$$\min\{ -t\mu'x + \frac{1}{2}x'\Sigma x \mid l'x = 1,\ x \geq 0 \},$$

where l is an $n-$ vector of 1's. Σ is a diagonal covariance matrix an μ is an $n-$ vector of expected returns. Assume the expected returns are ordered from smallest to largest (this is not restrictive). Best and Hlouskova [4] have shown the following. For $t = 0$, all assets will be held positively. As t is increased, the first asset will be the first to be reduced to zero and it will remain at zero for all t beyond this critical value. The next asset to be reduced to zero will be asset 2 and it will remain at zero for all greater t. As t is further increased further, asset 3 will be reduced to zero and so on.

A Matlab program is shown in Figure 8.13 which formulates this problem. The dimension is set to $n = 10$ but may be easily changed if one is interested in higher dimensional problems. The components of Σ and μ are set using random numbers and the components of μ are sorted in increasing order. Then the problem is solved for $t = 0$ using the Simplex Method for QP (Alg8). Then the parametric problem is solved for all $t \geq 0$. The output is shown in Figure 8.14. $\Diamond$

As part of the initialization, the model parametric quadratic programming problem is solved by the simplex method for quadratic programming with $t = \underline{t}$. Since that algorithm can only terminate when H_B is nonsingular, the H_B used as initial data for the simplex method for parametric quadratic programming is necessarily nonsingular.

Let $(x_B,\ x_N)$ denote the optimal solution for $t = \underline{t}$ and let $(u,\ v)$ denote the associated multipliers. Let $v = (v_B,\ v_N)$. In order to formulate the termination properties of the parametric method, we require

Assumption 8.1.
$$v_N \equiv v_N(\underline{t}) > 0.$$

From our previous discussion, Assumption 8.1 implies Assumption 7.2 and since the simplex method for parametric quadratic programming is a specialization of Algorithm 7, Theorem 7.2 applies and we have

Theorem 8.2 *Let Assumption 8.1 be satisfied and let* $x_{B_j}(t)$, $x_{N_j} = 0$, $u_j(t)$, $v_j(t)$, *and* t_j, $j = 1,\ \dots$ *be obtained by applying the simplex method*

for parametric quadratic programming to the model problem

$$\min \{ (c + tq)'x + \tfrac{1}{2}x'Cx \mid Ax = b + tp , \quad x \geq 0 \} , \qquad (8.19)$$

for $\underline{t} \leq t \leq \bar{t}$. Assume that for each j, the indices k and l are uniquely determined and that $\hat{t}_j \neq \tilde{t}_j$. Then the algorithm terminates after $\nu < +\infty$ iterations and for $j = 1, \ldots, \nu - 1$, $x_{B_j}(t)$, $x_{N_j} = 0$ is optimal for (8.19) with multipliers $u_j(t)$ and $v_j(t)$ for all t with $t_j \leq t \leq t_{j+1}$ and either $t_\nu = +\infty$, $t_\nu < +\infty$ and (8.19) has no feasible solution for $t > t_\nu$, or, $t_\nu < +\infty$ and (8.19) is unbounded from below for $t > t_\nu$.

8.3   Computer Programs

```
1  %      Alg8.m
2  function [x,xB,nB,IB,fB,HBinv,msg,dual,iter,alg] = ...
3                          Alg8(c,C,n,m,tol,xB,IB,nB,A,b)
4  %   Alg8 implements the Simplex Method for Quadratic
5  %   Programming.  It uses Procedures Psi1, Psi2 and Psi3
6  %   to do the matrix upating.
7  %   Assume xB,IB and nB all have the correct values
8  %   on input.
9  %   Initialzation
10 [ B,cB,CBB,gB,gN,fB,IN ] = simpobj(A,C,c,n,xB,IB,nB);
11 HB = vertcat([CBB,B'],[B,zeros(m,m)]);
12 HBinv = inv(HB);
13 alg = 'Algorithm 8 (Simplex Method for QP)';
14 next = 'Step1';
15 singular = 'false';
16 for iter=0:10000
17      iteration = iter;
18      if strcmp(next,'Step1')
19                              % Step 1
20          if strcmp(singular,'false')
21              next = 'Step1p1';
22          else
23              next = 'Step1p2';
24          end
25      end
26      if strcmp(next,'Step1p1')
27                              % Step 1.1
28          rhs = vertcat(gB,zeros(m,1));
29          sBu = HBinv*rhs;
30          sB = sBu(1:nB);
31          u = sBu(nB+1:nB+m);
```

```
32          gammaB = 0;
33          next = 'Step2';
34      end
35      if strcmp(next,'Step1p2')
36                                  % Step 1.2
37          sB =  w(1:nB);
38          gammaB = 1;
39          next = 'Step2';
40      end
41      if strcmp(next,'Step2')
42                                  % Step 2
43          if gammaB == 0
44              sigBtilde = 1;
45          else
46              sigBtilde = inf;
47          end
48          sigBhat = inf;
49          for i=1:nB
50              if sB(i) > 0.
51                  test = xB(i)/sB(i);
52                  if test < sigBhat
53                      sigBhat = test;
54                      ell = i;
55                  end
56              end
57          end
58          if sigBtilde == inf & sigBhat == inf
59              msg = 'unbounded from below';
60              return
61          else
62              sigB = min(sigBtilde,sigBhat);
63          end
64      next = 'Step3';
65      end
66      if strcmp(next,'Step3')
67                                  % Step 3
68          if sigB == sigBhat
69              next = 'Step3p1';
70          else
71              next = 'Step3p2';
72          end
73          if strcmp(next,'Step3p1')
74                                  % Step 3.1
75              if strcmp(singular,'false')
76                  xB = xB - sigB*sB;
77                  xB(ell) = [];
78                  [ HBinv,singular,w ] = Psi2(HBinv,m+nB,ell);
79                  IB(ell) = [];
80                  nB = nB - 1;
81                  [ B,cB,CBB,gB,gN,fB,IN ] = ...
82                      simpobj(A,C,c,n,xB,IB,nB);
```

```
83                           next = 'Step1';
84                   end
85
86               if strcmp(singular,'true')
87                   xB = xB -sigB*sB;
88                   xB(ell) = sigB;
89                   d = [];
90                   for i=1:ell-1
91                       d(i) = C(IB(i),kay) - C(IB(i),IB(ell));
92                   end
93                   d(ell) = (C(kay,kay) - C(IB(ell),IB(ell)))/2.;
94                   for i=ell+1:nB
95                       d(i) = C(IB(i),kay) - C(IB(i),IB(ell));
96                   end
97                   for i=1:m
98                       d(i+nB) = A(i,kay) - A(i,IB(ell));
99                   end
100                  d  = d';
101                  [ HBinv,singular,w ] = ...
102                          Psi3( HBinv,d,nB+m,ell );
103                  IB(ell) = kay;
104                  [ B,cB,CBB,gB,gN,fB,IN ] = ...
105                          simpobj(A,C,c,n,xB,IB,nB);
106                  next = 'Step1';
107              end
108          end
109          if strcmp(next,'Step3p2')
110                              % Step 3.2
111              xB = xB - sigB*sB;
112              [ B,cB,CBB,gB,gN,fB,IN ] = ...
113                      simpobj(A,C,c,n,xB,IB,nB);
114              u = - u;
115              for i=1:nB
116                  v(IB(i)) = 0.;
117              end
118              for i=1:n-nB
119                  j = IN(i);
120                  v(j) = A(:,j)'*u + gN(i);
121              end
122              smallv = min(v);
123              kay = find(smallv==v,1);
124              if smallv >= - tol;
125                  msg = 'Optimal solution obtained';
126                  x = zeros(n,1);
127                  for i=1:nB
128                      x(IB(i)) = xB(i);
129                  end
130                  dual = vertcat(v',u);
131                  return
132              end
133              d1 = [];
```

```
134                    for i=1:nB
135                        d1(i) = C(IB(i),kay);
136                    end
137                    d1 = d1';
138                    [HBinv,singular,w] = ...
139                            Psi1(HBinv,d1,C(kay,kay),A(:,kay),m+nB,nB);
140                     if strcmp(singular,'false')
141                        xB = vertcat(xB,0);
142             %          xB(nB+1) =0.;
143                        IB(nB+1) = kay;
144                        nB = nB + 1;
145                        kpos = find(kay==IN,1);
146                        gB = vertcat(gB,gN(kpos));
147                        gN(kpos) = [];
148                        next = 'Step1';
149                     end
150                     if strcmp(singular,'true')
151                        next = 'Step1';
152                     end
153                end
154            end
155 end            % end of for iter loop
```

Figure 8.2.   Simplex Method for Quadratic Programming, Alg8.m.

Discussion of Alg8.m

```
1  %      simpobj.m
2  function [ B,cB,CBB,gB,gN,fB,IN ] = simpobj(A,C,c,n,xB,IB,nB)
3
4  for  i=1:nB
5      B(:,i) = A(:,IB(i));
6      cB(i) = c(IB(i));
7      for j=1:nB
8          CBB(i,j) = C(IB(i),IB(j));
9      end
10 end
11 cB = cB';
12 fB = cB'*xB + 0.5*xB'*CBB*xB;
13 gB = cB + CBB*xB;
14 for i=1:n
15     temp(i) = i;
16 end
17 IN = setdiff(temp,IB);
18 for i=1:n-nB
19     for j=1:nB
20         CNB(i,j) = C(IN(i),IB(j));
21     end
22 end
```

```
23  for  i=1:n−nB
24      gN(i) = c(IN(i));
25      for j=1:nB
26          gN(i) = gN(i) + CNB(i,j)*xB(j);
27      end
28  end
29  return
```

Figure 8.3.   Simplex Objective Evaluation, simpobj.m.

Discussion of simpobj.m

```
1  %     eg8p1.m
2  n = 4;  m = 2;  tol = 1.e−6;
3  A = [1. 0. 1. 0. ; 0. 1. 0. 1.];  b = [2.  2. ]';
4  c = [ −3. −1. 0. 0.]';
5  nB = 2;  xB = [2  2]';  IB = [2 3];
6  C = [1 0 0 0 ; 0 1 0 0 ; 0 0 0 0 ; 0 0 0 0 ];
7  [x,IB,fB,HBinv,msg,dual,iter,alg] = ...
8                       Alg8(c,C,n,m,tol,xB,IB,nB,A,b);
9  AA = vertcat(−eye(n,n),A);  bb = vertcat(zeros(n,1),b);
10 checkKKT(AA,bb,x,n,n,m,dual,c,C,fB,iter,msg,alg);
```

Figure 8.4.   Example 8.1, eg8p1.m.

Discussion of Eg8p1.m

```
1  eg8p1
2  Optimization Summary
3  Optimal solution obtained
4  Algorithm: Algorithm 8 (Simplex Method for QP)
5  Optimal objective value   −4.50000000
6  Iterations    3
7  Maximum Primal Error      0.00000000
8  Maximum Dual Error        0.00000000
9
10
11   Optimal Primal Solution
12      1    2.00000000
13      2    1.00000000
14      3    0.00000000
15      4    1.00000000
16
17
18   Optimal Dual Solution
19      1    0.00000000
20      2    0.00000000
21      3    1.00000000
```

```
22     4    0.00000000
23     5    1.00000000
24     6   -0.00000000
```

Figure 8.5.   Example 8.1 (output), eg8p1.m.

Discussion of Eg8p1.m

```
1  %      eg8p2.m
2  n = 5;   m = 3;   tol = 1.e-6;
3  A = [-2 1 1 0 0;1 2 0 1 0; 4 3 0 0 1];   b = [ 2 14 36]';
4  c = [-5 2 0 0 0]';   nB = 3;   xB = [2 6 10]';
5  IB = [1 2 5];   C = zeros(5);
6  [x,IB,fB,HBinv,msg,dual,iter,alg] = ...
7                          Alg8(c,C,n,m,tol,xB,IB,nB,A,b);
8  AA = vertcat(-eye(n,n),A);   bb = vertcat(zeros(n,1),b);
9  checkKKT(AA,bb,x,n,n,m,dual,c,C,fB,iter,msg,alg)
```

Figure 8.6.   Example 8.2, eg8p2.m.

Discussion of Eg8p2.m

```
1  eg8p2
2  Optimization Summary
3  Optimal solution obtained
4  Algorithm: Algorithm 8 (Simplex Method for QP)
5  Optimal objective value    -45.00000000
6  Iterations    4
7  Maximum Primal Error         0.00000000
8  Maximum Dual Error           0.00000000
9
10
11   Optimal Primal Solution
12     1     9.00000000
13     2     0.00000000
14     3    20.00000000
15     4     5.00000000
16     5     0.00000000
17
18
19   Optimal Dual Solution
20     1     0.00000000
21     2     5.75000000
22     3     0.00000000
23     4     0.00000000
24     5     1.25000000
25     6    -0.00000000
26     7    -0.00000000
```

```
27      8    1.25000000
```

Figure 8.7. Example 8.2 (output), eg8p2.m.

```
1   %      Alg9.m
2   function [H1,H2,T,IB,fB,HBinv,msg,alg] = ...
3            Alg9(c,q,p,C,n,m,tol,xB,IB,nB,A,b,HBinv,tlow,thigh)
4   %    Alg9 implements the parametric Simplex Method for
5   %    Quadratic Programming.  It uses Procedures Psi1, Psi2
6   %    and Psi3 to do the matrix upating.  It is assumed xB, nB,
7   %    HBinv and IB contain the optimal solution for t = tlow.
8   %    Initialization
9   msg = [];
10  next = 'Step1';
11  alg = 'Algorithm 9 (Simplex Method for Parametric QP)';
12  singular = 'false';
13  T(1) = tlow;
14  interval = 0;
15  for iter=0:10000
16      if strcmp(next,'Step1')
17                                  % Step 1
18          [ B,cB,CBB,gB,gN,fB,IN ] = simpobj(A,C,c,n,xB,IB,nB);
19          soln1 = HBinv*vertcat(-cB,b);
20          qB = q(IB);
21          soln2 = HBinv*vertcat(-qB,p);
22          h1B = soln1(1:nB);
23          g1B = soln1(nB+1:nB+m);
24          h2B = soln2(1:nB);
25          g2B = soln2(nB+1:nB+m);
26          w1 = zeros(n,1);
27          w2 = zeros(n,1);
28          for j=1:n-nB
29              i = IN(j);
30              sum1 = c(i);
31              sum2 = q(i);
32              for k=1:nB
33                  sum1 = sum1 + C(i,IB(k))*h1B(k);
34                  sum2 = sum2 + C(i,IB(k))*h2B(k);
35              end
36              sum1 = sum1 + A(:,i)'*g1B;
37              sum2 = sum2 + A(:,i)'*g2B;
38              w1(i) = sum1;
39              w2(i) = sum2;
40          end
41          interval = interval + 1;
42          H1(:,interval)  = zeros(n,1);
43          H1(IB,interval) = h1B(:);
44          H2(:,interval)  = zeros(n,1);
```

```
45              H2(IB,interval) = h2B(:);
46
47
48
49  %                                     Interval limits
50          that =  Inf;
51          ell = 0;
52          for i=1:nB
53              if h2B(i) ≤ - tol
54                  temp = - h1B(i)/h2B(i);
55                  if temp < that
56                      that = temp;
57                      ell = i;
58                  end
59              end
60          end
61          ttilde = Inf;
62          kay = 0;
63          for i=1:n
64              if w2(i) ≤ - tol
65                  temp = - w1(i)/w2(i);
66                  if temp ≤ ttilde
67                      ttilde = temp;
68                      kay = i;
69                  end
70              end
71          end
72          t2 = min(that,ttilde);
73          t2 = min(t2,thigh);
74          T(interval+1) = t2;
75          if t2 == thigh
76              return
77          end
78          next = 'Step2';
79      end                           % end of Step 1
80
81      if strcmp(next,'Step2')
82                                    % Step 2
83          if t2 == that
84              next = 'Step2p1';
85          else
86              next = 'Step2p2';
87          end
88      end
89      if strcmp(next,'Step2p1')
90          [HBinvtemp,singular,w] = Psi2(HBinv,m+nB,ell);
91          if strcmp(singular,'false')
92              next = 'Step2p1p1';
93          else
94              next = 'Step2p1p2';
95          end
```

```
96          end
97          if strcmp(next,'Step2p1p1');
98              HBinv = HBinvtemp;
99              IB(ell) = [];
100             xB(ell) = [];
101             nB = nB - 1;
102             next = 'Done';
103         end
104         if strcmp(next,'Step2p1p2')
105             for i=1:m
106                 xi(i) = -w(nB-1+i);
107             end
108             kay = 0;
109             xmin = Inf;
110             for i=1:n-nB
111                 temp1 = A(:,i)'*xi';
112                 if temp1 >= tol
113                     test = (w1(i) + t2*w2(i))/temp1;
114                     if test <  xmin
115                         xmin = test;
116                         kay = i;
117                     end
118                 end
119             end
120             if kay == 0
121                 msg = 'No feasible solution for further t';
122                 return
123             end
124             d = [];
125             for i=1:ell-1
126                 d(i) = C(IB(i),kay) - C(IB(i),IB(ell));
127             end
128             d(ell) = (C(kay,kay) - C(IB(ell),IB(ell)))/2.;
129             for i=ell+1:nB
130                 d(i) = C(IB(i),kay) - C(IB(i),IB(ell));
131             end
132             for i=1:m
133                 d(i+nB) = A(i,kay) - A(i,IB(ell));
134             end
135             d  = d';
136             [ HBinv,singular,w ] = ...
137                         Psi3( HBinv,d,nB+m,ell );
138             IB(ell) = kay;
139             next = 'Done';
140         end
141         if strcmp(next,'Step2p2')
142             d1 = [];
143             for i=1:nB
144                 d1(i) = C(IB(i),kay);
145             end
146             d1 = d1';
```

```
147          [HBinvtemp,singular,w] = ...
148                     Psi1(HBinv,d1,C(kay,kay),A(:,kay),m+nB,nB);
149          if strcmp(singular,'false')
150              next = 'Step2p2p1';
151          else
152              next = 'Step2p2p2';
153          end
154     end
155     if strcmp(next,'Step2p2p1')
156          HBinv = HBinvtemp;
157          IB(nB+1) = kay;
158          nB = nB + 1;
159          next = 'Done';
160     end
161     if strcmp(next,'Step2p2p2')
162          for i=1:nB
163              sB(i) = w(i);
164          end
165          ell = 0;
166          sigB = Inf;
167          for i=1:nB
168              if sB(i) >= tol
169                  test = (h1B(i) + t2*h2B(i))/sB(i);
170                  if test < sigB
171                      sigB = test;
172                      ell = i;
173                  end
174              end
175          end
176          if ell == 0
177              msg = 'objective is unbounded from below';
178              return
179          end
180          d = [];
181          for i=1:ell-1
182              d(i) = C(IB(i),kay) - C(IB(i),IB(ell));
183          end
184          d(ell) = (C(kay,kay) - C(IB(ell),IB(ell)))/2.;
185          for i=ell+1:nB
186              d(i) = C(IB(i),kay) - C(IB(i),IB(ell));
187          end
188          for i=1:m
189              d(i+nB) = A(i,kay) - A(i,IB(ell));
190          end
191          d  = d';
192          [ HBinv,singular,w ] = ...
193                          Psi3( HBinv,d,nB+m,ell );
194          IB(ell) = kay;
195          next = 'Done';
196     end
197     if strcmp(next,'Done')
```

```
198          next = 'Step1';
199       end
200  end                              % end of "for iter=0:10000
```

Figure 8.8.   Simplex Method for Parametric Quadratic Programming, Alg9.m.

Discussion of Alg9.m. The output from Alg9.m includes T, H1 and H2 which contain the optimal primal solution and the end points of each parametric interval. The optimal dual variables for the equality constraints and nonnegativity constraints could be made available in a manner similar to H1 and H2.

```
1  %      eg8p3.m
2  n = 4; m = 2; c = [ 0 -1 0 0 ]; q = [ 1 1 0 0 ]';
3  C = [ 4 2 0 0; 2 1 0 0; 0 0 0 0; 0 0 0 0 ];
4  A = [ -2 -1 1 0; 0 1 0 1 ]; b = [ -2 3 ]'; p = [ 0 0 ]';
5  HBinv = [ 0 0 -1 0; 0 0 1 1; -1 1 -1 0; 0 1 0 0 ];
6  tlow = 0.; thigh = inf; IB = [ 2 4 ]; nB = 2;
7  xB = [ 2 1 ]'; tol = 1.e-6;
8  [H1,H2,T,IB,fB,HBinv,msg,alg] = ...
9          Alg9(c,q,p,C,n,m,tol,xB,IB,nB,A,b,HBinv,tlow,thigh);
10 T
11 H1
12 H2
```

Figure 8.9.   Example 8.3, eg8p3.m.

```
1  eg8p3
2  T =
3          0      2    Inf
4  H1 =
5          0      1
6          2      0
7          0      0
8          1      3
9  H2 =
10         0      0
11         0      0
12         0      0
13         0      0
```

Figure 8.10.   Example 8.3 (output), eg8p3.m.

Discussion of Eg8p3.m. The optimal solution for this example is a bit unusual in that $x(t) = (0, 2, 0, 1)'$ for all t with $0 \le t \le 2$ and $x(t) = (1, 0, 0, 3)'$ for all t with $2 \le t \le \infty$ so that the parametric parts are both zero.

```
1  %     eg8p4.m
2  n = 4; m = 2;
3  c = [ 0 -1 0 0 ]'; q = [ 1 1 0 0 ]';
4  C = [ 4 2 0 0; 2 1 0 0; 0 0 0 0; 0 0 0 0 ];
5  A = [ -2 -1 1 0 ; 4 1 0 1 ]; b = [ - 2 6 ]';
6  p = [ 0 -.5 ]';
7  HBinv = [ 0 0 -1 0; 0 0 1 1; -1 1 -1 0; 0 1 0 0 ];
8  tlow = 0.; thigh = inf; tol = 1.e-6;
9  IB = [ 2 4 ]; nB = 2; xB = [2 4 ]';
10 [H1,H2,T,IB,fB,HBinv,msg,alg] = ...
11        Alg9(c,q,p,C,n,m,tol,xB,IB,nB,A,b,HBinv,tlow,thigh);
12 T
13 H1
14 H2
```

Figure 8.11.   Example 8.4, eg8p4.m.

```
1  eg8p4
2  T =
3        0     2     4     8
4  H1 =
5        0     1     2
6        2     0    -2
7        0     0     0
8        4     2     0
9  H2 =
10            0          0    -0.2500
11            0          0     0.5000
12            0          0          0
13      -0.5000    -0.5000          0
```

Figure 8.12.   Example 8.4 (output), eg8p4.m.

```
1  %     eg8p5.m
2  n = 10; means = rand(n,1)*1.3; means = sort(means);
3  var = rand(n,1)/10.; A = []; b = [];
4  tol = 1.e-6; m = 1; c = zeros(n,1); C = zeros(n,n);
5  for i=1:n
6      C(i,i) = var(i);
7      A(1,i) = 1.;
8  end
9  b(1) = 1.; p = [0.]; IB = [1]; nB = 1;
10 xB = [1.]; HBinv = [1.];
11 [x,xB,nB,IB,fB,HBinv,msg,dual,iter,alg] = ...
12                      Alg8(c,C,n,m,tol,xB,IB,nB,A,b);
13 AA = vertcat(-eye(n,n),A);  bb = vertcat(zeros(n,1),b);
```

```
14  checkKKT(AA,bb,x,n,n,m,dual,c,C,fB,iter,msg,alg);
15  q = - means; tlow = 0.; thigh = inf;
16  [H1,H2,T,IB,fB,HBinv,msg,alg] = ...
17         Alg9(c,q,p,C,n,m,tol,xB,IB,nB,A,b,HBinv,tlow,thigh);
18  T
19  H1
20  H2
```

Figure 8.13.   Example 8.5, eg8p5.m.

Discussion of Eg8p5.m

```
1   eg8p5
2   Optimization Summary
3   Optimal solution obtained
4   Algorithm: Algorithm 8 (Simplex Method for QP)
5   Optimal objective value    0.00083085
6   Iterations    9
7   Maximum Primal Error      0.00000000
8   Maximum Dual Error        0.00000000
9
10
11   Optimal Primal Solution
12      1    0.09084192
13      2    0.06925724
14      3    0.01874428
15      4    0.57951243
16      5    0.03391913
17      6    0.09895379
18      7    0.01697901
19      8    0.02331578
20      9    0.03320274
21     10    0.03527370
22
23
24   Optimal Dual Solution
25      1    0.00000000
26      2    0.00000000
27      3    0.00000000
28      4    0.00000000
29      5    0.00000000
30      6    0.00000000
31      7    0.00000000
32      8    0.00000000
33      9    0.00000000
34     10    0.00000000
35     11   -0.00166170
36  T =
37     Columns 1 through 6
```

| 38 | 0 | 0.0024 | 0.0112 | 0.0131 | 0.0373 | 0.0486 |

39   Columns 7 through 11

| 40 | 0.0518 | 0.1170 | 0.2416 | 0.8003 | Inf |

41   H1 =

42   Columns 1 through 6

43	0.0908	0	0	0	0	0
44	0.0693	0.0762	0	0	0	0
45	0.0187	0.0206	0.0223	0	0	0
46	0.5795	0.6374	0.6900	0.7057	0	0
47	0.0339	0.0373	0.0404	0.0413	0.1404	0
48	0.0990	0.1088	0.1178	0.1205	0.4095	0.4764
49	0.0170	0.0187	0.0202	0.0207	0.0703	0.0817
50	0.0233	0.0256	0.0278	0.0284	0.0965	0.1122
51	0.0332	0.0365	0.0395	0.0404	0.1374	0.1598
52	0.0353	0.0388	0.0420	0.0430	0.1460	0.1698

53   Columns 7 through 10

54	0	0	0	0
55	0	0	0	0
56	0	0	0	0
57	0	0	0	0
58	0	0	0	0
59	0	0	0	0
60	0.1561	0	0	0
61	0.2144	0.2540	0	0
62	0.3053	0.3617	0.4849	0
63	0.3243	0.3843	0.5151	1.0000

64   H2 =

65   Columns 1 through 6

66	−37.1496	0	0	0	0	0
67	−3.9742	−6.8042	0	0	0	0
68	−0.7809	−1.5468	−1.6986	0	0	0
69	10.6314	−13.0484	−17.7432	−18.9419	0	0
70	1.5018	0.1158	−0.1590	−0.2291	−2.8880	0
71	4.9851	0.9417	0.1400	−0.0647	−7.8214	−9.1972
72	2.5345	1.8407	1.7031	1.6680	0.3371	0.1010
73	4.6006	3.6479	3.4590	3.4108	1.5831	1.2589
74	7.9529	6.5962	6.3272	6.2585	3.6559	3.1942
75	9.6985	8.2572	7.9714	7.8984	5.1334	4.6430

76   Columns 7 through 10

77	0	0	0	0
78	0	0	0	0
79	0	0	0	0
80	0	0	0	0
81	0	0	0	0
82	0	0	0	0
83	−1.3347	0	0	0
84	−0.7125	−1.0515	0	0
85	0.3868	−0.0960	−0.6059	0
86	1.6604	1.1475	0.6059	−0.0000

Figure 8.14.   Example 8.5 (output), eg8p5.m.

Discussion of Example 8.5. Figure 8.14 verifies the theoretical result that the holdings in each asset are reduced to zero and remain there in the order of increasing expected returns.

8.4 Exercises

8.1 Show that the dimension of H_B in the simplex method for quadratic programming can be at most $m + \operatorname{rank} C$.

8.2 Suppose that the simplex method for quadratic programming is applied to

$$\min \{ c'x \mid Ax = b, \ x \geq 0 \}.$$

By appropriate simplification, show that the method reduces to the revised simplex method for linear programming. Are all of the updating Procedures Ψ_1, Ψ_2 and Ψ_3 required? Explain. Compare with Exercise 7.4.

8.3 Consider the problem of Example 8.5 further. Instead of using random means, assume that each mean has value 1. Solve the resulting problem for all $t \geq 0$. How does this result generalize?

8.4 Complete the proof of Lemma 8.1.

8.5 Suppose that when using Procedure Ψ_1 in the Simplex Method for QP, Ψ_1 returns with singular = "true" and w. Show that the last m components of w are zero.

8.6 Repeat Exercise 5.8 with Algorithm 3 replaced with Algorithm 8.

Chapter 9

Nonconvex Quadratic Programming

We have seen examples of nonconvex quadratic programming problems in Section 1.4. Such problems are characterized by the possibility of many local minima each satisfying the usual optimality conditions. In Section 9.1, we verify that the optimality conditions are necessary for a strong local minimum and formulate conditions which are sufficient for a strong local minimum. In Section 9.2, we formulate a variation of Algorithm 4 which will determine a strong local minimum for a nonconvex quadratic program. Algorithm 4, by itself, will determine a point which satisfies the necessary conditions for a strong local minimum. When such a point is reached, tests are introduced to see if this point is a strong local minimum. If it is, the algorithm terminates. If it is not, the modified algorithm will determine a search direction along which the objective function can be further reduced.

A major difference between convex and nonconvex quadratic programming problems is that for the former any local minimizer is also a global minimizer whereas the latter may have many local minimizers. Algorithms that are designed for the nonconvex case will in general find only a strong local minimum, with no guarantee that is a global minimizer.

9.1   Optimality Conditions

Here we derive necessary and sufficient conditions for a strong local minimum for a nonconvex quadratic programming problem.

We consider the model problem

$$\min \left\{ \, c'x \, + \, \tfrac{1}{2}x'Cx \mid a_i'x \, \leq \, b_i \,, \quad i \, = \, 1,\ldots,m \, \right\} , \tag{9.1}$$

or equivalently,

$$\min \{ c'x + \tfrac{1}{2}x'Cx \mid Ax \le b \},$$

where C is symmetric. No other assumptions are made concerning C so that in general the objective function for (9.1) must be considered nonconvex. Let $f(x)$ and R denote the objective function and feasible region, respectively, for (9.1). A point x_0 is a strong <u>local</u> <u>minimum</u> for (9.1) if $x_0 \in R$ and for every $s_0 \ne 0$ such that $x_0 - s_0 \in R$, it follows that $f(x_0 - \sigma s_0) > f(x_0)$ for all positive σ sufficiently small. A strong local minimum is thus a feasible point for which other feasible points near x_0 give a strictly increased objective function value. A <u>global</u> <u>minimum</u> is a point x_0 such that $x_0 \in R$ and $f(x_0) \le f(x)$ for all $x \in R$.

Theorem 4.3 asserts that if C is positive semidefinite and $r = 0$, then x_4 is an optimal solution for (9.1) if and only if the optimality conditions for it are satisfied. A review of the proof of that theorem shows that the sufficiency part relies critically on C being positive semidefinite. However, provided that the term "optimal solution" is replaced with "strong local minimum," the arguments for the necessity part of the proof still apply. Thus we have

Theorem 9.1 *Assume only that C is symmetric. If x_0 is a strong local minimum for (9.1), then there is a vector u satisfying*

(a) $Ax_0 \le b$,

(b) $-g(x_0) = A'u$, $u \ge 0$,

(c) $u'(Ax_0 - b) = 0$.

Throughout this chapter, we define a point satisfying conditions (a), (b) and (c) of Theorem 9.1 to be a <u>Karush-Kuhn-Tucker point</u>.

We have seen examples for strong local minima in Example 1.7. Some others are shown in

Example 9.1

$$
\begin{aligned}
\text{minimize:} \quad & 2x_1 + x_2 - \tfrac{1}{2}(x_1^2 + x_2^2) \\
\text{subject to:} \quad -3x_1 + x_2 & \le 0, \quad (1) \\
x_2 & \le 4, \quad (2) \\
x_1 & \le 4, \quad (3) \\
-x_1 & \le 0. \quad (4)
\end{aligned}
$$

The feasible region and level sets for the objective function are shown in Figure 9.1. The level sets are circles centered at $\hat{x}_0 = (2, 1)'$. The objective function decreases away from $\hat{x}_0$.

Because $\nabla f(\hat{x}_0) = 0$, and no constraints are active at $\hat{x}_0$ it follows that $\hat{x}_0$ satisfies the necessary conditions for a strong local minimum. However, Figure 9.1 shows that points close to $\hat{x}_0$ give a strictly reduced objective value so that $\hat{x}_0$ is not a strong local minimum.

Next consider the point $\hat{x}_4 = (4/3, 4)$. There are two constraints active at $\hat{x}_4$, namely (1) and (2). For simplicity, let a_1 and a_2 denote their gradients so that $a_1 = (-3, 1)'$ and $a_2 = (0, 1)'$. Defining $u_1 = 2/9$ and $u_2 = 25/9$, it is easy to show that $\hat{x}_4$ satisfies $-\nabla f(\hat{x}_4) = u_1 a_1 + u_2 a_2$. Thus $\hat{x}_4$ satisfies the necessary conditions for a strong local minimum. Furthermore, in Figure 9.1 the level set for the objective function passing through $\hat{x}_4$ shows that feasible points close to $\hat{x}_4$ give a strictly larger objective function value. Therefore $\hat{x}_4$ is a strong local minimizer for this problem.

It can be shown that $\hat{x}_2$ is a strong local minimizer and that although $\hat{x}_1$ and $\hat{x}_5$ satisfy the necessary conditions for a strong local minimum, they do not satisfy the sufficient conditions (see Exercise 9.1).

Note that this problem is unbounded from below.

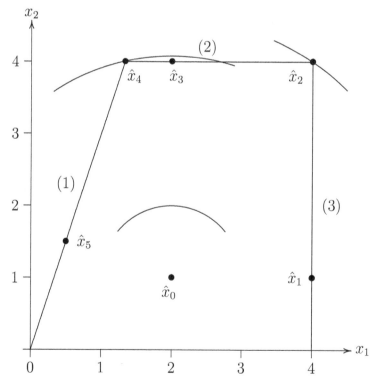

Figure 9.1. Geometry of Example 9.1.

For nonconvex quadratic programming problems, the optimality conditions of Theorem 9.2 are necessary for a strong local minimum but they are not in general, sufficient. This is illustrated in

Example 9.2
Consider the problem of Example 9.1 further. Observe that for $\hat{x}_3 = (2, 4)'$

$$-g(\hat{x}_3) = \begin{bmatrix} 0 \\ 5 \end{bmatrix} = u_2 a_2 = 5 \begin{bmatrix} 0 \\ 1 \end{bmatrix},$$

so that $\hat{x}_3$ satisfies the optimality conditions with $u_2 = 5$. However, with $s = (-1, 0)'$, $\hat{x}_3 - \sigma s$ is feasible for all σ with $0 \le \sigma \le 2$ and

$$f(\hat{x}_3 - \sigma s) = f(\hat{x}_3) - \tfrac{1}{2}\sigma^2$$

which is a strictly decreasing function of σ. Since close feasible points give a strictly decreasing function value, $\hat{x}_3$ is not a strong local minimum.   ◇

In the previous example, s satisfied $a_2's = 0$ and $s'Cs = -1 < 0$. If the generalization of this condition is precluded, then under the assumptions of nondegeneracy and strict complementary slackness the optimality conditions of Theorem 9.1 are also sufficient for a strong local minimum as shown in

Theorem 9.2 *For the model problem (9.1), assume only that C is symmetric and suppose that x_0 satisfies the following conditions:*

(a) $Ax_0 \le b$,

(b) $-g(x_0) = A'u$, $u \ge 0$,

(c) $u'(Ax_0 - b) = 0$,

(d) x_0 is nondegenerate,

(e) x_0 satisfies the strict complementary slackness condition,

(f) $s'Cs > 0$ for all i with $a_i's = 0$ and $a_i'x_0 = b_i$, $i = 1,\ldots,m$.

Then x_0 is a strong local minimum for (9.1).

Proof: Suppose without loss of generality that the constraints have been renumbered if necessary so that the first p constraints are active. Since x_0 is nongenerate, $a_1,\ldots,a_p$ are linearly independent. Let $d_{p+1},\ldots,d_n$ be any $n - p$ vectors such that $a_1,\ldots,a_p,d_{p+1},\ldots,d_n$ are linearly independent. Let $g_0 = g(x_0)$,

$$D' = [\, a_1, \ldots, a_p, d_{p+1}, \ldots, d_n \,],$$

and

$$D^{-1} = [\, c_1, \ldots, c_p, c_{p+1}, \ldots, c_n \,].$$

Let s_0 be such that $s_0 \neq 0$ and $x_0 - s_0 \in R$. Since $c_1, \ldots, c_n$ are linearly independent, there are numbers $\lambda_1, \ldots, \lambda_n$ such that

$$s_0 = \lambda_1 c_1 + \cdots + \lambda_n c_n . \tag{9.2}$$

Let i be such that $1 \leq i \leq p$. By definition of the inverse matrix, $a_i' s_0 = \lambda_i a_i' c_i = \lambda_i$. But because $a_i' x_0 = b_i$ and $a_i'(x_0 - s_0) \leq b_i$, we have

$$\lambda_i \geq 0 , \quad i = 1, \ldots, p . \tag{9.3}$$

Because x_0 satisfies the optimality condition

$$-g_0 = u_1 a_1 + \cdots + u_p a_p ,$$

it follows from (9.2) that

$$-g_0' s_0 = (u_1 a_1 + \cdots + u_p a_p)'(\lambda_1 c_1 + \cdots + \lambda_n c_n) .$$

Thus by definition of the inverse matrix

$$-g_0' s_0 = u_1 \lambda_1 + \cdots + u_p \lambda_p , \tag{9.4}$$

Equation (9.3) asserts that $\lambda_1, \ldots, \lambda_p$ are all nonnegative. There are now two cases to consider. Either at least one of the λ_i is strictly positive, or, all of them are zero. In the first case, let i be such that $\lambda_i > 0$. Then from (9.4) and the hypothesis of strict complementary slackness,

$$g_0' s_0 < 0 . \tag{9.5}$$

From (9.5) it is straightforward to show (Exercise 9.3) that $f(x_0 - \sigma s_0) > f(x_0)$ for all positive σ sufficiently small so that in the first case x_0 is indeed a strong local minimum.

In the second case ($\lambda_1 = \cdots = \lambda_p = 0$) (9.2) implies that

$$s_0 = \lambda_{p+1} c_{p+1} + \cdots + \lambda_n c_n . \tag{9.6}$$

By definition of the inverse matrix, each of $c_{p+1}, \ldots, c_n$ is orthogonal to each of $a_1, \ldots, a_p$. With (9.6), this implies that

$$a_i' s_0 = 0 , \quad i = 1, \ldots, p .$$

It now follows from the hypothesis of the theorem that

$$s_0' C s_0 > 0 . \tag{9.7}$$

Because $\lambda_1, \ldots, \lambda_p$ are all zero, it follows from (9.4) that $g'_0 s_0 = 0$. Taylor's Theorem implies that

$$f(x_0 - \sigma s_0) = f(x_0) + \tfrac{1}{2}\sigma^2 s'_0 C s_0 ,$$

which, with (9.7), shows that $f(x_0 - \sigma s_0) > f(x_0)$ for all nonzero σ and in particular all positive σ sufficiently small. Thus in the second case x_0 is also a strong local minimum. $\qquad\square$

Theorem 9.2 can be readily extended to problems having both linear inequality and equality constraints as follows. The proof is analogous to that of Theorem 9.2 and is left as an exercise (Exercise 9.4). Our model problem is

$$\left.\begin{array}{ll} minimize: & c'x + \tfrac{1}{2}x'Cx \\ subject\ to: & a'_i x \leq b_i, \qquad\qquad i = 1, \ldots, m, \\ & a'_i x = b_i, \quad i = m + 1, \ldots, m + r. \end{array}\right\} \qquad (9.8)$$

Theorem 9.3 *For the model problem (9.8), assume only that C is symmetric and suppose that x_0 satisfies the optimality conditions:*

(a) $a'_i x_0 \leq b_i$, $i = 1, \ldots, m$; $a'_i x_0 = b_i$, $i = m + 1, \ldots, m + r$,

(b) $-g(x_0) = a'_1 u_1 + \cdots + a'_{m+r} u_{m+r}$, $u_i \geq 0$, $i = 1, \ldots, m$,

(c) $u_i(a'_i x_0 - b_i) = 0$, $i = 1, \ldots, m$,

(d) x_0 *is nondegenerate,*

(e) x_0 *satisfies the strictly complementary slackness condition, and*

(f) $s'Cs > 0$ *for all* $s \neq 0$ *with* $a'_i s = 0$ *and* $a'_i x_0 = b_i$, $1 \leq i \leq m$, *and all s such that* $a'_{m+1} s = \cdots = a'_{m+r} s = 0$.

Then x_0 is a strong local minimum.

9.2   Finding a Strong Local Min: Algorithm 10

In this section we present an algorithm which will find a strong local minimum for (9.1) assuming only that C is symmetric. The analysis will be primarily based on satisfying condition (f) of Theorem 9.2. The nondegeneracy requirement as well as the strict complementary slackness requirement are very difficult to control so we shall just assume they are satisfied.

We first introduce the most important ideas of the algorithm in an informal setting and with some simplifying ordering assumptions. Suppose Algorithm 4 has been applied to (9.1) to obtain a Karush-Kuhn-Tucker point (i.e., a point satisfying the conditions of Theorem 9.1). Let this point be denoted by x_j and suppose only the first ρ constraints are active at x_j. Let

$$D'_j = [\, a_1, a_2, \ldots, a_\rho, d_{\rho+1}, \ldots, d_n \,] \tag{9.9}$$

and

$$D_j^{-1} = [\, c_1, c_2 \ldots c_\rho, c_{\rho+1} \ldots c_n \,]$$

be the final data obtained from Algorithm 4. In order to further assess the situation, we need an s satisfying the requirements of Theorem 9.2 (f). Since the columns of D_j^{-1} are linearly independent, we can represent s as

$$s = w_1 c_1 + w_2 c_2 + \ldots + w_n c_n, \tag{9.10}$$

for some scalars $w_1, w_2, \ldots, w_n$. But then we require $a'_i s = 0$ for $i = 1, 2, \ldots, \rho$. Using (9.10) and the definition of the inverse matrix, it follows that $w_i = 0$ for $i = 1, 2, \ldots, \rho$ and so (9.10) becomes

$$s = w_{\rho+1} c_{\rho+1}, \ldots, + w_n c_n. \tag{9.11}$$

Suppose first that in (9.9), $d_i = C c_i$ for $i = \rho+1, \ldots, n$ so that $c_{\rho+!}, \ldots, c_n$ are a set of normalized conjugate directions orthogonal to $a_1, \ldots, a_\rho$. Using (9.11), forming $s'Cs$ and using the fact that $c_{\rho+1}, \ldots, c_n$ are a normalized set of conjugate directions gives

$$s'Cs = w_{\rho+1}^2 + \ldots + w_n^2. \tag{9.12}$$

Not all of the w_i's in (9.11) can be zero for otherwise $s = 0$. Consequently we have $s'Cs > 0$ and assuming nondegeneracy does not occur and that strict complementarity is satisfied, x_j is a strong local minimum.

It may be that there is an $i > \rho$ for which $d_i \neq C c_i$ and $c'_i C c_i > 0$. In this case, column i can be updated to a conjugate direction using Procedure Ψ_1. This transforms column i of D_j^{-1} into a conjugate direction column. This is to be repeated as many times as possible.

Suppose secondly that there is a k with $k \geq \rho$, $d_k \neq C c_k$ and $c'_k C c_k < 0$. Setting the search direction $s_j = c_k$, it follows that

$$f(x_j - \sigma s) = f(x_j) - \sigma g'_j s + \tfrac{1}{2}\sigma^2 s'_j C s_j. \tag{9.13}$$

Because x_j is a Karush-Kuhn-Tucker point, $g'_j s_j = 0$ and (9.13) becomes

$$f(x_j - \sigma s_j) = f(x_j) + \tfrac{1}{2}\sigma^2 s'_j C s_j. \tag{9.14}$$

Because $s'_j C s_j < 0$, (9.14) shows that $f(x_j - \sigma s)$ is a strictly decreasing function of σ, from which we conclude that x_j is not a strong local minimum.

Furthermore, we can restart Algorithm 4, bypassing Step 1 and beginning at Step 2 with s_j and the column number k. Note that since $f(x_j - \sigma s_j)$ is a strictly decreasing function of σ, the first iteration of Algorithm 4 will use the maximum feasible stepsize and the updating will be done through Step 3.2 of Algorithm 4.

After some iterations, Algorithm 4 will complete with a new Karush-Kuhn-Tucker point with a strictly reduced objective function value.

Repeat these two steps until $c_i C c_i = 0$ for all i with $d_i \neq C c_i$.

Suppose thirdly that there are indices ρ and k with $d_\rho \neq C c_\rho$, $d_k \neq C c_k$, $c'_\rho C c_\rho = 0$, $c'_k C c_k = 0$, and $c'_\rho C c_k \neq 0$. Now define

$$s_j = \begin{cases} c_\rho + c_k & \text{if } c'_\rho C c_k < 0, \\ c_\rho - c_k & \text{otherwise.} \end{cases}$$

Then $s'_j C s_j = -2|c'_\rho C c_k| < 0$. As in (9.14),

$$f(x_j - \sigma s) = f(x_j) + \tfrac{1}{2}\sigma^2 s'_j C s_j, \tag{9.15}$$

so that $f(x_j - \sigma s_j)$ is a strictly decreasing function of σ. From this, we conclude that x_j is not a strong local minimum. Furthermore, we can restart Algorithm 4, bypassing Step 1 and beginning at Step 2 with s_j. Note that since $f(x_j - \sigma s_j)$ is a strictly decreasing function of σ, the first iteration of Algorithm 4 will use the maximum feasible stepsize and the updating will be performed through Step 3.2 of Algorithm 4. After some iterations, Algorithm 4 will complete with a new Karush-Kuhn-Tucker point with a strictly reduced objective function value.

We next present a detailed formulation of the algorithm using the index set $J_j = \{\alpha_{1j}, \alpha_{2j}, \cdots, \alpha_{nj}\}$.

ALGORITHM 10

Model Problem:

$$\min \left\{ c'x + \tfrac{1}{2}x'Cx \mid a'_i x \leq b_i, \ i = 1, \ldots, m \right\},$$

where C is symmetric but not necessarily positive semidefinite.

Initialization:
Start with any feasible point x_0, $J_0 = \{\alpha_{10}, \ldots, \alpha_{n0}\}$, and D_0^{-1}, where $D'_0 = [d_1, \ldots, d_n]$ is nonsingular, $d_i = a_{\alpha_{i0}}$ for all i with $1 \leq \alpha_{i0} \leq m$. Compute $f(x_0) = c'x_0 + \tfrac{1}{2}x'_0 C x_0$, $g_0 = c + Cx_0$, and set $j = 0$.

Step 1:   **Computation of Search Direction s_j.**
Same as Step 1 of Algorithm 4.

Step 2:   **Computation of Step Size σ_j .**

Same as Step 2 of Algorithm 4.

Step 3: **Update.**

Set $x_{j+1} = x_j - \sigma_j s_j$, $g_{j+1} = c + C x_{j+1}$, and $f(x_{j+1}) = c' x_{j+1} + \frac{1}{2} x'_{j+1} C x_{j+1}$. If $\tilde{\sigma}_j \le \hat{\sigma}_j$, go to Step 3.1. Otherwise, go to Step 3.2.

Step 3.1:

Set $d_j = (s'_j C s_j)^{-1/2} C s_j$, $D_{j+1}^{-1} = \Phi_1(D_j^{-1}, d_j, k, J_j)$, and $J_{j+1} = \{ \alpha_{1,j+1}, \ldots, \alpha_{n,j+1} \}$, where

$$\alpha_{i,j+1} = \alpha_{ij}, \quad i = 1, \ldots, n, \; i \ne k,$$
$$\alpha_{k,j+1} = -1.$$

Replace j with $j + 1$ and go to Step 1.

Step 3.2:

If there is at least one i with $\alpha_{ij} = -1$ and $a'_l c_{ij} \ne 0$, go to Step 3.3. Otherwise, set $D_{j+1}^{-1} = \Phi_1(D_j^{-1}, a_l, k, J_j)$ and $J_{j+1} = \{ \alpha_{1,j+1}, \ldots, \alpha_{n,j+1} \}$, where

$$\alpha_{i,j+1} = \alpha_{ij}, \quad i = 1, \ldots, n, \; i \ne k,$$
$$\alpha_{k,j+1} = l.$$

Replace j with $j + 1$ and go to Step 1.

Step 3.3:

Compute the smallest index ν such that

$$| a'_l c_{\nu j} | = \max \{ | a'_l c_{ij} | \; | \text{ all } i \text{ with } \alpha_{ij} = -1 \}.$$

Set

$$\hat{D}_{j+1}^{-1} = \Phi_2(D_j^{-1}, a_l, \nu, J_j),$$
$$D_{j+1}^{-1} = \Phi_1(\hat{D}_{j+1}^{-1}, a_l, \nu, J_j),$$

and $J_{j+1} = \{ \alpha_{1,j+1}, \ldots, \alpha_{n,j+1} \}$, where

$$\alpha_{i,j+1} = \alpha_{ij}, \quad i = 1, \ldots, n, \; i \ne \nu, \; i \ne k,$$
$$\alpha_{\nu,j+1} = l,$$
$$\alpha_{k,j+1} = 0.$$

Replace j with $j + 1$ and go to Step 1.

Step 4:

Step 4.1:

For all k satisfying $\alpha_{kj} = 0$ and $c'_{kj} C c_{kj} > 0$, update using Step 3.1. Repeat until $c'_{ij} C c_{ij} \le 0$ for all i with $\alpha_{ij} = 0$. Go to Step 4.2.

Step 4.2:
If there is a k with $\alpha_{kj} = 0$ and $c'_{kj}Cc_{kj} < 0$, set $s_j = c_{kj}$ proceed with Algorithm 4 by going to Step 2.

Step 4.3:
If there are ρ and k with $\alpha_{\rho j} = \alpha_{kj} = 0$, $c'_{\rho j}Cc_{\rho j} = c'_{kj}Cc_{kj} = 0$ and $c'_{\rho j}Cc_{kj} \neq 0$, set

$$s_j = \begin{cases} c_{\rho j} + c_{kj} & \text{if } c'_{\rho j}Cc_{kj} < 0, \\ c_{\rho j} - c_{kj} & \text{otherwise.} \end{cases}$$

Set $J_{j+1} = J_j$, and proceed to Step 2 using s_j as above.

Note: After each repetition of Step 4.1 and then Step 4.2, a new Karush-Kuhn-Tucker point is obtained. Eventually, we will have $c'_{ij}Cc_{ij} = 0$ for all i with $\alpha_{ij} = 0$.

Example 9.3
We consider the problem of Example 9.1 further. A computer program implementing Algorithm 10 is shown in Figure 9.4. A calling program for it with the data for Example 9.1 is shown in Figure 9.2 below. Table 9.1 shows the results of using different starting points for Algorithm 10. This shows that different starting points may (or may not) result in different strong local minima or the information that the problem is unbounded from below.

Starting Point	Conclusion	Final Point	Objective
$(1.8,\ 4)'$	strong local min at	$(1.333,\ 4)'$	-2.2222
$(4.,\ 1.)'$	strong local min at	$(4.,\ 4.)'$	-4.
$(4.,\ .5)'$	unbounded from below		
$(2.,\ 1.)'$	strong local min at	$(4.,\ 4.)'$	-4.

Table 9.1 Results of Different Starting Points for Algorithm 10

```
1  %       eg9p1.m
2  n = 2; m =4 ; r = 0; tol = 1.e-6; c = [ 2 1 ]';
3  C = [-1 0; 0 -1 ]'; A = [ -3 1; 0 1; 1 0; -1 0 ];
4  b = [0 4 4 0 ]'; ; Dinv = eye(2);
5  J = [ 0 0 ]
6  %x = [ 3 3 ]'
7  %x = [ 1.8   4 ]'          % strong local min at (1.333, 4.0)'
8  %x = [ 4 1 ]'              % strong local min at [4,4]'
9  %x = [ 4 .5 ]'            % unbounded from below
10 x = [ 2 1 ]'
11 [x,J,fx,Dinv,msg,dual,j,alg] = ...
12                         Alg10(c,C,n,m,r,tol,x,J,Dinv,A,b)
13 checkKKT(A,b,x,n,m,r,dual,c,C,fx,j,msg,alg);
```

Figure 9.2. Example 9p3: eg9p1.m.

Discussion of eg9p1.m

```
1  Algorithm 10
2  Optimization Summary
3  strong local min {extreme point}
4  Algorithm: Algorithm 10
5  Optimal objective value    -4.00000000
6  Iterations    2
7  Maximum Primal Error       0.00000000
8  Maximum Dual Error         0.00000000
9
10
11   Optimal Primal Solution
12     1     4.00000000
13     2     4.00000000
14
15
16   Optimal Dual Solution
17     1     0.00000000
18     2     3.00000000
19     3     2.00000000
20     4     0.00000000
```

Figure 9.3. Example 9p3: eg9p3(a).m.

Discussion of eg9p3(a).m

We next establish the termination properties of Algorithm 10 as follows.

Theorem 9.4 *Let Algorithm 10 be applied to its model problem assuming only that C is symmetric. Then Algorithm 10 terminates in a finite number of steps with either the information that the problem is unbounded from below, or with* a Karush-Kuhn-Tucker point x_j. In the latter case, assume that x_j, satisfies the strict complementary condition and the nondegeneracy assumption. Then

(a) if $\alpha_{ij} \neq 0$ for $i = 1, \cdots, n$ then x_j is a strong local minimizer and

(b) if $\alpha_{ij} = 0$ for at least one i then x_j a weak local minimizer.

Proof: If Algorithm 10 terminates with the message "the problem is unbounded from below," then it exhibits a search direction s_j such that $x_j - \sigma s_j$ is feasible for all nonnegative σ and $x_j - \sigma s_j \to -\infty$ as $\sigma \to \infty$ so that the problem is indeed unbounded from below.

Algorithm 10 must terminate in a finite number of steps since even though Steps 4.1, 4.2 and 4.3 may be repeated several times, the number of α_{ij}'s having value zero is decreased by at least one. In addition, application of Algorithm 4 cannot increase the number of such α_{ij}'s.

If Algorithm 10 does not terminate with the message "the problem is unbounded from below," suppose it terminates at iteration j with

$$D'_j = [d_{1j}, d_{2j}, \cdots, d_{nj}], \quad D_j^{-1} = [c_{1j}, c_{2j}, \cdots, c_{nj}], \qquad (9.16)$$

and

$$J_j = \{ \alpha_{1j}, \alpha_{2j}, \cdots, \alpha_{nj} \}. \qquad (9.17)$$

Then either

$$\alpha_{ij} \neq 0, \quad \text{for } i = 1, 2, \ldots, n, \qquad (9.18)$$

or

$$c'_{ij} C c_{ij} = 0, \quad \text{for all } i \text{ with } \alpha_{ij} = 0, \qquad (9.19)$$

and

$$c'_{kj} C c_{\rho j} = 0, \quad \text{for each pair } k, \rho \text{ with } \alpha_{kj} = \alpha_{\rho j} = 0. \qquad (9.20)$$

Now assume x_j satisfies the strict complementary slackness condition. Let s be any vector satisfying $a'_i s = 0$ for all i with $a'_i x_j = b_i$. From Lemma 3.1 (with D and D^{-1} interchanged)

$$s = (s' d_{1j}) c_{1j} + (s' d_{2j}) c_{2j} + \ldots + (s' d_{nj}) c_{nj}. \qquad (9.21)$$

But $d_{ij} = a_{\alpha_{ij}}$ for all i with $1 \leq \alpha_{ij} \leq m$. This implies $s' d_{ij} = d'_{ij} s = a'_{\alpha_{ij}} s = 0$ for all i with $1 \leq \alpha_{ij} \leq m$. With these results, (9.21) becomes

$$s = \sum_{\alpha_{ij}=-1} (s' d_{ij}) c_{ij} + \sum_{\alpha_{ij}=0} (s' d_{ij}) c_{ij}. \qquad (9.22)$$

In the case of *(a)*, (9.22) reduces to

$$s = \sum_{\alpha_{ij}=-1} (s'd_{ij})c_{ij}. \tag{9.23}$$

Because $\{\, c_{ij} \mid$ all i with $\alpha_{ij} = -1 \,\}$ is a set of normalized conjugate directions,

$$s'Cs = \sum_{\alpha_{ij}=-1} (s'd_{ij})^2. \tag{9.24}$$

If $s'd_{ij} = 0$ for all i with $\alpha_{ij} = -1$, then from (9.23), $s = 0$. Because we are only considering $s \neq 0$, there must be at least one i with $s'd_{ij} \neq 0$ and from (9.24), $s'Cs > 0$. This completes the proof of part *(a)*.

For the case of *(b)*, (9.18), (9.19) and (9.21) all hold and from these it follows that

$$s'Cs = \sum_{\alpha_{ij}=-1} (s'd_{ij})^2 \geq 0,$$

from which the case of *(b)* follows. □

9.3 Computer Programs

```
1  %      Alg10.m
2  function [x,J,fx,Dinv,msg,dual,j,alg] = ...
3                            Alg10(c,C,n,m,r,tol,x,J,Dinv,A,b)
4  %    Alg10 implements Algorithm 10 to find a strong local
5  %    minimizer (if one exists).
6
7  %    What follows is Algorithm 4 (down to the cut line.
8  %    When Alg4 determines 'Optimal solution obtained',
9  %    we interpret this as a Karush-Kuhn-Tucker point has
10 %    been located and we have to transfer to Step 4 for
11 %    further testing.
12 %
13 %  Initialization
14 alg = 'Algorithm 10';
15 fx = c'*x + 0.5*( x'*C*x);
16 g = c + C*x;
17 next = 'Step1';
18 for j=0:10000;
19     if strcmp(next,'Step1')
20         [next,k,s,msg,dual,big] = ...
```

```
21                              Step1(J,next,g,Dinv,n,m,r,tol);
22            if strcmp(msg,'Optimal solution obtained')
23                lista = find( J ≥ 1 );
24                if length(lista) == n
25                    msg = 'strong local min {extreme point}'
26                    return
27                end
28                next = 'Step4';
29            end
30        end
31
32        if strcmp(next,'Step2')
33            [next,sigma,sigtil,sighat,ell,stCs,msg] = ...
34                         Step2(J,m,A,b,C,s,tol,big,x);
35            if strcmp(msg,'Unbounded from below');
36                return
37            end
38        end
39
40        if strcmp(next,'Step3')
41                                              %  Step 3: Update
42            x = x − sigma*s;
43            g = c + C*x;
44            fx = c'*x + 0.5*( x'*C*x);
45            if sigtil ≤ sighat
46                next = 'Step3p1';
47            else
48                next = 'Step3p2';
49            end
50        end
51        if strcmp(next,'Step3p1')
52                                              %  Step 3.1
53            d = stCs^(−.5) * C * s;
54            [Dinv] = Phi1(Dinv,d,k,J,n);      % k is from Step 1
55            J(k) = −1;
56            next = 'Step1';                   % end of Step 3.1
57        end
58
59        if strcmp(next,'Step3p2')
60                                              % Step 3.2
61            list = find( J == −1 );
62            big = −1.;
63            for i=1:length(list)
64                dotprod = abs( A(ell,:)*Dinv(:,list(i)) );
65                if dotprod > big
66                    big = dotprod;
67                end
68            end
69            if big ≥ 1.e−6
70                next = 'Step3p3';
71            end
```

```
72          if strcmp(next,'Step3p2')
73              [Dinv] =  Phi1(Dinv,A(ell,:)',k,J,n);
74              J(k) = ell;
75              next = 'Step1';
76          end
77      end
78
79      if strcmp(next,'Step3p3')
80                                          %  Step 3.3
81          big = -1;
82          nu = 0;
83          for i=1:length(list)
84              dotprod = abs( A(ell,:)*Dinv(:,list(i)) ) ;
85              if dotprod > big
86                  big = dotprod;
87                  nu = list(i);
88              end
89          end
90          [Dhatinv] = Phi2(Dinv,A(ell,:)',nu,J,n);
91          [Dinv] = Phi1(Dhatinv,A(ell,:)',nu,J,n);
92          J(nu) = ell;
93          J(k) = 0;
94          next = 'Step1';
95      end
96
97
98  % ───────────────── end of Alg 4 ─────────────────
99
100     if strcmp(next,'Step4')
101         next = 'Step4p1'
102     end
103     if strcmp(next,'Step4p1')                % Step 4.1
104         list = find(J == 0);
105         if length(list) ≥ 1;
106             for i=1:length(list)
107                 s = Dinv(:,list(i))
108                 stCs = s'*C*s
109                 if stCs ≥ tol
110                     d = stCs^(-.5) * C * s;
111                     [Dinv] = Phi1(Dinv,d,list(i),J,n);
112                     J(list(i)) = -1;
113                 end
114             end
115         end
116         next = 'Step4p2'
117     end
118     if strcmp(next,'Step4p2')
119                                          % Step 4.2
120         number = 0
121         list = find( J==0 )
122         if length(list) ≥ 1
```

```
123              for i=1:length(list)
124                  test = Dinv(:,list(i))
125                  stCs = test'*C*test
126                  if stCs ≤ -tol
127                      number = number + 1
128                      k = list(i)
129                      s = test
130                  end
131              end
132          end
133          if number ≥ 1
134              % here we have s = Dinv(:,k) and s'Cs < 0
135              next = 'Step2'
136          else
137              next = 'Step4p3'
138          end
139      end
140      if strcmp(next,'Step4p3')
141
142      %  test to see if c_{ij}'*C*c_{ij} = 0 for all
143      %  i with alpha_{ij} = 0   (which should be true)
144          for i=1:n
145              if J(i) == 0
146                  check = Dinv(:,i)' * C * Dinv(:,i)
147              end
148          end
149
150                                          % Step 4.3
151          rho = 0
152          k = 0
153          for i=1:n
154              for j=i+1:n
155                  if J(i) == 0 & J(j) == 0
156                      check1 =  Dinv(:,i)' * C * Dinv(:,j)
157                      if abs(check1) ≥ tol
158                          rho = i
159                          k = j
160                      end
161                  end
162              end
163          end
164          if k ≥ 1 & rho ≥ 1
165              check2 =  Dinv(:,rho)' * C * Dinv(:,k)
166              if check2 < 0
167                  s = Dinv(:,rho) + Dinv(:,k)
168              else
169                  s = Dinv(:,rho) - Dinv(:,k)
170              end
171              next = 'Step2'
172          else                       % optimal solution
173              list = find(J ≠ 0)
```

```
174              if length(list) == n
175                  msg = 'Strong local min'
176                  return
177              end
178              if length(list) == n-1
179                  msg = 'Weak local min'
180                  return
181              end
182           end
183        end
184  end     % iteration loop for j
```

Figure 9.4.   Algorithm10: Alg10.m.

Discussion of Alg10.m. Alg10.m implements Algorithm 10 to find a strong local minimum for a QP (if one exists). The first part of the algorithm uses Alg4 to find a Karush–Kuhn–Tucker point. Lines 1 to line 98 are identical to those for Alg4 with one exception. When Step1 returns with the message "Optimal solution obtained", Alg 10 (lines 23 to 29) counts the number of active constraints in J_j. If this is n then the current point is an extreme point and assuming strict complementary slackness it is also a strong local minimum and the algorithm terminates with this information. If the current point is not an extreme point, control transfers to Step 4 for further analysis.

Step 4 (line 100) immediately transfers to Step 4.1 (line 103). Step 4.1 is implemented in lines 103–117 and constructs a list of those i for which $\alpha_{ij} = 0$ (line 104). If there are more than one such elements, each element, list(i), is checked as follows. We set s to be column list(i) of D_j^{-1}, compute "stCs" $= s'Cs$ and if this bigger than "tol", we update according to "Step 3.1" by setting "d" $= s'Cs^{-1/2}Cs$, "Dinv" $= D_{j+1}^{-1} = \Phi_1(D_j^{-1}, d, list(i), J, n)$ and "J(list(i))" $= J_{j+1}(list(i)) = -1$.

Step4p2 is similar to Step4p1 in the sense that it constructs a list of those α_{ij} having value 0, but for Step4.1 the additional requirement is $c'_{ij}Cc_{ij} > 0$ and with Step4p2 this additional requirement is $c'_{ij}Cc_{ij} < 0$. When the construction of the list is complete, "number" is the number of elements on the list. In line 133, if "mber" ≥ 1 we have an index k with $\alpha_{kj} = 0$, $s = c_{kj}$ and $s'Cs < 0$. Control now transfers to Step2 where a new Karush–Kuhn–Tucker point will eventually be determined by Alg4. If "number" $= 0$, then we must have $c'_{ij}Cc_{ij} = 0$ for all i with $\alpha_{ij} = 0$ and control transfers to Step4p3. Note that this last condition is tested numerically in lines 142–148. There is no ";" at the end of line 146, so the value of "check" is printed.

After performing this test, the coding proceeds with Step4p3 on line 151. Lines 153–163 look for a pair "i" and "j" with "abs(Dinv(:,i)'C Dinv(:,j) $\geq tol$", where "Dinv(:,i)" denotes the i-th column of D_j^{-1}. When such an "i" and "j" are determined, they are recorded in "rho" and "k" respectively. If "rho" and "k" so determined are both larger than or equal to 1 (line 164), then "s" is determined as in Step 4.3 of Algorithm 10 (lines 165–171) and control passes to Step 2 with the "s" just obtained.

If at least one of "k" or "rho" is zero then control passes to line 173. If the number of nonzero elements in J_j is "n," then assuming strict complementary slackness is satisfied, the current point is a strong local minimum. If the number of nonzero elements in J_j is "n-1," then assuming strict complementary slackness is satisfied, the current point is a weak local minimum.

9.4   Exercises

9.1 Continue Example 9.1 by showing $\hat{x}_2$ is a strong local minimum and both $\hat{x}_1$ and $\hat{x}_5$ satisfy the necessary conditions for a strong local minimum but not the sufficient conditions.

9.2 Suppose the feasible region for Example 9.1 is changed by the addition of the constraint $-x_2 \leq 0$. How does this change the analysis for the modified problem?

9.3 Suppose $g_0 s_0 < 0$. Show that $f(x_0 - \sigma s_0) > f(x_0)$ for all positive σ sufficiently small.

9.4 Prove Theorem 9.3.

9.5 Formulate an algorithm to determine if a given symmetric matrix C is either positive definite, positive semidefinite or indefinite. For the case that C is indefinite, your algorithm should prove this conclusion by determining a vector s such that $s'Cs < 0$. It may be helpful to review Step 4 of Algorithm 10.

Bibliography

[1] M.J. Best, "Equivalence of some quadratic programming algorithms", *Mathematical Programming* **30**, 1984, 71–87.

[2] M.J. Best, "A compact formulation of an elastoplastic analysis problem", *Journal of Optimization Theory and Applications*, **37**, 1982, 343–353.

[3] M.J. Best, *Portfolio Optimization*, CRC Press, Taylor and Francis Group, 2010.

[4] M.J. Best and J. Hlouskova, "The efficient frontier for bounded assets", *Mathematical Methods of Operations Research*, **52**, 2000, 2, 195–212.

[5] M.J. Best and K. Ritter, "A quadratic programming algorithm", *Zeitschrift für Operations Research* , Vol. 32, No. 5 (1988) 271–297.

[6] M.J. Best and K. Ritter, *Linear Programming: Active Set Analyis and Computer Programs*, Prentice–Hall Inc., Englewood Cliffs, New Jersey, 1985.

[7] R.G. Bland, "New finite pivoting rules for the simplex method", *Mathematics of Operations Research*, Vol. 2, (1977), 103–107.

[8] G.B. Dantzig, *Linear Programming and Extensions*, Princeton University Press, Princeton, NJ, 1963.

[9] R. Fletcher, "A general quadratic programming algorithm", *Journal of the Institute of Mathematics and its Applications*, **7**, 1971, 76–91.

[10] P.E. Gill and W. Murray, "Numerically stable methods for quadratic programming", *Mathematical Programming* **15**, 1978, 349–372.

[11] H.M. Markowitz, *Portfolio Selection: Efficient Diversification of Investments*, Coyles Foundation Monograph, Yale University Press, New Haven and London, 1959. This has more recently been reprinted as, H.M. Markowitz, *Portfolio Selection*, Blackwell Publishers Inc., Oxford, UK, 1991.

[12] G. Maier, D.E. Grierson, M.J. Best, "Mathematical programming methods for deformation analysis at plastic collapse", *Computers and Structures* **7**, 1977, 499–612.

[13] Ben Noble, *Applied Linear Algebra*, Prentice-Hall, Inc., Englewood Cliffs, New Jersey, 1969.

[14] W.F. Sharpe, *Portfolio Theory and Capital Markets*, McGraw–Hill, New York, 1970.

[15] J. Sherman and W.J. Morrison, "Adjustment of an inverse matrix corresponding to changes in the elements of a given column or a given row of the original matrix", *The Annals of Mathematical Statistics*, **20**, 1949, 621.

[16] C. van de Panne and A. Whinston, "The symmetric formulation of the simplex method for quadratic programming", *Econometrica* **37**, 1969, 507–527.

[17] M. Woodbury, "Inverting modified matrices", Memorandum Report 42, Statistical Research Group, Princeton University, Princeton, 1950

Index